SYSTEMS ANALYSIS AND DESIGN
Method and Invention

Robert E. Leslie

Reader's Digest Association

Prentice-Hall
Englewood Cliffs, N.J. 07632

Library of Congress Cataloging-in-Publication Data

Leslie, Robert E.   (date)
   Systems analysis and design.

   Bibliography: p.
   Includes index.
   1. System analysis.   2. System design.
I. Title.
T57.6.L47 1986   003   85–12447
ISBN 0–13–880311–0

**Editorial/production supervision
and interior design: Tracey L. Orbine
Cover design: Photo Plus Art
Manufacturing buyer: Gordon Osbourne**

To Alice Angier, Tom, and Susan

Printed in the United States of America

10 9 8 7 6 5 4 3 2 1

ISBN 0-13-880311-0   01

Prentice-Hall International (UK) Limited, *London*
Prentice-Hall of Australia Pty. Limited, *Sydney*
Prentice-Hall Canada Inc., *Toronto*
Prentice-Hall Hispanoamericana, S.A., *Mexico*
Prentice-Hall of India Private Limited, *New Delhi*
Prentice-Hall of Japan, Inc., *Tokyo*
Prentice-Hall of Southeast Asia Pte. Ltd., *Singapore*
Editora Prentice-Hall do Brasil, Ltda., *Rio de Janeiro*
Whitehall Books Limited, *Wellington, New Zealand*

CONTENTS

Preface v

Introduction 1

Chapter 1 *The Systems Development Life Cycle 7*
Chapter 2 *The Systems Development Life Cycle Concluded 23*
Chapter 3 *Charting Process Flow 36*
Chapter 4 *Charting Data Flow 60*
Chapter 5 *Requirements Analysis 81*
Chapter 6 *An Introduction to Data Analysis 85*
Chapter 7 *Data-in-Motion 98*
Chapter 8 *Data Transmission 125*
Chapter 9 *Data-at-Rest 146*
Chapter 10 *Modern Database Design 169*
Chapter 11 *Strategic Planning for Information Systems 205*

Chapter 12 *Data Administration, the Data Dictionary, and Documentation 237*
Chapter 13 *Research Techniques 264*
Chapter 14 *Getting and Using Software 286*
Chapter 15 *Estimating Benefits, Costs, and Time 308*
Chapter 16 *Information-System Design 345*
Chapter 17 *The Bakery Problem—a Case Study in Design 389*
Chapter 18 *Invention 421*

Questions and Exercises 442

Glossary 452

Recommended Reading List 463

Appendix A *A Case Study Showing a New System Specification for a Large Business Enterprise 465*

Index 483

PREFACE

This book has been a joyful burden to me, and now this preface is the last part to be written. After I finish it, my household is off to Maine to walk the beaches without a care and to eat a little too much lobster and corn-on-the-cob.

The book was written while I was also working full time as a systems analyst. The writing was done in the evening, late at night, at daybreak, on weekends, holidays, and "vacation" days taken one day at a time, over a period of more than a year. No work project in my life has so consumed me and forced me so close to the edge of my capacities. Yet, it was a joy. I did it my way—all alone in front of my IBM PC in the solitude of my home office. Well, not quite all alone—there were a few people who became my special support network.

Before mentioning them, I would like to identify the expected readers of this book. In the bull's-eye are graduate school or advanced undergraduate school students of Information Science, graduate business school students, and information systems professionals already practicing. Outside the bull's-eye but still very much on the target are the managers and planners in the user community who have to live with the end products of systems change and participate in the analysis at all stages. Also on the target are the large value-forming community of those who opt to be well informed. And then there are small users of information systems who are finding in increasing numbers that the proper use of information may be the difference between profit and bankruptcy. Indeed, graduate school students may find themselves, at some time in their future, small-scale entrepreneurs making vital use of information, perhaps on microcomputers.

ACKNOWLEDGEMENTS

I would like to thank the following friends and colleagues for their support and review:

Robert Hawkins   Reader's Digest
Paul Chasinov   Pace University and Reader's Digest
Neil Hennessy   Reader's Digest
Charles Del Priore   Reader's Digest
Blake Ives   Dartmouth Graduate School of Information Science
Howard Jacobson   Bridgeport University
Tom Schricker   Reader's Digest
James Fletcher   International Business Machines
Maury Ross   International Business Machines
Marilyn D'Allura   Reader's Digest
Mary Bondi   Reader's Digest
Otto Ehrenberg   Pyschologist and Author
Miriam Ehrenberg   Psychologist and Author
Dana Raphael   Anthropologist, Teacher, and Author
Philip Young   Association for Computing Machinery and Philip
Lawrence Group, Inc.
William Fisher   Reader's Digest
Ralph Bondi   Reader's Digest
A. A. Newcomb   Union Carbide
Donald Kutler   Sterling Lord Agency

I am particularly grateful to Robert ("RJ") Hawkins, Professor Paul Chasinov, and Dr. Howard Jacobson for their generosity, helpful criticism, and the use of their extensive libraries.

Finally, I want to express my gratitude to the hard-working editors at Prentice-Hall for their guidance. Particularly I thank Charles Iossi, Patricia Henry, Tracey Orbine, and Jim Fegen for their patience and help.

Robert E. Leslie

Sherman, Connecticut

INTRODUCTION

This is a book about doing systems analysis and design. We call this specifying systems. Specifying a complex system effectively is a significant challenge. What is required are bright analysts who have appropriate methods, common sense, and a flair for invention. We now have a group of hard methodologies which can be taught and which provide the ground for the inventive process.

When we speak of *systems* we mean *natural systems* as opposed to devices (which are artificial systems of relative simplicity) and as opposed to random unstructured systems of enormous complexity (such as junk thrown together for removal at the town dump). A natural system falls in between these two extremes because it exhibits organized relationships of an unvarying nature, has hierarchical form, exhibits evolutionary potential, and can replicate itself. Since such a natural system has these organizational qualities, it is more predictive than a heap of junk.

A human being is a natural whole system. It has suprasystems of which it is a subsystem, such as a family, a work organization, or a political unit such as a country. The human has subsystems, such as the heart, lungs, muscles, nerves, eyes, ears, and so forth. These subsystems have complex relationships with each other for the purpose of maintaining the dynamic equilibrium of the total organism.

It is not easy to imagine a human with three eyes and no stomach or blood vessels. Such a *creature* is possible to imagine, but we would not call it a human being. In a system the components and their attributes and relationships to each

1

other are unvarying. At any given point in time a representative component may be replaced, but the component will remain. In the lining of a human stomach, cells die and are replaced by other cells, but the same *type* of stomach cells remain to form the stomach.

A baseball team is a system containing a subsystem called the pitcher. The pitcher may be replaced in the fifth inning by another pitcher, and the game of baseball can proceed. Remove the role of pitcher, however, and you no longer can play baseball.

The human being has major subsystems, which in turn have subsystems down to the cell level. Cells have molecular subsystems, which in turn have atomic and then subatomic subsystems. This hierarchical leveling is true of all natural systems. We can look at the levels in a work organization in the same way, seeing hierarchy.

All natural systems evolve. The little hamburger stand can become the giant fast-food chain. All natural systems are capable of replication. The well-run hamburger emporium in Boston can be replicated in San Francisco, once the organizational structure is mastered.

Therefore a natural system consists of subsystems of unvarying organization, arranged hierarchically and having the potential for evolution and replication.

In a natural system the whole system is *never* equal to the sum of its parts. The law of synergy is at work here. Natural systems must always be considered as having an element of the mysterious. The systems analyst who treats a natural system as a device is headed for a lot of suffering.

In the practical phenomenal world there are two reasons for systems to exist: first, to process raw inputs into refined outputs, and second, to use raw inputs to maintain the dynamic equilibrium of the system so that it survives. In either case we can speak of systems as doing the same thing—which is:

A system processes inputs into outputs using processes which in turn use system resources and system triggers.

In this narrow but useful construction even Beethoven's *Ninth Symphony* can be called an output—an exalted output but an output nonetheless. (In the last chapter of this book we will expand this definition to include the metaphysical aspects of natural systems.)

This book, and all books on systems analysis, are an expansion of this definition. To illustrate these different functions, take the case of baking bread. The inputs consist of such raw materials as flour, yeast, milk, and shortening. The processes consist of such activities as mixing, rolling, rising, baking, and cooling. The triggers are such events as the timer going off *x* minutes after baking begins. The outputs are bread loaves ready for consumption. The system resources consist of such things as the recipe and the bake oven.

In the case of information systems, where we make information much as we bake bread, the inputs to the system consist of collections of unrefined data called transactions, the outputs consist of reports and responses to queries, the function of the system is to process these unrefined transactions into a data storage resource from which we then process our refined outputs, the triggers tell the system when to initiate a process (e.g., produce a daily report at 7:00 A.M.), and the resources consist of such elements as stored data, people, and a physical support system including computers, data storage devices, terminals, and data transmission facilities.

The premise of this book, which is the basis for all research into general systems theory, is that all natural systems follow the same organizational rules and are subject to the same methods of analysis. This is quite an important concept, because once the systems analyst has mastered the technique of analysis in one subject area, this knowledge is transferable to the solution of other systems problems in other subject areas. Also, this is a concept which transcends any particular device used in problem solving. Systems analysis is a much broader concept than mapping manual operations onto a computer, as important an endeavor as that is. Most of this book will be concerned with specifying systems improvements using computers, but it would be wrong to leave it at that. Let us just say that systems analysis using computer resources is the logical place to develop our methodology.

A hard technique of analysis methodology exists and can be taught, and this book aims to present the several charting methods, each serving a different purpose, which together make for a coherent system for specifying analysis. All the methodologies discussed in this book are proven as effective, and the author will attempt to be faithful to the intent of each method, even while suggesting some improvements.

What is different about this book is that all the methods are integrated into one system, so that the reader will hopefully grasp a whole larger than the sum of each method taken separately. Each method will be dealt with rigorously, not as a summary but at the working level. Some methodologies will be seen to be alternatives to each other, and the author will present the pros and cons of each approach.

Books that give over their entire length to one methodology have certain advantages over a book that tries to teach the whole system of systems analysis, and the reader will be encouraged in this book to do further research into techniques. This will always be necessary as part of the learning process, and references to such books will be given here as the occasion arises (but always within a larger context). Those using this book in the classroom context are the most fortunate. They can work the exercises at the back of the book as a team and receive the directed guidance of the teacher as they seek to learn systems analysis using this book as a primary resource.

Even after this research the novice analyst will not be completely prepared to do a major specification for a complex system. At this point colleagues as mentors take over to further the learning process—which, of course, never stops.

We will be concerned with the class of systems dealing with human social organizations. Such an organization might be, for example, a business corporation, a government agency, or an educational institution. We will teach applied systems analysis based on a sound theory of systems which is widely accepted by those who specialize in general systems theory.

Applied systems analysis is the deliberative attempt to create a reductionist model of a complex social organization. One purpose of this exercise is to determine which operations in the model are similar enough to the actual dynamics of the organization to be mapped onto a device, which operations are suitable for a person-device dialogue, and which operations are left entirely in the human province with device support regarding information for decision making. This often very expensive endeavor is usually undertaken because the system under analysis is failing to meet existing or new requirements regarding its working mandate.

Other purposes for doing this exercise involve gaining a better understanding of the organization to improve various aspects of performance aside from automation, to prepare for possible future change, and to instruct.

Two models are involved, called the current system specification and the new system specification. The new system is either a major revision or an entirely new approach to handling the system processes.

These models, current and new, are both built with the same analytic tools. These tools, taught in this book as an integrated whole, consist of:

- process flow charts
- data flow diagrams
- decision tables
- physical system charts
- motion-data tables
- user-view diagrams
- database tables
- strategic plan architecture
- matrix charting
- dictionary definitions
- design structure charts
- system and program charts
- design flow charts
- panel flow diagrams

A glance at this list of tools or methodologies reveals that this approach is largely nonlinear. Text is not eliminated as a method of defining the model but is definitely assumed to play a supportive role, with the charting technique being the essential method.

The reader is particularly urged to consult Appendix A, which is a case study demonstrating the minimum specifications package needed to do the systems analysis for a new system. This appendix should be referred to only after finishing Chapter 4 which discusses tracking data flow with data flow diagrams. Thereafter, Appendix A can serve as a guideline as we discuss all the methodologies involved at different responsibility levels of systems analysis.

Aside from these charting methodologies, the chapters ahead will cover some of the main subjects that one must understand in order to do the systems analysis. These include:

- the systems development life cycle
- process analysis
- data analysis
  - the relationship between data, information, and knowledge
  - data in motion
  - physical systems for managing data
  - data at rest
  - modern database design
  - the data dictionary
- strategic planning for information system development
- implementation languages
- application packages
- database management systems
- research techniques
  - document analysis
  - interviewing
  - working with users
- cost-benefit analysis
- system design
  - design for batch and on-line systems
  - design charting techniques
  - charting the user-terminal dialogue flow
- invention, creativity, and insight

A glance at this list will show that computer systems design is included in this book on systems analysis. The author believes it is impossible to do even high-level analysis without having a strong background in systems design. The potential users of a new system rightly demand a conceptual glimpse at the final results and also want enough cost-benefit estimation to make an investment decision. Therefore a study of design is introduced in this book about specifying systems.

During the discussion of these various topics in systems analysis we will deal with the different methodologies that—taken together with a broad understanding of the concepts involved—form the tool box of the modern systems analyst.

The writer of a major undertaking such as this book on systems analysis needs something to sustain the soul during the long, lonely days and nights that pass as such a work struggles to unfold. The author's feeling of a particular readership has been most helpful. One such readership has been the author's own workmates and associates in the information field. This book, besides through the references, has been researched experientially. There have been days when the author struggled for clarity with others in business meetings on such subjects as the common database and then rushed home to write about the same thing.

The other readership which has always been present during writing sessions are those wonderful young men and women students at the Dartmouth Graduate School of Information Science with whom the author had the pleasure of sharing knowledge while giving a guest lecture shortly before starting this endeavor. In order to qualify for admission to this school each student had to already have work experience in the field in addition to the rigorous academic requirements. Most of these students had left good jobs and interrupted their lives and even gone into debt in order to undertake two years of graduate school learning at Dartmouth. Whether these particular valiant searchers after knowledge ever read this book or not, it is this community the author has in mind as one word follows the other onto the computer screen. They deserve a proper book on this subject.

1 | THE SYSTEMS DEVELOPMENT LIFE CYCLE

As we observed in the Introduction, the human social organization and every other natural system consists of the following classes of components, existing in a state of dynamic, balanced equilibrium:

- inputs in motion
- outputs in motion
- transforming processes
- triggers that start processes (e.g., do something at 8:00 A.M.)
- resources that enable processes (data, people, money, etc.)

Our aim is to describe the system at hand in terms of these components, evaluate the usefulness of this component relationship and develop better alternatives. Such a system has at least the following attributes:

- relationship between components—e.g., process 1 sends an input to process 2
- goals or mandates—the reason for the system
- equilibrium or balance—the internal forces that maintain the system identity in spite of all change

- evolutionary potential—the ability to grow to something new
- irreducibility—no pitcher, no baseball team
- replication—the ability to generate a duplicate, such as a branch store just like the first store

Figure 1-1 shows the generic model for all steady-state open-ended systems—which for our purposes means all systems.

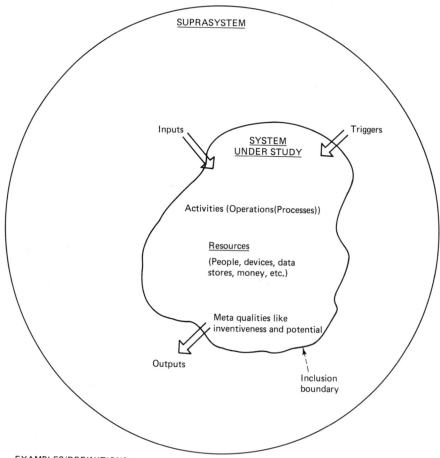

SUPRASYSTEM

Inputs   Triggers

SYSTEM
UNDER STUDY

Activities (Operations(Processes))

Resources

(People, devices, data
stores, money, etc.)

Meta qualities like
inventiveness and potential

Outputs

Inclusion
boundary

EXAMPLES/DEFINITIONS

Triggers: Perform a process 4 P.M. Friday.
Inclusion boundary: Arbitrary selection of system scope.
Meta quality: Thought is a meta quality of brain.
Activities are major subsystems that consist of operations
which consist of serial processes.

Figure 1-1   Generic System Model.

AN APPROACH TO SYSTEMS ANALYSIS

Systems analysis is the process by which we specify systems through decomposition and synthesis. Analysis results in the inductive (bottom-up) understanding of systems parts and the deductive (top-down) understanding of the whole system, which is greater than the sum of its parts. In terms of the specific, applied tasks of the systems analyst we will be dealing mainly with two subjects:

- specifying the current system model
- specifying a feasible future system satisfying new goals and requirements

Communicating Analysis

A major concern of the systems analyst is the interface role of communicating on the one hand with the user-operator of the system and on the other hand with the system designer. Figure 1-2 shows this interface role. The problem is that often the user is accustomed to sequential text with imbedded graphics while the designer needs nonlinear graphics with incidental text. As a consequence, often it is necessary to produce two sets of specifications identical in content but different in form.

The Cost of Systems Analysis

No matter how big or small an organization is, systems analysis often is the largest single cost factor in developing a new or revised system. Analysis and design often accounts for more than half the total cost of system development, while program-

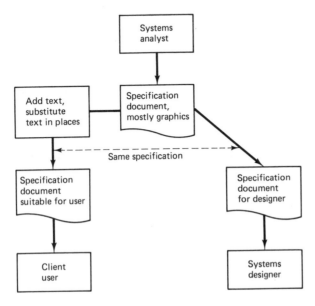

Figure 1-2 Interface Role of the Systems Analyst.

ming code may account for less than 15%. The reason is that correction of mistakes or omissions in the systems analysis phase is much less expensive than the reworking of solutions after implementation has commenced. The price of serious mistakes in analysis is systems failure. That is why the systems analysis methodologies mentioned in the Introduction are so very important.[1]

THE SYSTEMS DEVELOPMENT LIFE CYCLE

The systems development life cycle begins when someone perceives an opportunity or a problem or an intolerable situation, and it continues through to postimplementation maintenance and performance measurement. The steps in the cycle involve checkpoints and participants. Participants include such types as The Concerned User, The Executive Sponsor, The Analyst, The Designer, The Implementer and the Operator. One individual may play several parts. There could be upwards of one hundred checkpoints in a major system development effort.[2]

The basic steps of the system development life cycle are shown in Figure 1-3. Notice that this chart represents a one-level approach to describing a system. Another approach would be to chart a top level, as in Figure 1-4, then to show a second level of detail for each box, and so on through other levels as needed. The second level of detail for ANALYSIS (box 2) would show boxes 2-1, 2-2, 2-3, and so forth. This figure gives us a brief look ahead at *leveling*, which is of primary importance in systems analysis.

A quick look at almost any box in Figure 1-3 will reveal unanswered questions of detail. We can see that a full workup of tasks involves many levels, until we reach a level beyond which it is impossible to go. This bottom level is called the *base level*.

In describing a social organization it might be appropriate to describe the work of a single human organism as a base level. To descend further to a description of the human body's subsystems, such as the cardiovascular or nervous system, would probably be inappropriate, although there are always exceptions. For instance, we might indeed be interested in the workings of the cardiovascular system if we were designing an excercise program for employees that would improve health and work performance.

Let us review now each of the life-cycle elements shown in Figure 1-3. This will allow us to position ourselves as learners of systems analysis skills within the larger context of the development path.

[1] See Matt Flavin, *Fundamental Concepts of Information Modeling* (New York: Yourdon Press, 1981), and Michael Jackson, *System Development* (Englewood Cliffs, N.J.: Prentice-Hall, Inc., 1983), for two other recent methodologies that deserve attention.

[2] The record for system development checkpoints is probably held by the McAuto Division of McDonnell Douglas for their STRADIS package for analysis and design support. STRADIS includes automated charting software called STRADIS/DRAW.

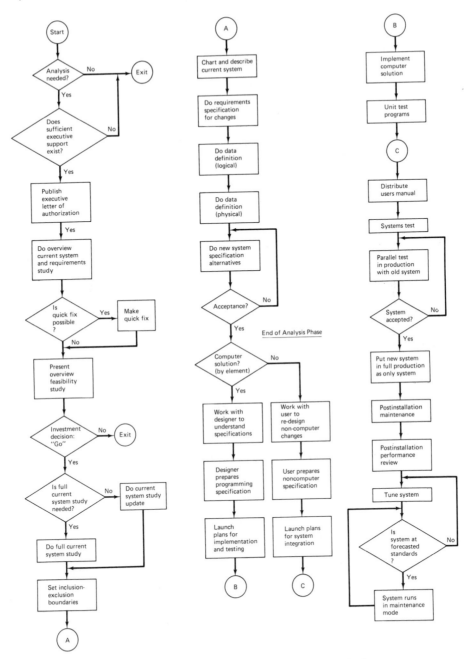

Figure 1-3 Highlights of the System Development Life Cycle.

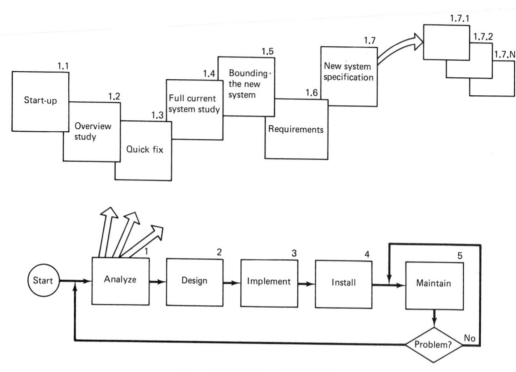

Figure 1-4 An Example of Life-Cycle Levels.

ELEMENTS OF THE LIFE CYCLE

The Problem: Is Analysis Needed?

Analysis is undertaken for a variety of reasons. Usually the system has failed to meet performance standards, in which case it must be repaired or replaced. Often new requirements force a change.

It is important for the analyst and operations manager (user) to have a performance standard for the critical operations in the system, showing (often in graphic form) the minimum, average, and maximum forecasted standards and a point on the chart that represents the current actual performance. Standards are established from the top down: for the system, for the major activities that make up the system, for the operations within each activity, and even for some of the processes in each operation. These performance standards are a required output of a new system specification. Figure 1-5 gives an example of a forecasted performance standards chart.

The first analysis of the problem often is the end of the matter, since a quick rule-of-thumb cost-benefit analysis results in a decision to make some minor changes and let the system limp along with the problem not fully resolved.

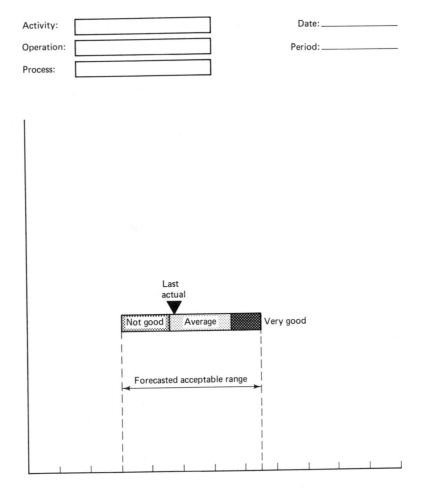

Figure 1-5   Performance Measurement.

Sometimes, however, it is clear that something must be done. This results in a "formal" project for the analyst, who must then prepare to do some extended work.

Does Sufficient Executive Support Exist?

To illustrate this element let us take the analogy of a used car. Let us say we are getting only 300 miles to a quart of oil, the radiator boils over on hot days, the brakes lock, and the front end is loose. We have already replaced the clutch, the starter motor, the battery, and the muffler (twice). The body is rusting out and most of the exterior panels are dented. Fixing up this old heap one more time seems like a poor solution to the problem. We would like to go out and buy a new car: that is the reasonable solution to the used-car problem.[3] The trouble is, we have other important priorities to consider.

This "used-car" dilemma now plagues most social organizations around the world. The executives responsible for the allocation of major resources, such as money and labor, often do not perceive the problem of reconstructing systems that are not in a state of collapse as having top priority. Therefore the analyst will many times find a system in need of major revision but be unable to gain executive support for such a change. Of course the systems analyst is not alone in this situation; the systems user shares it and has even more at stake. And, of course, when we speak of "the analyst," this always implies an analyst team with its own management. It is usually the Systems Development Manager who has to do the job of enlisting executive support.

The mark of a high-quality organization is the care it gives to the maintenance of its various subsystem activities. A major skill the systems analyst must cultivate is the ability to present to senior management the true nature of the problem at hand so that these executives can make a good investment decision.

The Executive Letter of Authorization

Executive support for any major analysis project must involve a letter from the manager who has responsibility for the bounded area of study. This is, in effect, a signed direct order. It states the scope of the analysis and briefly gives a reason. It invokes cooperation from all concerned parties. More than any other single act this letter can insure project success. There are powerful motivators for noncooperation (about which we will have more to say later). Cooperation involves sharing all information, even confidential information, with the analyst.

In some cases the signature to this authorization is symbolic. For a localized systems study the department manager may just walk the analyst over to a supervisor and say: "Fred, Jane Brown here needs to know all about how you people operate. Give her your full cooperation." This is certainly the equivalent of the letter of authorization.

In the case where a large study is to be conducted, involving the sales, manufacturing, and accounting functions, only the chief executive officer has responsibility for the bounded area.

The Overview: Current Systems Study Including New Requirements

A full study of a large system might involve a team of analysts for an extended period. Before that is undertaken (now that we have determined a major change is required and have official executive support), a fast overview analysis should be

---

[3] See Gerald M. Weinberg, *An Introduction to General Systems Thinking* (New York: John Wiley & Sons, 1975), pp. 253–256, for an entertaining and thorough discussion of the "Used Car Law." This is a recommended book for systems analysts.

done to establish the first fluid parameters of the job at hand. This is a task for a skilled senior analyst who has a thorough knowledge of the operations of the subject area (or the class of similar subject areas). We will not describe at this point in the book exactly how this analysis is to be done. It is enough to say that we will be getting a fix on the "chunks" of the system. We will be blocking out the system level minus one. We will be defining these subsystem "chunks" by defining their resources, inputs, outputs, triggers, and processes. We have two main objectives:

- Finding out enough about the problem so that we can make a guesstimate regarding the costs and benefits involved in the change.
- Finding out not only "where the gold is buried" but if there is any gold that can be quickly extracted. This is a metaphor for finding out if there is some quick fix that will produce a significant benefit. It is suprising how many times the quick fix that produces a significant portion of the possible benefit is overlooked by the operations management. Living with a situation every day seems to make it very difficult to see the simple solution to a problem.

Examples of quick fixes: Case 1.   The sales department is using a time-sharing service to run an on-line system because, two years earlier, they found that if they ran the system "in-house" they were getting a 30-second response time. This was intolerable. Using the outside service they are getting a 5-second response time. They are spending $300,000 a year for this outside service and have long ago abandoned the idea of coming "in-house". Meanwhile the in-house operating system and the hardware have been upgraded significantly. The analyst has come into this department to examine a number of problems. Perhaps the main complaint is the high cost of operations. A light flashes on in the analyst's head as he begins to do the general analysis and runs into this particular opportunity. A few phone calls and a visit to a couple of offices and the analyst determines that in-house operating costs for this system would now be about $30,000 a year and that a 2-second response time is a certainty. The high-level overview study reveals situations where benefits of $470,000 plus or minus $40,000 per year are possible following a one-time system revision cost of about $170,000. The quick fix can be done immediately for $2000 resulting in a benefit of $270,000 yearly (plus the benefit of the 2-second response time which may be worth at least another $20,000 yearly). This leaves a remaining benefit of about $200,000.

This happens all the time, everywhere, perhaps because events are moving so fast in the information industry. Very often the user, being a poor sport, cancels the rest of the job. However, next time these people need help they will certainly want to use this analyst.

Case 2.   The manufacturing operation is faced with shortages that show up only during product assembly. This has been going on since the system of parts delivery at the last minute was put into place to cut down on inventories. These last-minute deliveries go right from the receiving dock to the shop floor hold area. This is only one of a group of problems in the shop floor operation. There is, for unknown reasons, too much idle time on certain machines. There is a difference in productivity on the third shift (much worse), but it is not an overall problem, just specific areas. Destroys have gone up sharply for reasons not understood. But the most vexing problem is that of the parts shortages. Finished goods to be shipped to the customers are piling up because they are missing one part. Relationships with parts vendors are deteriorating.

Enter the analyst, always on the lookout for the quick fix. The analyst sees that this is a problem in the purchasing department and, upon interviewing in that department, finds out from one of the buyers that sometimes owing to the vendors' negligence the requested parts are not delivered by truck as scheduled. As a matter of fact it happens all the time! There are heroes and heroines in the purchasing department who in response to agonized calls from the shop perform great feats of expediting.

As the analyst walks out the door of the purchasing department, the hair begins rising on the back of the analyst's neck followed by that old familiar stab of joy that signals the arrival of the *Aha!* moment. Why not make a hot report of expected daily vendor deliveries? The information is available on the databases having to do with materials management—of that the analyst is certain. There is a purchase order involved and a vendor and a set of delivery dates to particular locations. It is all there already.

An expediter can call and confirm the expected deliveries the day before and then call the vendor again 7:30 A.M. on the day of the delivery. It would be wise to add to the vendor database the name and telephone number of the contact as well as the earliest time a call can be made. Someone is going to have to come to work early to logon the terminal and follow the menus to the screen that shows the deliveries due that day. And why not make another report showing late deliveries by vendor by week?

Does the Need for Analysis Still Exist?

Sometimes the development work ends here. The benefits no longer seem to warrant the effort, or there are other problems pending on the backlog list that now seem more important.

Do Feasibility Study

Often the brief cost-benefit analysis already mentioned is not enough to convince senior managers to make an investment in the project. A project with a systems development effort needing major financing goes to the top operating management of the organization. In a business corporation it goes to the executive committee or to some special committee appointed by and including key members of the executive committee. The executive committee acts as top resource manager for the company, handling the advancement of management personnel and making investment decisions. A large systems project is such an investment decision, and in some corporations the analyst may have twenty minutes once a year to advocate a proposal. John and Jane Analyst stand outside the board room with their box of slides waiting to go "on." When they go on their presentation will be a major input into the investment decision. Their estimate of benefits will be what the company expects to realize, and their reputations will ride on this commitment.

In addition to cost-benefit information the feasibility study should give a general understanding of the revised system (or new system, if something like an acquisition is involved).

Is a Full Current Systems Study Needed?

When dealing with a very stable system with much the same data and processing in place over a period of years and no new goals planned, and if there was already a current system study done, it may not be necessary to do it over again. Perhaps it would be sufficient to go through enough analysis to update the previous study and then publish a revised edition.

Under no circumstance should an analyst agree to undertake a revision to a system which is not understood. It may be that the analyst's contribution is very circumscribed, it is still necessary to understand that part of the system and the way it interfaces with the other parts of the whole system. It is still necessary to have a conceptual understanding of the whole system.

The analyst who is forced to make changes to a system without having the opportunity to develop a sufficient understanding should seriously consider finding another place to practice—on one level a career is involved, on another a planet is involved, since the chief sin of the current age is changing what is not fully understood, thus gambling with our future.

To conclude this subject on a humbler note—it is not probable that an analyst or analyst team will ever fully understand the current system. We can fully understand devices but not living systems. We must still do our best to approach this goal and at least pass the threshold where our ignorance causes grievous error. How we attempt to achieve the fullest understanding is the subject of many chapters in this book.

Do Current System Study—Charting and Describing

The current system study describes the existing system in both graphics and language. Resources such as stored data and finances are related to processes, and processes are related to inputs and exports. After the system has been so described, conclusions are drawn about performance. These conclusions are related to the problems that brought the analyst onto the scene in the first place. Both the good and the bad are mentioned. Systems problems are the subject of intense thought and discussion, and solutions emerge. These solutions are noted but are not usually a part of the current system documentation.

There are situations where nothing more is required from the analyst than such a current system study. Such situations usually involve very high level analysis, such as a study done before acquisition of a business or the making of a foundation grant, or a study to chart the potential future of a corporation preparatory to recommending an investment, or the study of a political unit such as a city. In the latter case the analyst is called a city planner, which is just another name for the same job of systems analysis. Systems analysts *don't* work exclusively with computer systems! Almost all large systems now involve computers. All natural systems involve more than computers alone.

Very high level systems analysis will be further examined in later chapters, which deal with advanced systems analysis.

The current system graphic may be done in current system specifics and converted to abstract after review with the user or directly into abstract. What this means is that "John matches the invoice to the receiving sheets and gives it to Mary who okays the invoice" (system specific) becomes "match up invoice-receipts and approve invoices" (abstract). Usually the analyst (working under a tight budget) does the abstracting as the interview notes are transferred to the charts.

BASIC PROBLEMS OF ANALYSIS

Special problems relating both to the current system study and the new system specification must be dealt with at this point in the overview.

Inclusion-Exclusion

This involves setting the boundaries between what is outside the system and what is to be considered inside it. What is outside the system will be the systems environment. In doing this bounding it is important to start from the top down. Buckminster Fuller, one of the great systems analysts, said to always start with the universe and work your way down to where you want to study.[4] So in doing systems analysis we will always resist for a while the user's desire to deal with the system at the level of the current complaints and first do our job of starting at a higher level, viewing the problems from there, and then work our way down to the appropriate level for problem solving. As we do this we will keep in mind the difference between jurisdictions and processes. We are interested primarily in process and information flow, regardless of department. This may affect what we include in the system to be studied.

It is important to know how to draw an arbitrary circle around a cohesive

---

[4] R. Buckminster Fuller, *Operating Manual for Spaceship Earth* (New York: E. P. Dutton, 1971), pp. 59–60.

cluster of business operations and decide that this is our area of analysis. If we include too much, we will be using up resources that might better be used elsewhere. If we do not include some vital system part, we will fail to come up with a good solution.

It is clear, for example, that if we take the human being as a system and decide what to include in this system, we will include the neural flow, the blood flow, the digestive flow, and so on. But what about the preprocessing that goes into baby food to simplify digestion? What about blood plasma stores? What about the human metaquality of self-consciousness, which is of prime importance in defining the human species? There is a group aspect to self-consciousness, or the idea of self. It is dependent as much on the group culture as it is on the functioning of the individual organism. Where do you then draw the perimeter between inside and outside? In problems of social organization it is even less clear, and so this business of inclusion-exclusion is quite a subtle thing. The best approach is probably to include more than you need and then work your way down to what you *must* include.

Douglas R. Hofstadter calls the ability to see elements of a system working in groups "chunking"[5] and points out that what differentiates the chess master from the novice is not so much the ability to see many moves ahead but rather the ability to see groups of pieces in "chunks." This entails an understanding of the relationships involved. This same approach is needed when deciding what organized operations make up our "chunk," which we call our system under study.

This inclusion-exclusion process occurs at other points in analysis. At each subsystem level we will have to partition the process flow, and sometimes it is not simple to decide where the partition is made. Another consideration is what part of a new system is to be automated and what is to remain nonautomated. In all these problems of inclusion-exclusion we struggle to arrive at some right point between the two polarities, taking into consideration all the forces at play.

Leveling

Leveling, as Fuller says, starts with the universe and moves vertically downward through hierarchies until we reach our study system; then it moves vertically downward through the different subsystem levels shown in Figure 1-4. At each level we will decompose the chunk from the previous level until we come to a level where it is meaningless to decompose further. This is called the base level, and the partitioned processes at this level are called base levels. These base levels are processes which are simple enough to be mapped onto a computer as software modules.

If a business corporation is what we call system, we can then decompose corporation, going down one level, into Sales, Manufacturing, Customer Service,

[5] Douglas R. Hofstadter, *Godel, Escher, Bach* (New York: Random House, 1979), pp. 285–288.

Material Management, Accounting, and so forth. Then we decompose these chunks one level down so that Accounting becomes General Ledger, Fixed Assets, Payables, Receivables, Costing, Profit and Loss Reconciliation, Tax Services, and so on. Any one of these chunks will go down through a number of further levels before it is reduced to a set of base levels.

Partitioning

Leveling takes us top-down vertically as we decompose chunks into lower levels; partitioning takes us horizontally as we take a subsystem chunk and decompose it into its transforms at that level. Open envelope; take out contents; remove money; ring up money; route rest of contents to order processing; separate orders to type— these are examples of horizontal transforms. In this case we have partitioned a higher-level chunk called "route order mail" into the lower-level processes or transforms.

We go from level to level vertically, then, until we need go no further, and at each level we partition into transforms suitable for that level. A transform called "write purchase order" may be suitable at one level, but at a lower level we would decompose (partition) that into a string of transform processes one of which would be called "write vendor name and address" (a base level—why decompose that?). Transforms are serial processes, a verb acting on a noun, with the output of Transform-*a* becoming the input to Transform-*b*.

Relating Process Flow to Information Flow

All life and all living things are a story of process flow. We are born, we learn to talk, we go to school, we get a job, we get married, we have children, we grow old, and we die. Information is marvelously developed by the human mind to record process. We were born in the year 1961 when John Kennedy was President of the United States. Our first words were "Mama, Dada." We went to school first at the Evelyn J. Bumpers Elementary School. We married a guy named Charlie or a gal named Hilda and so on. We know the process of our lives not only by the wrinkles around our eyes but even more significantly by the story of our lives. This story is, of course, data about process. Making the story relate to the process is the work of the analyst and the simple purpose of information theory. Doing this clearly is the work of data analysis, which is one hat the systems analyst wears. It is important that the student of systems analysis grasp down deep at the sticking point this subtle interplay between process and information.

Having considered some of the basic areas in analysis whence come our tools and methods, let us return to the system development life cycle, which is centered around the steps shown in Figure 1-3.

THE CURRENT SYSTEM STUDY CONCLUDED

Diagrams, Data, and Other Deliverables

Our current system study has been done using one of two diagramming methods: process flow charts and data flow diagrams. These are the two methods used by most analysts. We have defined the data resource as either data-in-motion or data-at-rest. Data-in-motion are transactions of information. Data-at-rest are stores of information. Hopefully, all data plus data attributes and data relationships have been stored in a data dictionary. A data dictionary is really a necessity in medium to large organizations, and some work on the dictionary should be a deliverable from the current system study. Other deliverables will include decision charts and tables, performance measurement charts like Figure 1-5, on-line versus batch definition, and text which supports the charts and provides study continuity and conclusions.

Get User Sign-Off

When we are dealing with the users (the managers and operators of the system), it would be nice to hand them the technical material described above in order to communicate our understanding of their system, which we have derived from communicating with them and examining the documents they have provided. This is often not possible, since most users will not read charted specifications at this level of detail. Suprisingly, the same is sometimes true of managers of the systems analyst's own department. They may set a standard by which system specification is to be done, using a rigorous charting methodology, but when it comes to going through thirty pages of charts as opposed to fifty pages of text with some charting imbedded, they and the user opt for the text. Therefore it very likely will be necessary for us to transfer our graphics to a "Victorian novel,"[6] preceded by a brief introduction in the terse Dashiell Hammett tradition. This allows users and managers to read the summary and browse the text at the very least. The graphics cannot be done away with, for it is the graphics of the current system that are revised to become the graphics of the new system, which becomes the vital input to the design and implementation process.

It is difficult to write the Victorian novel specification. This is one of many good reasons why systems analysts should be broadly educated, as opposed to a strictly "trade school" type education. The Victorian novel as written by Kipling, Dickens, or a Brontë is a suspenseful work of great clarity and not the dreadful opus that the narrowly educated technician might turn out. These remarks are, of

---

[6] Tom DeMarco, *Structured Analysis and System Specification* (New York: Yourdon Press, 1978), p. 13. This book is must reading for systems analysts. Fortunately, it is exceedingly well written in a lively, personal style.

course, a response to those who coined the expression "Victorian novel" in the pejorative sense in order to make the case for a purely graphic specification. It would be nice if that were always possible. At any rate, we must get a written response from the users stating that we have indeed accurately described their operation.

THE SYSTEMS DEVELOPMENT LIFE CYCLE CONCLUDED

FURTHER ELEMENTS OF THE LIFE CYCLE

Get Approval to Go Ahead with Requirements Analysis and New System Specification Alternatives

At this point in the development cycle we have a checkpoint where the users can decide whether to go ahead with further analysis. The current system study contained conclusions, made recommendations, and perhaps gave a rough cost-benefit analysis. These conclusions and recommendations are formally presented to the user. Depending on the scope of the changes (any major revision can be called a new system), the users at some level and at some point in time may give a written go-ahead.

Do Requirements Specification

This element entails listing as discussion points all the processs steps in each operation where changes are contemplated (allowing for additions to be made to this list). For instance: "Check Invoices against Purchase Orders and Receiving Slips in order to authorize payment of the Invoice. Yes–No." If the user circles Yes and then adds: "Also get the Receiving Slip signed by the person who made the

Original Requisition," we have a new requirement. So we end up with net requirements derived from the current study and modified to reflect new goals.

By using the current system study model, the analyst works out with the user-client a reformulation of those operations that present problems. Goals are reformulated for these problem operations, with sharply defined new goals replacing old fuzzy goals. Where formerly the goal was to respond to customer complaints regarding slow delivery of product with a form letter, now the user also wants "more timely delivery of product," and this must now be sharpened to an enforceable parameter such as "delivery of product within 48 hours of stamped receipt of order." This becomes a new measurement to be used later to judge the implementation. Involved here might be something as simple as a new data item: "stamped receipt date-time" or as complicated as going from a manual to an on-line computer environment.

Remember that the lexicon of our hierarchical approach to specifications will be: *subsystems* within systems (which we will also call *activities*, *operations* within activities (which are multileveled), and *processes* (which exist as a horizontal flow within each operation). The requirements will be at the level of processes within each operation.

Total requirements and partial requirements. The total requirements of a system need not be reviewed each time a change is made, although it should be available as background. When it is available, it should be maintained by updating it with the new requirements and purging obsolete requirements. This requirements list is a support document for the new system specification as well as a work order.

Figure 2-1 shows the flow of requirements from general goals to specific processing.

Data requirements. Data requirements are shown separate from process requirements. Here are several alternatives, depending on the circumstances.

1. Data can be shown in a data dictionary in a temporary category (not in production).
2. Data can be shown as lists or tables to be incorporated at a later development phase into the organizational data scheme.
3. Data can be shown as *user view charts*,[1] each chart answering a process question. This is a method of formal "data modeling" whereby the user views are then decomposed into *relations* in the "third normal form" and then synthesized into high-affinity subject data bases. Software packages are available that do this decomposition and synthesis for you.[2] Don't be put off by

[1] James Martin, *Computer Data-Base Organization*, 2d ed. (Englewood Cliffs, N.J.: Prentice-Hall, Inc., 1977), pp. 60–80.

[2] Database Design Inc., Ann Arbor, Michigan, and Holland Systems, Ann Arbor, Michigan, both offer automated data-modeling packages.

General Goal: increased profits

Specific Goal: 4%–5% increase in revenue next fiscal year

Requirement: same-day acknowledgement of order to speed up payments

General Goal: decreased costs

Specific Goal: 1% decrease in billing costs next fiscal year

Requirement: no billing of accounts where balance due is less than $1.00

General Goal: better service

Specific Goal: 18-day turnaround in portal-to-portal order fulfillment time-in-process

Requirement: optical scanning of all machinable orders

General Goal: improve quality of life within organization

Specific Goal: reduce noise levels

Requirement: install carpeting in all workplaces

Figure 2-1 Goals and Requirements Examples.

this terminology. Relations in the third normal form are nothing more or less than flat tables, where the columns are a key with a set of attributes which relate only to the whole key and nothing but the key and have only one value in each column-row box. Don't worry if that is not clear now. We will return to it. This approach is often taken by companies who are committed to a strategic plan based on the integrated corporate database, as opposed to lower-level subsystem application databases which are uploaded and matched to arrive at corporate data.

Get User Approval of Requirements.   At this point in time all the requirements on the user's wish list have been clarified, and agreement on this list is now formalized by a requirements document.

Do New System Specification

Current system study and approved new requirements at the ready, it is now time to specify the proposed new system. The same method used to model the current system is used to model the new system. This is a skilled task, because many detail processes, all of which have been charted, must be related to the new requirements. In order to be able to acknowledge an order on the same day as the stamped receipt of the order (a new requirement) it may be necessary to redo an entire batch system to on-line. Network communication enhancements may be involved if the receipt of orders involves a distributed operation. It may be necessary to provide exception processing for orders involving credit risks, which must be reviewed by human eyes. Credit rules may have to be relaxed. Some new requirements may involve a greater change to corporate business policy than was originally understood. This means meetings, possibly between people who have a different view of what is best. Jurisdictions may be disturbed and even threatened. There is an iterative tug and pull process here which no programmer ever had to contend with.

Nevertheless, and with a heroic application of *inventiveness*, the new system is specified (with flow charts, data flow diagrams, and other charting methods such as user view charts) in such a way that it can be turned over to a design and implement team and also be made understandable to the user.

Alternatives.   It will often be necessary to show user decision makers a number of alternatives, each demonstrating a different cost-benefit profile. One alternative is always the full study, which is by definition the one with all the bells and whistles (incorporating all the user wish list requirements). The other alternatives will usually be scaled down versions of the first.

Phasing.   The ideal solution is where the best possible system is accepted as a phased operation: "top down and left to right." In this solution, graphically depicted in Figure 2-2, the long-range strategic planning for integrated data and a process development architecture is put into place as subject data bases and a systems development plan, all related on an *extended* data dictionary. Then the applications under this umbrella that make up in total the best new system are prioritized left to right, with left done first and right done last, following the 80-20 rule. The 80-20 rule states that the first application is the one that provides 80% of the benefits for 20% of the total new system cost. More often than not it works out that way.

More phasing.   Now that the total new system has been specified left to right (a metaphor describing the appearance on Figure 2-2) in a horizontal series of development phases, there is one more necessary step in phasing, which goes on in the development of each application phase. This is the vertical top down movement of the system from the most passive state to the most dynamic state. This can be done after the system is installed and over time. In a materials management system, for instance, purchasing support is one phase. In the beginning the new system provides collected information about workload needs (bill of materials and projected inventory) to assist the people in the purchasing department prepare material requisitions, which are then synthesized into the purchase orders that represent the material requirements for a given time period, hopefully taking full advantage of vendor bulk-order discounts. In a later, more dynamic phase of

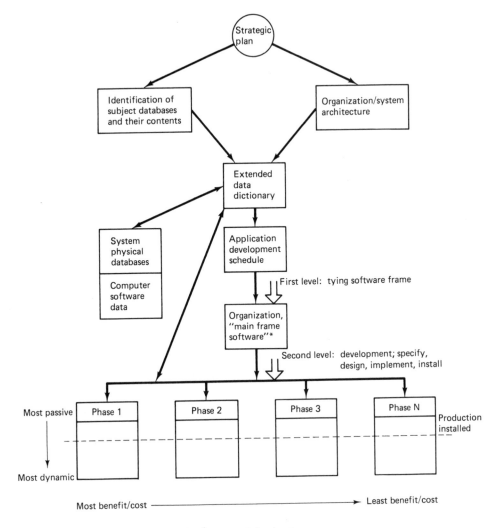

Figure 2-2 Phased Development Model.

the system the purchase orders are cut directly from the workload, the inventory, and the manufacturing time requirements (subject to human review and override).

The prototype.   We cannot leave the subject of phasing without a mention of the *prototype*. What is it? Is it a demo of the system? Is it a throwaway system that works (sort of) and keeps the wolves from the doorstep while the slower real work goes on? Or is it an extendable, viable first phase in implementation? In all three of these cases the prototype should be thought of as part of phasing.

If the first deliverable phase of a system based on thorough analysis must be quite a long way off into the future, it may be wise from the user's perspective to do a prototype. The developers of the full-service system and the users will learn from it a lot about what is really needed and what really works. Also, any prototype system put into production that performs functions better than before will yield benefits. Perhaps the prototype will have a less-than-desirable method of data entry with minimum data validation. It may have an inefficient data base with slow response time and may lack operational features beyond a simple store and retrieve of essential data. However, it will do a real job. In this case it is important not to build too much into the prototype, for the users may settle for less than they should and cancel out the more sophisticated system in favor of a "brassboard" system that doesn't work under stress and can't be further developed through phases.

Review of Figure 2-2

This graphic has already been used as a reference in the discussion on phasing but warrants a review in more general terms.

Downward control of the strategic plan.   The strategic plan is a plan for data and process development for the whole organization. It is a system development plan with the addition of a formal organizationwide review of operations and data, performed as a major effort by a team of users and analysts producing a long-range "strategic plan" document. This is done up front before the development plan is further implemented.

This plan defines whole-system data as entities of data (like the flat tables already mentioned) clustered into affinity groups which are clustered into subject data bases to be made available as corporatewide integrated data. Also defined at this time is the general strategy for *communicating* not only data but also voice, message (such as facsimile), and video information. This involves organizationwide *networking*, often through complex transmission systems. For this part of the strategic plan the systems analyst must develop enough expertise to relate successfully to the user on the one hand and the communications experts on the other. The costs and the benefits are significant enough in this area for us to focus on communications in a later chapter.

This plan also defines business functions (such as sales and manufacturing) and, related to these jurisdictions, defines the system in terms of activities such as advertising or purchasing or product assembly. Related to these activities are the subject databases and their entity groups. An excellent place to establish this relationship is in a full-service data dictionary, which allows the building of user-developed categories and relationships. This is shown at the top of Figure 2-2 as the data entities and the system architecture feeding the dictionary as inputs. This makes the dictionary the container for the strategic plan. As each phased application is implemented, the analyst models the data for it. The output of the data modeling as well as with the subject databases are converted into the physical databases by the database administrator (DBA). When these physical databases are loaded with actual values and software is put in place to use the data, you have a production system. While the development is going on, you have stored on the dictionary two kinds of things: (1) forecasted data and systems, and (2) actual data (databases containing segments containing data elements) and systems (links to programs).

The mainframe level.   Before going to the application level of subsystem specification the organization should implement the unifying mainframe structure that holds together the whole system. This includes a network structure (as mentioned above) to handle and integrate information of the following kind:

a. centralized information at headquarters

b. distributed information at branches

c. private information at desks

d. suprasystem information at desks (news, stock quotes, library browsing, ...)

The relationship beween items a, b, and c above brings up the subject of uploading and downloading, which makes local information available at higher system levels and corporate information available at lower system levels.

All phased application development must take place within the context of this mainframe. Another mainframe activity is the screen system, which controls on-line maintenance and display of applications. There should be a screen architecture, so that new users can intuit their way through the screen flow to the data they need.

A final mainframe consideration we will touch on in this chapter is the batch versus on-line transaction flow. An organization processing 500,000 simple transactions a day through an optical scanner with most of the information in the header card is not going to be much interested in on-line input. Another organization where the customer is standing in front of the input operator wanting verification of his or her share of a perishable inventory is certainly not going to be much interested in a batch system. The example just given might be an airline reservation system. The inventory is perishable because once the flight lifts off, all empty seats are gone forever.

The choice of method, batch or on-line, is critical to the cost-benefit ratio, since on-line systems are usually many times as expensive as batch systems in terms both of software needs and of additional hardware. The analyst should be careful not to recommend an on-line system when a modified batch system would suffice. The basic question is: does the user need the information in two seconds or is two days just fine?

The second level: phased application development.   As we do each phased application, we must decide whether to buy or build and how much to do at first cut. If we are lucky enough to be able to buy what we need without being overwhelmed by package modifications, we should buy, and that may simplify our top-down development from passive to dynamic. We must be wary, though, of installing a takeover dynamic package in an environment which is unprepared for this trauma. No package is going to work if the folks don't cooperate and there is a definite culture lag. If we build, we must use the highest-level language consistent with a solution to the problem.

New System Specification Particulars

The basic specification is the diagram or chart which is the model of the new system, showing manual-automated boundaries and decision support-operational boundaries. This diagram or chart is done top-down first, showing the whole system with its major inputs and outputs (data flows or transactions) and files or databases (data stores). On the next level down are shown the major subsystems along with data flows and stores, and so on down to the base level—the bottom operations level where the processes operate on the data. This is exactly the same as the current systems study particulars, except that the new requirements have been applied as modifications or gross changes.

Supporting this document are the text and charts and tables that clarify complicated problems. As we have said before, it may be necessary to write text to replace the charting for the user audience that has not been close to the development of the system. Those users who have been on the specification team with the analyst have already seen the charting and are comfortable with it.

Also identified in this specification are the transaction types. These include the standard maintenance transactions such as add, delete, and update and also the system special transactions such as payments, stop billing, orders, pressure readings, account withdrawals, and medical claims.

If formal data modeling was started during the requirements analysis, it is now finished off when the specification alternative has been selected. You will remember that *user views* are charts showing data needed to support processes, which are decomposed into entities and then synthesized into subject databases which become the model for the actual, physical corporate integrated database. If formal data modeling is not done because the organization is building application databases at a lower level of integration, then this is the time to finalize your physical data files.

When design begins, the database administrator (or someone most responsible for file organization) will define an *external database view* tailored to the application, which is not the same as the physical database. The external view given to the applications team may contain parts of three physical databases "stitched" together. The problem of getting data for the less complex database or one-level file is very much simpler, but then the data presents other problems.

And to repeat: all views of the data in a complex system belong on the extended data dictionary. If this is done right, applications data can be copied over into the programs that need the data if these programs are in second- or third-generation languages (Assembler or COBOL, for instance); or, if the data dictionary has dynamic input processing, the input data stream goes directly through the dictionary, and field validation is done at dictionary time.

Those analysts who will be working with full-service databases will be using what is known as a three-level storage scheme—as opposed to a file such as an address book or fixed-length tape records, which is a one-level scheme, or a VSAM file, which is a two-level scheme. Figure 2-3 shows the official three-level ANSI/X3/SPARC database as defined by the American National Standards Institute. Note the program interfaces between each level. This is the *database management system* (*DBMS*), which manages each level so that changes at one level do not affect system stability at another level.

Define Implementation Plan

There is a wide range of levels of complexity when it comes to defining the implementation plan. In a small organization this plan may involve a modest application revision; at the other extreme it may involve a major new system for a large government or business organization, where an automated Pert critical-path tracking system is needed to keep all the pieces of the puzzle on the table and on schedule.

Whatever the level of complexity and organization size, the following elements are involved:

- people to do what only people can do
- programs to drive the hardware
- computers and computer peripherals, such as terminals and other equipment
- facilities, such as office space for users
- training for developers and users
- training manuals
- development documentation
- performance measurements
- a communication network plan
- a project plan, putting all resources together to a schedule covering such topics as:
  1. launch meeting
  2. design plan due date
  3. designer's full implementation plan
  4. conversion plan
  5. systems testing plan
  6. parallel testing plan
  7. production turnover plan
  8. tune-up, maintenance, and postinstallation evaluation plan

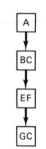

EXTERNAL SCHEMA (ONE APPLICATION VIEW)

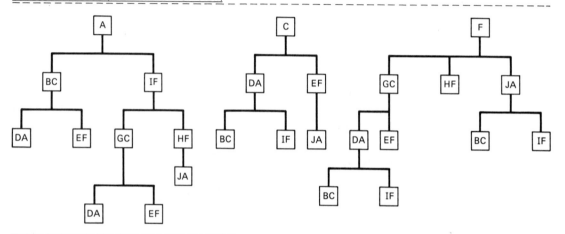

Note: The two letters combined show the pointer relationship.

CONCEPTUAL SCHEMA (LOGICAL DATABASE SYSTEM)

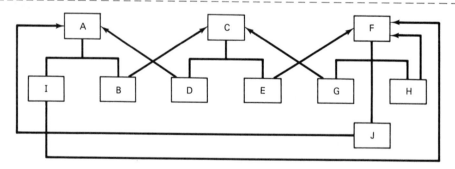

INTERNAL SCHEMA (PHYSICAL DATABASE SYSTEM)

Figure 2-3   The Three-Level ANSI X3 Database System Organization.

Basically, anything that we can forsee, involving implementation costs, should be covered by the analyst in the new system specification. Otherwise a decision to go ahead cannot be made on the basis of full knowledge of all the resources that would be involved.

Get Acceptances

Here, again, are the components of the new system specification package, which is now to be reviewed for acceptance by the user:

- the charts or diagrams specifying the most ambitious alternative
- accompanying text
- data definition from simple file to integrated databases
- system development plan and data definition on extended data dictionary for complex system development effort
- physical system requirements, including networking
- implementation plan, including batch versus on-line considerations
- cost-benefit analysis
- choice of several alternatives from broadest to narrowest, including cost-benefit analysis

This specification package may be a major undertaking involving a large team of analysts, or for a small job it may be done in two or three pages.

Before the specified system can be built, it must be accepted by all those users who have interests involved. Some will not have been intimately connected with the project during analysis, but these people will now make the investment decision. Of course by this time a very considerable portion of the budget needed for the entire project has already been spent, since analysis is a very large cost factor. At any rate, at this point the cards are on the table concerning what the new system would mean to all the jurisdictions, and we are in a very political situation. It is not enough for a proposed system to be cost-effective and technically feasible; it must also be politically feasible. For a big change this sometimes turns out to be something like a contract negotiation, where each "player" offers something for something. Here are the players:

- *the users*, who have two concerns: (1) that the change does not put their group at a disadvantage relative to where they are now, or conversely, puts them at an advantage, and (2) that, since they will have to participate heavily in all stages of design, implementation, and installation, they will be able to endure what is always a risky and disruptive experience
- *the management*, who must invest money and other resources and make choices as to what is most important. Their job is made more difficult because of the

technical nature of the investment decision. When it comes down to the decision, they often end up betting on people they know will deliver a quality product. Their confidence in the analyst team is therefore often a critical factor.

- *the design and implementation team*, who, once accepting the analyst's specifications as complete and reasonable, have the responsibility for actually seeing that "the machine goes out the door"—and on time. They have the right to refuse to accept the specifications.

It must be emphasized that the quality of the specification package for the new system, the clarity of the various presentations associated with it, the respect the various "players" have for the capabilities of the analyst, and the level of trust they place in the character of the analyst all go to achieving a quick acceptance. Programmers who become analysts often have trouble with these aspects of systems analysis, which are far removed from the solitary joys of constructing an elegant and nonpolitical program.

At this point in the system development life cycle we have reached the end of the system specification phase. The analyst will stay with the project through the rest of the cycle as a consultant—or, in a small organization, the analyst now puts on another hat: that of the designer.

Prepare Design for Computer Portion of New System

The following major tasks are involved:

- designing system structure
- designing program structure within system
- specifying processing at the base level if not done
- designing transactions (data-in-motion)
- acquiring application data stores (data-at-rest)
- specifying appropriate languages for various tasks
- evaluating and recommending off the shelf software
- giving customizing rules for acquired software
- preparing on-line screen flow and appearance
- preparing output reporting detail
- designing back-end query-language facility
- further development of project implementation plan
- further development of physical implementation plan
- further development of project documentation
- further development of operations manual

- further development of testing plan
- presentation of design specification to all concerned

Implement System

If the implementation team accepts the design specification, they now have the responsibility to implement the system according to the schedule. Here are the major tasks involved:

- finish detail charting if not already done by designer
- coding
- further development of project documentation
- further development of testing plan
- unit testing of individual programs
- systems testing of all programs together
- parallel testing against the old system
- assist in user training
- prepare system for production, including job control language and computer-center instructions
- turn system over to postinstallation maintenance group
- assist in postinstallation fine tuning
- participate in postinstallation review

As we have indicated before, in a small shop doing modest systems—or even in a larger organization where the system is state-of-the-art—the systems analyst may become the designer and then become the programmer. Having dealt with all the problems and group interactions in the analysis and design phase, perhaps the analyst deserves the treat of some quiet programming.

3 CHARTING PROCESS FLOW

PROCESS VERSUS INFORMATION

When we think of systems, we think first of process flow.[1] Cells grow and die. Humans grow, do tasks, have children, and pass on. Cities are started at a crossroads near a dock and become vast complexes of modern urban life. This is all clear. Yet as humans we find it difficult to experience process phenomena without also experiencing information about the process. This is a unique quality possessed by humanity—this self-awareness that allows us to organize and process information about the physical world as it is, as it was, and as it may be.

We cannot dwell at length here on the much-discussed issue of whether thought (information) or process is primary. However, we do have to develop a perspective on this matter in order to practice systems analysis in an effective way. For our discussion we can accept that we humans cannot deal with phenomena without also dealing with information except for brief innocent moments—those moments when the musician gets "lost" in the music and the runner breaks through a barrier into the "pure" experience of movement.

We will now entertain two methodologies for recording systems analysis, both of which accept process and information as being bound together. One method

[1] Fritjof Capra, *The Turning Point* (New York: Simon and Schuster, 1982), p. 267.

places emphasis on process tracking, the other places emphasis on data tracking. Each ends up defining both. This chapter will be concerned with placing the emphasis on process tracking.

FLOW CHARTS: THE UNIVERSAL CHARTING TOOL

We will now deal with flowcharting as the only all-purpose charting tool. Flow charts can handle the specification of the current system, the new system analysis, the new system design, and the new system implementation at any level of detail and with the use of the same few symbols. However, to do all this successfully we will need "enhanced" flow charts.

Process Flow Chart Symbols

Flow charts use only the symbols shown in Figure 3-1 to model everything except physical systems (in which case special device and connector symbols are needed). The symbols shown are the decision diamond, the operations box, the subroutine box or octagonal, the line arrow, the page or subroutine connector and the branching table. Note that these symbols can define a purely logical system (solution-independent) and that no symbols relating to physical implementation are included. Examples of physical symbols on commonly used templates are symbols for disk files, terminals, printed output, and the like.

The decision diamond.   Figure 3-2 shows the variations on use of the decision diamond. The decision diamond always shows a binary decision, such as Yes-No. It is best if the diamond need contain only a simple relational expression, although a compound relational expression is shown. A simple relational expression is: people $> 6'1''$. A compound relational expression is: (people $> 6'1''$ or people < 28 years) and (hair $=$ brown).

We see an example in Figure 3-2(a) of the *precise decision*, where a name is related by a colon to another name or literal value. The conditions on one arrow path include equal, not equal, greater than, less than, equal or less than, and equal or greater than. The complementary condition goes on the other line arrow.

Figure 3-2(b) shows the *switch*, where the name of the switch in the diamond always, as a convention, has a suffix of SW to identify a switch. A switch can be either *on* or *off*.

Figure 3-2(c) shows the *question* form, wherein a question is posed inside the diamond relating a name with a name or a literal.

Figures 3-2(d) and (e) show the multiple simple form and the compound form, which combine (a), (b), and (c) in various ways as shown. The multiple simple form gives the same results as a decision table or decision tree. Chapter 4 will show examples of these charts.

For the purposes of analysis, which is to specify current and new systems at

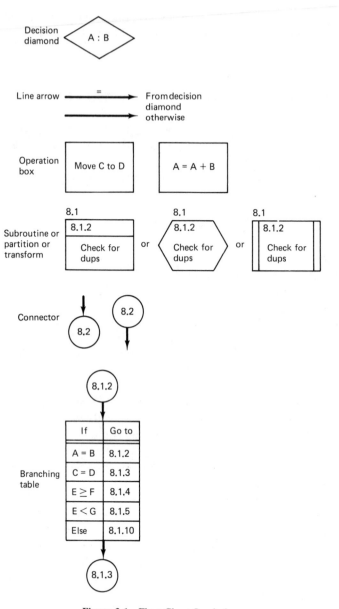

Figure 3-1 Flow Chart Symbols.

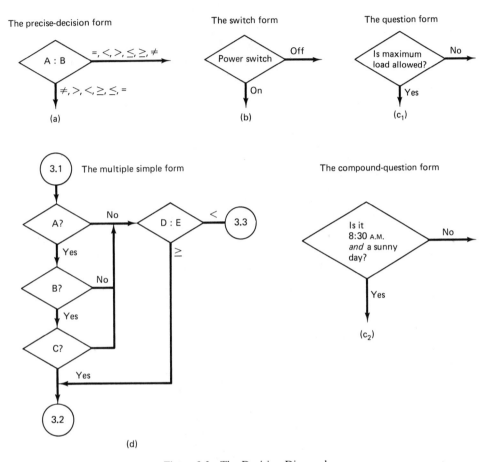

Figure 3-2   The Decision Diamond.

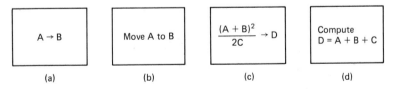

Figure 3-3   The Operations Box.

the logical level independent of physical solutions, we will rely on the question form of the decision diamond but not exclude the other forms if applicable.

The operations box.   Figure 3-3 shows some of the varieties of operations boxes and the conventions associated with this form. Figures 3-3(a) and (b) show the same *move operation* with different conventions. Figure 3-3(c) shows a *com-*

putation (the arrow means "store result in D"). Figure 3-3(d) shows a *logical action*.
Most often we will be using Figure 3-3(d) in system specification.

Figure 3-4 shows decision and operation chart forms combined into a system

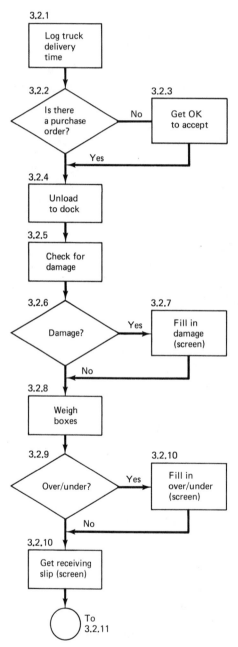

Figure 3-4 Receiving Goods Specification.

specification. Note the ubiquitous line arrow, which in this charting convention shows process flow but not data flow. This is the third in our cast of symbols. As a convention we will flow our charts from primarily top down and secondarily from left to right. It is smart to draw on a large surface using small symbols, because the object is to see the complete picture at one viewing of the operation we are modeling. If you draw on 11 × 14 paper you can usually find a copy machine which will shrink your charts down to $8\frac{1}{2}$ × 11. It is good to keep all the chart symbols uniform in size to the extent possible, as the appearance of the finished chart will be more pleasing.

Now that we have the decision and operation symbols in our repertoire, we can also perform a DO loop, as shown in Figure 3-5. The DO loop shows an iterative operation (such as a table lookup) and always has a bust-out diamond at the top or bottom of the loop.

The bust-out in Figure 3-5 is the decision diamond that contains the relational expression: Counter : 100. When the counter is greater than 100, we bust out. The operation "do something important" might be letting the telephone ring ten times before we discontinue ringing, or doing a lookup to a telephone directory, or one of many other similar loops. Current systems are full of manual DO loops just waiting to be automated.

In the flow chart form of specification, operations and decisions can exist at

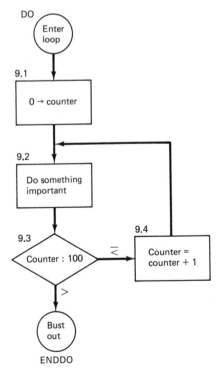

Figure 3-5   The DO Loop.

all levels of charting, not just the base level. However, we will never show operations at a higher level if they can be deferred to a lower level. For instance, the following flow is inappropriate: (1) publish sales plan, (2) identify author of sales plan, (3) plan raw materials acquisitions. Clearly the explicit operation "identify author of sales plan" is inappropriately leveled in this flow and should be pushed down to a much lower level.

Partitioning was mentioned in Chapters 1 and 2 as a major factor in successfully charting specifications. In order to understand a large activity, we need to partition this activity into comprehensible transform modules (each module or partition operating on inputs to make transforms into outputs). It is very hard to "make dinner for guests" until we divide this process up into (1) before dinner munchies and drinks, (2) soup course, (3) main course, (4) salad, (5) dessert and coffee. Now we can allocate chores, figure out timings, and make a decent meal for the guests. This (one more time) is *partitioning*. [Note that our flow is also properly leveled. One level up would be (1) work, (2) drive home, (3) shop for food, (4) make dinner for guests.]

The sub-routine box or octagonal.   This form is used to do both partitioning and leveling and is the compleat analyst's secret to successful specification modeling. This symbol stands for all of the following:

1. A pointer to a lower level in the system hierarchy (therefore used for leveling).
2. A common operation used many times in the system which we want to specify only one time.
3. A "black box" imported from the suprasystem environment, the lower levels of which we do not care to specify. An example is off-the-shelf software to do accounting or shop floor control or a square root routine. We may or may not want to specify the internals, settling in some cases for control of the inputs needed and the outputs expected in much the same way as we operate our car without knowing or caring much about the detailed operation of the voltage regulator. A variant on this is the use of a fourth-generation non-procedural language (see glossary reference to languages), where an input syntax generates an appropriate output such as a query, report, or input screen.

We will be dealing in this chapter with the subroutine symbol in the first case: used for leveling.

Let us look at another familiar example which shows the importance of leveling and partitioning as a method to arrive at understanding of a process (and therefore makes the subroutine symbol the most important of analytical tools): the manufacture of a car. We would not have much chance of understanding car assembly if we simply described, at the lowest level of detail, starting at the front of the car and working to the rear, how the parts were put together. It is possible to understand even simple devices and organizations only by working top-down. In the case of the car, which is not simple at all, we need to deal with the engine and clutch, the electrical system, the ignition system, the braking system, and so forth. Even then we would have to divide such subsystems as ignition down to alternator, voltage regulator, battery, starter motor, spark plugs, and so on to begin to get close to a real comprehension. When we reach the base level, such as the battery components, we finally have a chance to understand the process at the technical level. We then know we have to "add water" and "charge" certain batteries. Meanwhile we have the responsibility to relate the partitions at each level to other partitions on the same level. We must know how the alternator relates to the battery, so that we know that if the alternator is not charging, the battery will soon go dead.

Figure 3-6 shows the subroutine charted on two levels. In Figure 3-6(a) in the chart DO DISPLAYS two subroutines DISPLAY E and DISPLAY F are drawn as boxes with a line dividing the box into two parts. Above the line is the numeric address where the one-level-down detail of the subroutine can be found. Below the line is a brief description of the subroutine, consisting of a verb and a noun. This is the way we will be showing most charting above the base level: as just such a subroutine transform box. The ubiquitous IBM template shows the subroutine box as an octagonal figure, which is almost impossible to draw freehand. Since most analysts do several freehand revisions before getting around to making respectable-looking charts with a template, we will not show examples of the octagonal form here.

At addresses 2.1 and 3.1 we see the actual processing of DISPLAY E and DISPLAY F. Note that DISPLAY F calls on another subroutine one level further down: 3.3.1 FORMAT F.

In DO DISPLAYS, DISPLAY E and DISPLAY F are horizontal partitioning. The relationship between DISPLAY F and FORMAT F is an example of vertical leveling. The former calls the latter.

Let us accept that, no matter how we define a whole system, there is always a higher calling level. Let us say for the purposes of our study that DO DISPLAYS is our whole system. It calls on two subsystems, one of which calls on a sub-subsystem. DISPLAY E does not call on any further subroutines and is therefore clearly a base level. DISPLAY F does call on a subroutine, and we will have to exercise judgement to decide whether to call it a base level. If we had called FORMAT F DO SQUARE ROOT or FORMAT DATE, it would be clearly a base level, since these are "black boxes" about which we care little.

The process flow definition of base levels.   The base level is always shown as a subroutine. It may include other subroutines, but they are uninteresting. Further decomposition of the subroutine by either leveling or partitioning would be illogical (like dividing a simple sentence into two sentences) or inappropriate (like breaking up a perfectly good compound sentence).

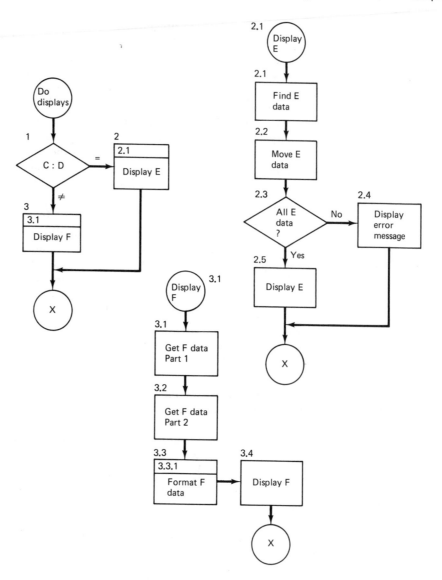

Figure 3-6 Subroutine Example.

Any subroutine always begins with a connector and ends with a connector. You always go in the first connector and exit from the second connector.

Base-level subroutines consist mostly of decision diamonds and operation boxes. Higher-level subroutines consist mostly of subroutine boxes with as few decision diamonds as possible and only rarely operations boxes, but we would hardly want to bury the operation "press missile launch button" in the middle of a base-level transform. Or would we? The next chapter will state the "structured" case for doing just that.

The connector.   The connector is used to show the entrance and exit of a process subroutine. It is also used to continue a process onto another page or view.

The line arrow.   The line arrow shows serial flow. Out of a diamond it shows the binary decision.

The branching table.   The last symbol in our tool kit for tracking process is the branching table, which is a large rectangle showing rows of branching decisions under the column headers IF. . .GO TO. This replaces a series of decision diamonds. *Branching* refers to a departure to another place from a serial flow. This gives us a one-arrow-in, many-implied-arrows-out capability. One decision diamond gives us only one in, two out.

Figure 3-7 shows an example of an operation specification using all the flowcharting tools we have covered, including the branching table. Note that we can chart this process quite easily with flow charts (although we have not as yet shown the data being processed). If this processing is within the domain of what we intend to automate, we can easily convert this system flow chart into a solution-oriented design flow chart, ready to program into computer software. We do not change the charting method except to become more precise.

INCORPORATING PROCESS FLOW CHARTS INTO A COMPLETE METHODOLOGY

We now have covered all the process symbols needed to do a system specification (and a system design and implementation coding or specifying after bounding off the operations to be automated). But in order to proceed, we need more information than process flow. We need to configure around the process flow the input-data flow, the output-data flow, the triggers, and the resources, including the stored data resource in the form of files and databases.

This will be our *core* specification. Other system resources that we will need to specify are the network communications support and other aspects of the physical system, including equipment and facilities, financial support, and personnel. We will be using the extended data dictionary to relate the entire configuration, not just to define the data.

Elements of the Core Specification

The trigger.   If we are specifying an operation such as 8.5 in Figure 3-7, where some event sets off a process (something is done at 4:00 P.M. daily), we have a *trigger*. Triggers are significant catalysts that must be incorporated into the specification.

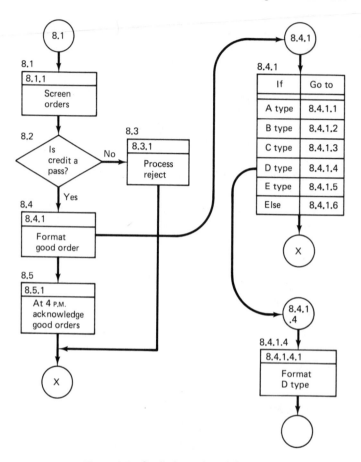

Figure 3-7   Credit Screening of Orders.

The input and output data.   This is the transactional information; it includes, in addition to digital transactions, video, message, and voice transactions.

Transactions are data-in-motion through whatever physical carrier. To the extent that this data-in-motion requires processing, it forces the system to assemble a series of transforms to do the job. For instance, order transactions come into the system and demand processing through an accept-reject process transform. Accepted and rejected orders are outputs of that process and become inputs to another process.

The process.   This is our flow chart detail that transforms the inputs into the outputs. It is our transaction.

The data storage resource.   In the case of the credit screen processing shown in Figure 3-7 there is probably a data store for past customer history and for credit rules. In the case of simple logical record files the logical record must be identified with the process. As you go down levels, the data is identified more precisely in the same way as the process flow becomes more precise, as in the example of preparing a dinner for guests. The same is true of a data base resource, except that in this case the database involves several segment entities in a hierarchical relationship—that is, parent segments with multiple children segments.

In either case, once you get down below the highest charting levels it is difficult and sometimes impossible to show the data resource in and out of each transform partition. That is just too much freight for the line arrow to carry in a real systems analysis solution, and too difficult to arrange through a data dictionary.

Our approach to that, in this method, will be to show the data in sufficiency for the set of transforms at any level but not in-line. This comes down in most cases to saying that we will show the data once per page for the process flow shown on that page.

In the instance of the database as resource, the application view of the database is shown—not the physical database, which may involve parts of several physical databases. Figure 3-8 shows one application view of a set of conceptualized physical databases. In Figure 3-8 we have three relations or segments (3, 4, 5) as the applications view of the data store. For a given set of processes (an operation) the three shaded data elements and their keys may be all we need to show. At a higher level we may need to show only the name descriptions of the segments or some one phrase alias for that. Whatever we need to show in the way of stored data to understand the process we do show. For those smart organizations that are maintaining a data dictionary it is an axiom that what is named on the chart is available for access on the data dictionary, so that if we use a summarized alias on the chart, that alias when accessed on the dictionary will point to the segments and

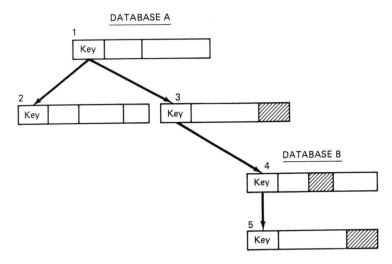

Figure 3-8 Physical vs. Application Databases.

data elements involved. (Instead of "point to," however, we should use the phrase "has relationship to" in dictionary parlance.)

It must be clear that what we are discussing here concerning transactions and data stores does not refer only to information that has been machined into a computer. It also refers to that phone call which enters a transform and that list of vendor phone numbers on a well-worn sheet of paper in a desk drawer (with a less-than-two-second access time). Concerning voice transactions, the voice message stored in a public database waiting for the recipient to dial the voice mailbox and get the message is certainly a data store and certainly the concern of the systems analyst specifying a best system.

Resources other than data stores.   Data is not the only resource we need to incorporate in our charts on process flow. It is important to remember that we are not specifying computer systems; we are systems analysts, and therefore we specify whole systems, unless some unwise manager shears off our wings. Computer systems are only part of the organization's task flow. Here are some of the more important resources other than data:

The Human Resource.   There are people at many points in the process flow: People preparing input and handling output. People interpreting information. People overviewing a dynamic computer system. People using information from a passive computer system. People interacting on-line with an automated system—asking questions, answering questions from the software, modifying the system. People working with minimum need for computer support (such as finalizing a deal or acquiring a major talent or painting a portrait which will become a book illustration). Social organizations without humans as the primary resource would actually be devices rather than natural systems.

Financial Resources.   Many systems can be expressed to some degree in financial terms such as labor, machine, facilities, and materials costs or investment and return on investment. To operate a system involves ongoing costs, which are subject to accountability. Changing a system involves a development cost, which also must be accounted for. A new system will have certain forecasted operating costs, which must also be accounted for. Senior managers do nothing more important than make the best financial investments possible. Our job is to accurately state the business case. This type of microeconomics relates all social values to the money index. It works for the system but sometimes results in grave consequences beyond the system.

Equipment, Facilities, and Material Resources.   The workplace, the big machines and little pencils, and the raw materials that go into the products (in many cases nothing but information) make up the rest of the resources that may be required in order to specify a system. Some of these resources that we probably will not be asked to include are such things as parking lots, rest rooms and lounges, and dining facilities, and beyond these such resources as a healthy pleasant environment, a safe workplace, reasonable access to the workplace from the homeplace, and the other amenities of a reasonable civilization which certainly determine the future course of any given organization subsystem.

Some of these considerations may be outside the scope of our analysis, but the resources which must always be part of a systems analysis are:

- data stores
- people
- finances
- equipment, facilities, and materials

THE FULL METHODOLOGY FOR SPECIFYING SYSTEMS BY PROCESS FLOW

Finally we are prepared to deal with a full definition of this methodology. Starting as high up in the scheme of things as we dare to think (see footnote 4 in Chapter 1 for B. Fuller's definition of "as high up"), we work our way down through suprasystems through a process of inclusion-exclusion to the level where we know we should begin studying. When we have included the relevant and excluded the rest at this level, this becomes our whole system. Having dropped vertically to our proper level and drawn our circle of horizontal inclusions, we now declare this to be the domain of our analysis.

This process is conceptualized in Figure 3-9 and made explicit as an example in Figure 3-10. Figure 3-9 shows graphically how the charting will follow our vertical-leveling, horizontal-partitioning concept of how to decompose a system for the purpose of synthesizing a new system which might include more automated machine functions.

The example in Figure 3-10 concerns a major league baseball team. We decide to start as high as we dare think, which in this case is the suprasystem: sports teams. Maybe, our reasoning goes, the stadium facility with its operating costs could be shared with another sports team. We decide to work our way down several levels to the level: a major league baseball team. At level 1.n we partition sports teams into 1.1 amateur and 1.2 professional. At level 1.2.n we partition professional sports teams into component sports. Selecting professional baseball, we level and partition down in a similar fashion through leagues to divisions. We now level down to one selected team in one division, having satisfied ourselves along the way that for the purposes of our study we have nothing to pick up from a higher level. We are at 1.2.4.1.2.1.1, which defines the domain of a particular partition at a particular level.

This being our whole system, we reset our level-partition numbering system to 1 and begin the same process just described but in full detail in order to specify our whole system—in order to get the change requirements—in order to specify

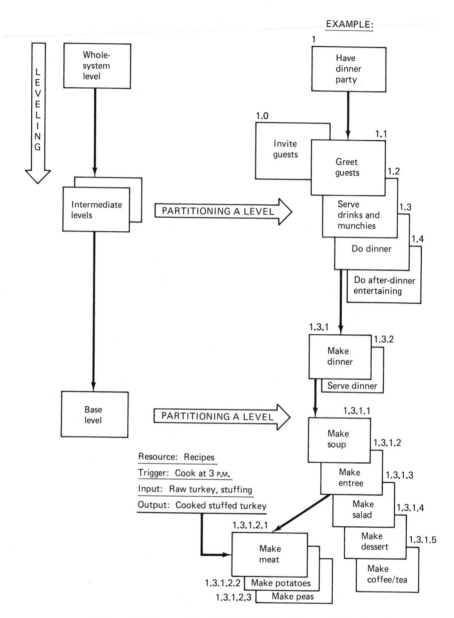

Figure 3-9   Process Charting Methodology Conceptualized.

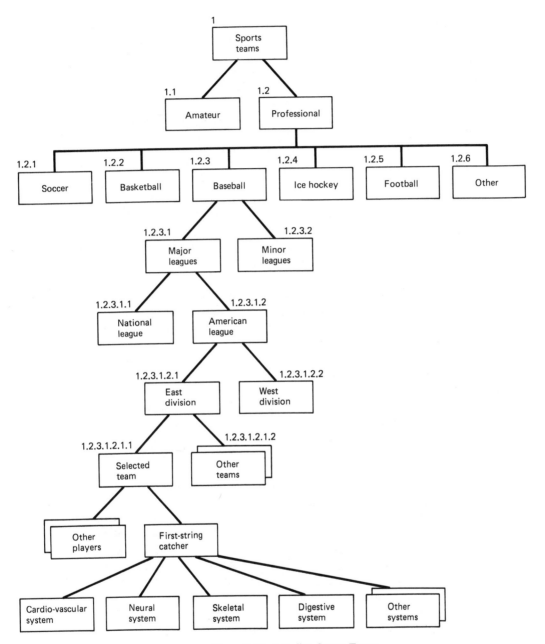

Figure 3-10   Leveling Sports Teams.

the new system after the requirements have been applied. That is all there is to it (the catch is all the new computer parts, perhaps numbering in the many thousands, having to work at the zero-defect level—one loose screw and we could have a most unpleasant crunch).

Notice also in Figure 3-10 that we have decomposed one major league baseball team three further levels down: play area (such as infield, outfield) and, within play area, position (such as second base), and, within second baseman, the major digestive subsystems of the human organism. The final level is blatantly absurd, taking us out of baseball and into the first year of medical school. Infield play and second baseman also are wrong, since our goal is to study a whole major league baseball team. At each of these levels we are presented with a completely different set of processes, as we shall see in Figure 3-11.

When we actually do the applied work of systems analysis we will be reading documents (in a new way—decomposing as we read each document) and interviewing the operators and users of the system. Regarding our baseball team, we will be interviewing the customers and TV watchers as well as the management and the players and the support people. We are always interested in the same thing: the process which we will chart according to our method and the surrounding information about inputs, outputs, triggers, and resources. This is what we are writing down on our yellow sheets of ruled paper as we talk to the people. As we chart the process, these are the things we will surround our process charts with: the data-in-motion, the stored data, the little whistles and time-bombs that start processes, and the rest of the resource information as needed. This total is what we present as our core system specification.

We do this iteratively: first the big picture at 100 miles up, then at 40,000 feet, and so on down to the base-level view, which may be eyeball to eyeball.

At each level the information as detailed above is appropriate to the level of the process we are dealing with. At all levels the naming conventions from broad to precise are all documented, and *relationships are defined* in the data dictionary. This means that if we describe a data store only as "transportation costs" at a high level, we will be able to go to the data dictionary and find not only a definition of transportation costs but also all the relationships implied. What precise data elements are *owned* by transportation costs? What is the relationship between spring training costs and transportation costs? Does spring training costs *own* transportation costs—but only from February to April? Is this relationship exclusive or shared with another owner? If the relationship is shared, then the person who is interested only in spring training costs had better not start changing the data definitions regarding transportation costs. This is even more serious if the data is nonredundant shared data.

Case Study: Major League Baseball Team

Figure 3-11 is an example to tie together the discussion in this chapter regarding process flow charting methodology. Figure 3-11(a) sets the context at the top level of the playing cycle for major league baseball (*Note:* The author has never done

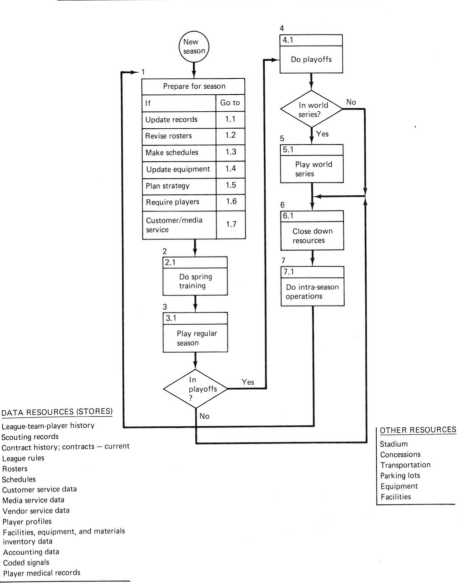

(a)

Figure 3-11   Baseball Team System Specifications.

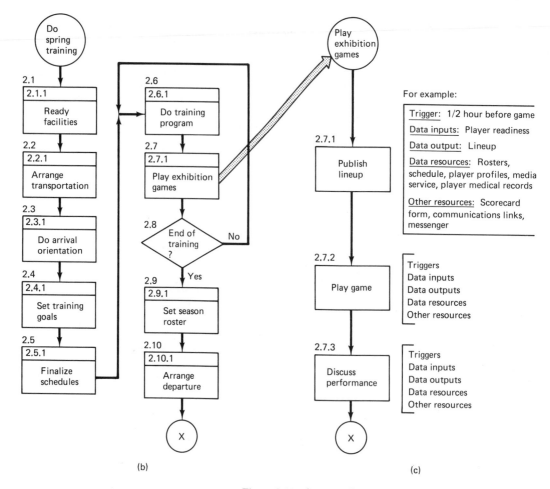

Figure 3-11 (*continued*)

systems analysis for a major league baseball team but wanted to pick a process which might be easily understood and even have some pleasurable associations for a large number of readers; if you dislike big league baseball, please accept the author's apology for this choice.)

Observe, looking over Figure 3-11(a), how the various elements of analysis that we have discussed are arranged around the process charts. At this high level we have partitioned "play major league baseball" into the following subroutines:

- prepare for season
- perform spring training
- perform regular season
- do playoffs

- play World Series
- close down resources
- execute intraseason operations

These are the processes. Also shown in association are the data flow, such obvious triggers as today's game time, and such data-at-rest stores as rules, rosters, schedules, and league-team-player history. These data stores must become more specific as we descend levels.

In Figure 3-11(b) we decompose subroutine "perform spring training" to partitions such as "arrival orientation" and "play exhibition games." When we descend one level in Figure 3-11(c) and decompose "play exhibition games" into such partitions as "publish game roster," we are at the base level and can ask ourselves such questions as: What data is needed to publish the game roster and who needs to see it? And how quickly do we need to send the roster? Should we print it on a micro in the dugout or transmit over a local area network? In the current system the manager writes it out with a pencil and the batboy runs around with it to the sports reporters, the TV and radio announcers, the umpires, and the person who keys it up on the big board. The trouble is, anyone not in the ball park is out of luck.

Note that each of these partitioned boxes in Figure 3-11 describes a process in terms of a cogent action verb acting on a noun. This form should be observed diligently.

Further comments on the baseball team case study.   Before going on, we should pause a moment to marvel at the extent to which information systems are involved in the game of baseball as played in the majors. Particularly extensive is the need for stored data as shown in Figure 3-11(a). Information control seems to be the difference between a tax writeoff and a profitable enterprise. It appears to also be the difference between winning and losing ball games.

Such data as customer service and media service could mean happy season ticket holders and a well supplied media. Information regarding other teams is vital.

Now that we have defined a current system using a charting method, we can use the same approach to specifying the new system. Something—the requirements analysis—must come in between. We have already begun some speculations about requirements changes during the current system study. Now let us continue these speculations in a more formal way.

Let us say, using the major league baseball team case, that we found a decision support system was in place using a fourth-generation language and database management system. At the conclusion of last season the management of the team discussed improving the information/process along the following lines:

- a programmed automated pitching machine—a ball-throwing machine under computer control that can throw to a batter's weaknesses and also simulate particular pitching styles; the machine must be controllable from a micro in the dugout.

- an airline and bus scheduler to track team travel in spring training and on the road
- detail information on exhibition-game revenue
- on-line facilities and equipment inventory system
- owners-players contract database
- season-ticket-holders database expansion to include demographics, year of enrollment, seat preference, complaint history and response to other promotions

New system specification.   New requirements in hand, we proceed to rewrite those portions of the current system affected. We have a real deadline, since the system must be revised and ready to go in April. We find that the first requirement, an automated ball-throwing machine that can train batters and get them out of slumps, is so state-of-the-art, so costly to develop, as to be hardly justifiable. However, just finding out that it may be feasible has so excited the field manager that he is willing to fight for it. Much the same is true about expanding the ticket-holders database: while not innovative, it is costly and may not be justifiable.

Under these circumstances the analyst charts and specifies the most costly and complete of the new specifications, giving everybody everything they want. Then the analyst does two variations: sans ball-thrower and sans ball-thrower and ticket-holders database expansion. Now the general manager and the owners have three alternatives.

To make the new specification we will cut open the current system specification and, using exactly the same methodology, chart the new specification. Of course the requirements analysis is just touched on here, and it would be much more detailed than what has been mentioned above. What equipment is inventoried? What are the details of travel schedules? How are we to arrive at all the season-ticket-holder information requested?

A NAME FOR OUR CHARTING METHOD

The charting method presented in this chapter traces back to the early days of government and commercial computing in the mid-1950s. At that time two charting systems emerged as contenders. One, associated with Univac, was a monolevel system of charting, where a transaction was carved out of a flat plain using a method known as setting variable connectors. These flow charts looked like present-day global network topology flows, where the variable connector was like a node—or a switch on a train track—always set in one direction until programmed in another direction. The other method, associated with IBM, was based on top-down sub-routining through levels and across transforms.

Sometimes the author wonders whether IBM's eventual dominance in computing wasn't somehow involved with this analysis methodology. As more complex systems went into development, it became clear that the variable-connector charting approach could not handle the complexity.

In 1963 a full-fledged methodology was developed at IBM called the Study Organization Plan (SOP). This was the brainchild of three brilliant IBM analysts: Thomas Glans, Burton Grad, and David Holstein. This plan, which was on the wordy side in terms of the delivered specification document and short on good charting technique, was the father and mother of both Business Systems Planning (BSP) and Hierarchical Input Process Output (HIPO) charting. The IBM manuals for SOP,[2] although out of print, are given as a reference out of respect for the past.

SOP, or rather the children of SOP, underwent criticism in the late seventies and early eighties from a remarkable new group of theorists and business educators, who, working backward from a vision of standardized COBOL and forward from data transaction analysis, mixed together an eclectic Granola of techniques and hygienes into what became famous as the "structured" method. Structured actually has no hard meaning (although it is far more meaningful than "SOP"!).

Those innovators who applied the structured approach to the practice of systems analysis and design[3] made a major contribution. Their charting system is discussed in the next chapter as the other hard systems methodology credible enough to be presented in this book.

Unfortunately, in their zeal to advance their systems theories, some in the structured group took an adversary position in regard to the older existing method based on SOP, which they called the *classical method*. What is unfortunate is that they distorted the nature of the method they hoped to replace, virtually ending flowcharting practice. They convinced data processing managers that the "classical" method was machine- and solution-dependent, perhaps because IBM stamped onto the same template as the basic charting symbols just covered other symbols referring to terminals, storage devices, and printers.

Some analysts who were not in control of their technique did take a wrong approach to system specification, particularly in regard to specifying solutions prematurely. But certainly, during the "pre-structured" years, there were many examples of top-down logical system specification.

A more relevant charge was that the classical method was wordy and therefore

---

[2] IBM Study Organization Plan, SF20-8135, 8136, 8137, 8138.

[3] For instance: (a) Victor Weinberg, *Structured Analysis* (New York: Yourdon Press, 1978); (b) Tom DeMarco, *Structured Analysis and System Specification* (New York: Yourdon Press, 1978); (c) Chris Gane and Trish Sarson, *Structured Systems Analysis: Tools and Techniques* (Englewood Cliffs, N.J.: Prentice-Hall, Inc., 1979).

imprecise. There are two approaches to handling a problem like this, where a basically sound approach with a good charting method is flawed. One, you improve the method and reduce the impact of the flaws; two, you throw away the method and try something else. Those who live in glass houses should not throw stones. Structured people use the phrase "Victorian novel" to describe the wordiness of the classical method. Those who have attempted to wade through "structured English," the COBOLese answer to literacy, may be forgiven if they long for the pure water of *David Copperfield*.[4]

Let us accept the word "classical" as a good label for the flowcharting method. That being the case, the author proposes that the method put forth in this chapter based on the classical tradition be called the "neoclassical" method. What, after all, is a revived method without a neoname?

WHICH IS BETTER—NEOCLASSICAL OR STRUCTURED?

To conclude this chapter we will list the advantages of the neoclassical method. At the end of the next chapter we will list the advantages, as we see them, of the structured method. The smart analyst will carry more than one tool in the toolbox and will learn which tool is proper for the task at hand. The chapter exercises will facilitate arriving at a feeling for this selection process.

Advantages Gained Using the Neoclassical Charting Method of Tracking Process with Augmented Flow Charts

1. The same charting tools can be used to show analysis, design, and implementation.

2. The same charting tools can be used to describe manual processes and computerized processes. This is important, because more and more we are specifying not just manual systems but a combination of programs and manual operations.

3. Process and information flows can be mixed using the same charting approach. "Process" in this case means such things as the movement of materials in a warehouse-manufacturing operation or flows of liquid in a refinery operation.

4. It is easier to follow vertical leveling through subroutines in flowcharting than to follow leveling in other methods. In a flowchart the appearance of the subroutine box tips you off that this particular transform has a lower level.

5. With the flowcharting approach it is simple to mix in decision diamonds and operations boxes at any level. Who is going to guarantee that there will not be some high level decision that must be shown? The data-flow-diagram technique does not allow for this, as we will see next chapter.

[4] Charles Dickens' famous novel *David Copperfield*.

6. Flowcharting as a universal all-purpose specification, design, and (with extenders) physical system description tool can be easily automated into very satisfactory graphics, so that such actions as changing a third generation language (COBOL) program can automatically reproduce new flow chart documentation. This can also be used by the analyst as a productivity aid.

Structured English as a method to show process is limited and strictly a linear approach. It is related best to a single programming language, COBOL. We have in the United States a multibillion-dollar investment in COBOL, and COBOL will not go away during the next two decades. But COBOL, at the third-generation level, often is not the language of first choice anymore.

For new system specification we need a flexible charting tool for use with fourth-generation nonprocedural languages, such as RAMIS II,[5] and fifth generation languages, such as LISP,[6] which through artificial intelligence (AI) may handle situations where the users simply cannot lead you to an up-front stable specification of the transactions that drive the system; they can give only relationships and parameters. More of the problems we will be facing in the future will require such an approach—problems such as providing really good health care and government services. In both of these cases we are still just beginning to learn something about what we should be doing—so how could we give a complete specification?

Above and beyond building new systems through languages is the increasing use of prebuilt software, which we will be hooking onto our basic organizational information system as customized black boxes. Flow chart subroutines will handle this problem just fine.

7. Finally, it is clumsy to show the data flow along the line arrows that move between partitions. This works out all right for a small example, but when it comes to the real thing there just isn't enough bandwidth there to handle all the communication needed. This is doubly true when specifying design flow.

[5] RAMIS II Data-Base Management System and Language. A product of Mathematica Products, a subsidiary of Martin-Marietta Corp.

[6] LISP—a language constantly being improved, based on tables or lists of relationships; suitable for self-modifying artificial intelligence systems development. See Laurent Siklossy, *Let's Talk Lisp* (Englewood Cliffs, N.J.: Prentice-Hall, Inc., 1976).

4 | CHARTING DATA FLOW

Having made the best case we know how for the charting of process flow, we will now change hats and do our utmost to present the best case for data flow charting—or, as it is commonly called, charting with data flow diagrams (DFD). This is a so-called "structured" methodology.

Outside of the charting technique itself and differences regarding the tracking of data rather than the tracking of process, the major theories about doing systems analysis which have concerned us in Chapters 1–3 remain the ground upon which the discussion in this chapter is based. We are still discussing system specification by top-down decomposition through leveling and partitioning. We are still dealing with the relationship between the system specification and the data storage technique. We are still concerned with knowing the current system and knowing the new requirements and changing the current system into the new system. We are still dealing with the same checkpoints regarding the system development life cycle.

THE RATIONALE FOR TRACKING DATA TO ARRIVE AT PROCESS

The average organization has many fewer transactions than it has processes. One important transaction will go through many operations, each operation containing a series of processes, before the transaction is resolved. The perspective of the business from the point of view of the transaction is often more comprehensive than that from the operating departments that process the transaction.

With regard to difficulties in handling software packages and fourth-generation language systems, this is solution dependence rearing its head again. Not until design time should we concern ourselves with software solutions—not even if the current system is entirely written in fourth-generation and/or packages. Do we or do we not want to know our business regardless of past implementations?

The problem of the users' inability to supply up-front requirements, because the requirements simply cannot be known in advance, remains perplexing no matter what method is used. In this case—which is the case for the knowledge-based, expert system—requirements are both evolutionary and interactive. The expert system that performs diagnosis in internal medicine better than any one doctor and is still growing in power through each use is, after all, still part of some health-care system that can be charted.

The purpose, then, of tracking vital chunks of data is to arrive at the many processes at the base level that transform this data from inputs to outputs. Only at this level (as we have demonstrated in earlier chapters) can we, by describing the processes in each operating bubble (or box), map the process logic onto the computer software, or change the process through any other means, with full understanding of what is involved.

CHARTING TOOLS USED FOR DATA FLOW DIAGRAMS

The *data flow diagram (DFD)* is the core specification in this method. Figure 4-1 shows the very few charting forms that are necessary.

Shown in Figure 4-1(a) as a square box is the external source or receiver of data. This might, for instance, be the customer who sends in orders and receives bills.

Shown in Figure 4-1(b) is the *transform bubble.* Two variations of this have been put forth. The circle or real bubble is the better-known and is used by both Victor Weinberg[1] and Tom DeMarco.[2] The rectangular bubble shown beneath the circle is the form used by Chris Gane and Trish Sarson.[3] The reason for the rectangular bubble is the perceived need to enter more information than can be contained in the bubble.[4] Tom DeMarco, who is the purist in this group of four writers and educators in the structured method, holds that the data content of the bubble must be just the bare bones of one verb and a noun, since our objective is

[1] Victor Weinberg, *Structured Analysis* (New York: Yourdon Inc., 1978), p. 45.

[2] Tom Demarco, *Structured Analysis and System Specification*, rev. ed (New York: Yourdon Inc., 1978), pp. 39, 97.

[3] Chris Gane and Trish Sarson, *Structured Systems Analysis* (Englewood Cliffs, N.J.: Prentice-Hall, Inc., 1979), p. 27.

[4] Gane and Sarson, *Structured Systems Analysis*, p. 32.

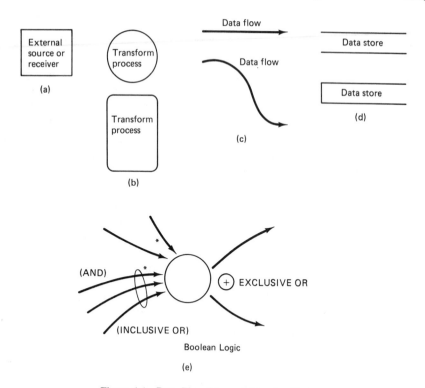

Figure 4-1   Data Flow Diagram Charting Forms.

not to explain the process but to partition into leveled transforms.[5] The author suggests that the rectangular bubble is hard to draw freehand and the template containing this special form is not one you are likely to have around the shop, it must be specially ordered.

The line arrow is much more important in this method because it carries the data flow: the data into the transform and the data out of the transform. All lines must be identified by their data. Figure 4-1(c) shows the line arrows. The variations on the use of the line arrow are significant, because again they answer needs that different proponents of this method have perceived. Three different points of view are advanced by the practitioners mentioned above regarding line arrows.

1. When multiple lines go into or leave a transform, Weinberg offers the ability to use Boolean logic describing symbols to represent AND, INCLUSIVE OR, and EXCLUSIVE OR.[6] This clearly indicates a felt need for decision logic above the base level. DeMarco advises against using Boolean decision

[5] DeMarco, *Structured Analysis and System Specification*, p. 67.

[6] Victon Weinberg, *Structured Analysis* (New York: Yourdon Inc., 1978), p. 57.

logic.[7] Our examples will not use Boolean logic beyond showing examples in Figure 4-1, since this seems to the author to represent a consensus view of those who use the DFD approach.

2. DeMarco shows the line arrow as a curved line giving a different "feeling" than the straight lines and right angles of Weinberg and Gane–Sarson. In the Demarco approach there is more of a sense of flow—a sense of data in motion. Our examples will use the curved line.

3. Regarding the data flow along the line arrow, this would be a serious problem, like the problem of expressing the process within the confines of the bubble. The answer here is to take advantage of the fact that the method allows multiple lines in and out of a bubble and to break up the wordy data flow into several briefly named data flows. It is either that or lengthen the line.

This concentration on detail of form, worrying about whether to use a circle or a square or a curved or straight line, may read as petty to the nonchartist. To those who mean to use this method to specify systems it is just as serious a matter as the concern of the professional tennis player with the type of racket to be used in a tournament. What we are trying to do with these forms is to invent solutions to problems as we move from the fluid to the concrete and from the tentative to the certain. We need a method for all seasons, but especially to communicate the fluid and the tentative new idea.

The final diagramming element shown in Figure 4-1 is the open rectangle or two parallel lines, which indicates the data store (such as a database, file, Kardex, or phone book). Gane and Sarson, unlike the others, show the data store as the two parallel lines joined at one side to make an open rectangle. We will use their approach.

These are all the charting forms we need to use this methodology. Again, as in the flowcharting forms of Chapter 3, much can be developed out of a few tools. Essentially a system of any complexity whatever is shown with the bubble, line, data store rectangle, and external box.

Figure 4-2 shows an abstract example of the way these forms relate to each other in a one-level data flow diagram. Four external sources-receivers are represented by the square boxes. On the lower path, data flow B_1 is transformed into B_2 and then meets with data flow A_2 and data flow C_1 out of data store-2, and so on. Note how data stores are shown in three ways in this example:

1. data flowing one way out of the store (1.2)
2. data flowing to and from the store (1.3)
3. A store between two transforms (1.4–1.6), implying a time delay.

(If there is a time delay, there must be some trigger to start the flow going again.

---

[7] DeMarco, *Structured Analysis and System Specification*, p. 61.

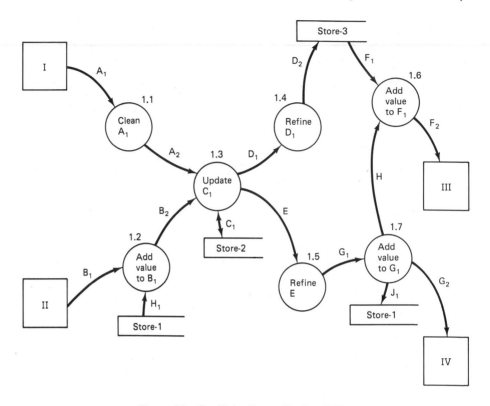

Figure 4-2   Combining Forms Use in a DFD.

In this methodology we will have to dig into the base-level process logic of 1.6 to find the trigger.)

CASE EXAMPLE: USE OF THE DATA FLOW DIAGRAM

Let us now turn to a more realistic example of data flow diagramming to show the methodology. Consider a medium-size bakery operation. This bakery makes the following products:

1. white bread in several sizes
2. whole wheat bread
3. rye bread in two sizes
4. assorted types of rolls
5. assorted types of cakes
6. doughnuts
7. breadcrumbs, stuffing, and croutons (from stale returns)

The bakery has four types of customers:

1. supermarkets and other retail stores
2. airlines (demanding fresh high-quality baked goods)
3. schools
4. other institutional accounts, such as hospitals and nursing homes

The school orders are taken over the telephone each afternoon from 3:00 P.M. until 4:00 P.M. All the other orders are taken in person by the route salespeople, who are the contact with the customer. There are many salespeople, each with a delivery truck. They pick up today's baked goods at the plant by 3:00 A.M., make deliveries, and return to the plant the next afternoon. As they deliver today's order, they take the written order for tomorrow on a form called the route order. It is in effect a bill of materials for finished goods to be baked overnight and loaded on the trucks by 3:00 A.M. In order to accomplish this baking there must be a material requirements plan which transforms the forty bills of materials plus the phoned-in school orders into the nightly raw material and baking requirements. A raw materials inventory must be maintained, in which no shortages are allowed and ability to store perishable items is limited. The raw materials include:

1. white flour
2. whole wheat flour
3. cake flour
4. granulated sugar
5. confectioners' sugar
6. waxed paper
7. honey
8. salt
9. milk
10. yeast
11. vitamins
12. flavors
13. sourdough
14. plastic bags
15. plastic bag ties
16. and so on

One main transaction drives the whole system: the route order. In addition there are return-of-stale and receipt-of-payment or credits transactions. Other transactions in the system include a raw materials purchase order, a raw materials receiving slip, and puts and takes to and from stock. There is also the bake order which is the sum of the route orders expressed in terms of finished goods rather than customers on routes. Finally, there are transactions to accounting and transportation (for truck servicing), which we have arbitrarily not bounded in the system.

There are a number of high-activity data stores in this fast-moving business, where each night is a major production effort. Here are some of them:

1. raw materials inventory
2. stale statistics
3. transaction copy files
4. sales performance statistics
5. monies received sales record
6. pending customer orders
7. ring up tapes
8. standards and methods
9. long-range materials planning
10. daily requirements
11. recipes file
12. route-customer requirements
13. customer file

The process flow observed by tracking the data is:

1. The route salespeople deliver the finished product to the customer and pick up the next day's order—except for school orders, which the bakery order clerk takes on the telephone.
2. The route salespeople come in one at a time and turn over to the order manager the money received, the stale counts logged in, and the new route orders.
3. The route orders combined are converted to the bake order by product.
4. Additional raw materials are acquired if needed (raw materials are ordered in bulk from sales forecasts).
5. The baking production is done to the bake order requirements.
6. The bake order is converted into the next-day set of customer orders by route.
7. The route trucks, having been serviced, are loaded according to the route-customer orders.
8. The route salesperson checks the order and goes out to start the process all over again.

Figure 4-3 shows the data flow diagram at the whole-system level, decomposing RUN BAKERY into ten partitions.

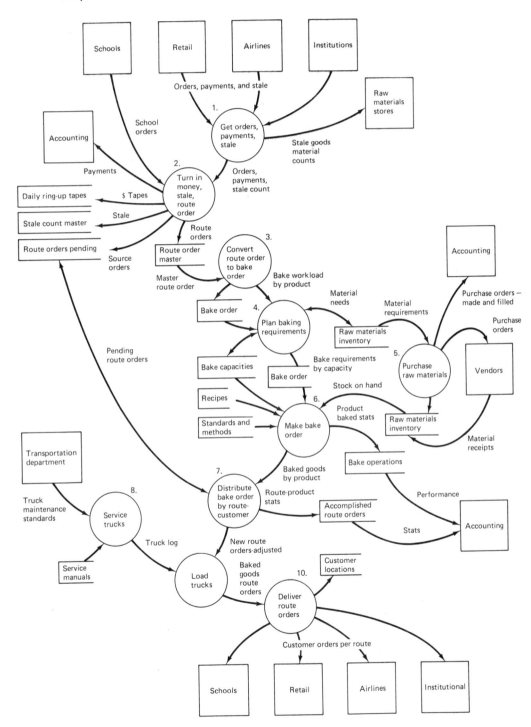

Figure 4-3 RUN BAKERY Data Flow Diagram.

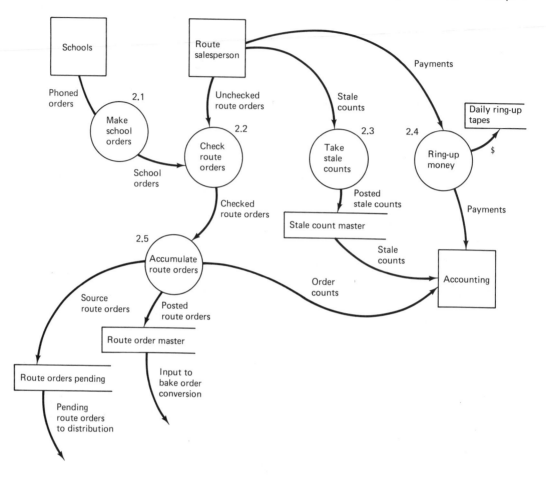

Figure 4-4   Turn in Money, Stale, Route Order.

Figure 4-4 shows the second partition, TURN IN MONEY, STALE, ROUTE ORDER down-leveled to the next level, which looks like a suitable base level.

Figure 4-5 shows the sixth partition, MAKE BAKE ORDER, down-leveled and partitioned to what may be a suitable base level.

Notice how the information from this bakery operation can be picked up from the DFD at a glance. The relationships between the data-in-motion, the transforming processes, and the data-at-rest can be observed as in a picture. Compare this to the written information that preceded it. Compare it in terms of the leveling and partitioning decomposition that is at the heart of understanding this system.

Is MAKE BAKE ORDER really at the base level? In terms of building an information system it is, but if our requirements were process automation, we would want to decompose further. For instance, partition 6.3 in Figure 4-5 might

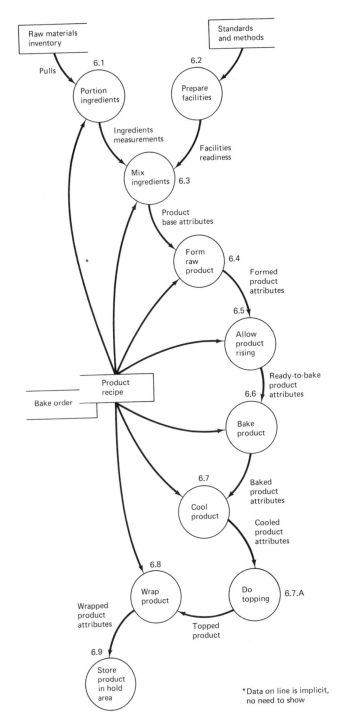

*Data on line is implicit,
no need to show

Figure 4-5 Make Bake Order.

then be decomposed from MIX INGREDIENTS to PORTION FLOUR, POR-
TION YEAST, PORTION WATER, PORTION FAT SUBSTANCE, PORTION
MILK, and so on. That would be our third level, which assumes that the mixing
can all be done in one step.

Even in this somewhat simplified diagramming of a bakery system we can
see that this method is both enhanced and made difficult by carrying so much
information about flow data and store data in the transform flow. For those who
for this reason opt for the square bubble we give a reference regarding obtaining
the Gane–Sarson template[8] with its maxi-rectangular bubble. The other option is
to break up each wordy bubble into several not-wordy bubbles. Sometimes this is
helpful, but it may just overcomplicate the flow.

COMPLETING THE STRUCTURED SYSTEM SPECIFICATION

Having covered the substance of data flow diagramming, we are still not ready to
do a current system specification or a new system specification—any more than
we were in the last chapter after we had learned process flow charting. Several
important additional deliverables must be added to the data flow diagrams to
complete the system specification package.

1. *Specification of the system data.* Chapters 6 through 12 cover this subject in
   detail. This involves the tie to the data dictionary or whatever data description
   method is in place.

   Any organization with a significant investment in information should include
   a data dictionary in their system planning.

2. *Specification of the physical system and terminal dialogue.* This is covered
   primarily in Chapter 8 and includes the relationship among people, terminals,
   computers, storage devices, and data transmission systems.

3. *Specification of the process logic at the base level.* Except for tracking the
   decomposition of the system through levels and partitions, we have really
   shown very little of how the system works! *All* the process logic of the system
   is inside the bubbles, and we have been able to describe process only as a
   single verb acting on a single noun. In the bakery problem are we really ready
   to improve the operation of the bakery, perhaps through the extensive use
   of computers, based on what we know from the data flow diagrams? No, we
   are not ready, but yes, we would be ready if we had a detailed process flow
   covering all decisions and operations within all base-level bubbles.

[8] DATAFLOW/STRUCTURE CHART TEMPLATE. Improved System Technologies, 888 Sev-
enth Ave., New York, N.Y. 10019, 212-586-1098.

Specifying System Data

Bypassing until Chapter 6 the differences among words such as information, data, message, transaction, communication, and knowledge, we will for now chunk all this overlapping semantics into the word DATA. We have discussed two kinds of data: data-at-rest in stores (which we have shown, in our DFDs, as an open-ended rectangle) and data-in-motion (which we have shown flowing along the line arrows between bubbles).

In both these cases we can further subdivide data into that which is defined to the computer system and that which is not. For instance, a physical database containing segments made of data elements, or a file containing records made of data fields, has been defined to the computer system. In this case the data names shown on the DFDs must be linked to the already defined physical data. Either the same name must appear in both places or the *alias* appearing on the DFD must be linked via a dictionary to the program data name—for instance, WHITE-FLOUR-INVENTORY-IN-POUNDS as the alias of BAK-WF-LB-CNT.

That data which does not yet exist on the physical databases or files of the computer system but which is in the process of data analysis for automation is best defined to a data dictionary as a different category of data. This will be data in the logical modeling pipeline, waiting to be synthesized into the physical database or file. This data is certainly shown on the DFD.

That data which is not involved with processes that are to be automated but which is to stay in the "manual" part of the system (such as LIST-OF-CHRONIC-COMPLAINERS) should also be linked to the data dictionary. Some organizations don't keep noncomputer data in the dictionary. This omission cannot be afforded. Again, we pay dearly when we do not take a whole-system approach to our resources. Confusion over data is a major cause of system malfunction. Sometimes the real benefit from automation has been not the increased speed of the system but rather the improved data definition.

Decision support systems are based almost entirely on the capture and retrieval of well-defined data—as opposed to operational systems, which are heavily process oriented.

The other main category of data is that which is not input to a process but rather describes the process. This other category also includes data about data, which is called ATTRIBUTES and RELATIONSHIPS. Here are some examples:

1. data describing process: COOLING-TIME
2. data *attributes* of WHITE-BREAD: WEIGHT; CALORIES
3. data *relationships* between WHITE-BREAD and MILK
   - white-bread <u>CONTAINS</u> milk
   - milk <u>IS-INCLUDED-IN</u> white-bread

This last category of information, which relates indirectly to the data shown on the data flow diagram, is necessary for a full understanding of the specification. We have to know such facts as the loaf-weight to understand the process.

We can sum up by saying that the system specification, whether current or future, is not complete unless the data requirement is linked successfully to the data flow diagrams. When we make reference to the data requirement, we are dealing in the broadest of terms concerning the meaning and scope of data.

Base-Level Process Specification

What we have been doing, while decomposing whole systems through various levels, is deferring the process logic until the decomposition has been completed and we are left with dozens or hundreds of base level process chunks. It has been essential in this structured specification not to show the process until now. What we now have is a network map of the system showing these base levels in proper relationship to each other. This has been done with data flow diagrams.

A new charting methodology.   Now we must go to a new charting methodology to show the base-level process logic. What is proposed is that we go to *structured English*. Structured English is nothing else but modules of go-to-less COBOL using *nested-if* logic. This is a technique used in structured programming. The reason given for its use is the precision and lack of ambiguity believed to result only from this method. This structured English is to be supported by decision tables and decision trees to show complex decision logic that is beyond the reach of the nested-if. The main teachers of this method do advise making the effort to dress it up so that it appears to be just a tightly written outline not tied into COBOL. This approach seldom works in the field. Most data names are written in pure COBOL. Some analysts may not have the writing skills—or, more to the point, the time—to edit their COBOLese into readable language.

Flow charting: one method for all situations.   In the neoclassical flow-charting method we would now continue to flowchart the serial process flows of the base levels. This option can work here, too. No harm will come to those who wish to take an eclectic position and flowchart the base levels. The flow charts could be turned over to the programmer for reworking into technical, solution-dependent flow charts. Depending on the language level, it might even be possible to program directly from the analysts' charts.

Tight English and decent outlines.   Another solution is advanced by Gane and Sarson, who concede that structured English is hard for users to read. They advance the term "tight English"[9] as a substitute for structured English. In this method we first do a private structured-English specification and then convert it into a readable tight-English specification. The author believes that "tight English"

[9] Gane and Sarson, *Structured Systems Analysis*, pp. 104–105.

is our old friend "decent outline use" and suggests as another alternative that we just go right to a decent outline in the language of our choice without going to the COBOL language for structure.

It is desirable to use tables to support comprehension—all kinds of tables, including decision tables. Other tables would include lookup tables and branching tables, as shown in the preceding chapter.

Passing the buck.   In one case the structured-English approach becomes interesting— when there is a shop standard that in almost all applications COBOL (and, furthermore, structured Cobol) is *the* implementation language. In this case it is easier to convert a structured specification into a structure chart into a COBOL module. The great dangling carrot here is that coding COBOL becomes a trivial operation to be performed by trained monkeys who will work for bananas. What may be lost is a sense of responsibility. The analyst knows the specification will be reworked by the designer. The designer knows that the programming and unit testing will be done by the monkey. The monkey isn't all that smart or involved, and we lose that last best chance to catch logical bugs before the machine goes out the door. That last chance was the sharp eyes of the better-than-average, professional programmer who delights in finding design flaws and is intelligently absorbed with the process at the level where the thought model becomes the physical reality. It is all too easy for the designers to get sloppy when they don't have to do the coding and testing.

Another comment on structured COBOL as part of the system specification is that it is skewed toward a COBOL solution, and we are not ready to be committed to language solutions. When it comes to solutions, we may be working at all language levels and also making software package purchases.

To sum up our progress in the discussion of process specification, we have up to this point mentioned several alternatives:

1. alternatives advanced by the developers of the method
   - structured English
   - tight English
2. other alternatives
   - a decent outline
   - flowcharting

Examples showing the different alternatives.   We will now take a base-level process fragment and develop it in each of the alternatives mentioned. The author's point of view is that all methods will work but not equally well. Structured English—not as advanced by the structured teacher but as practiced in many shops in order to simplify design and implementation in COBOL—is not recommended, although an example will be shown here. Coding in COBOL is the smallest part of the major costs that go into the system development cycle. The major costs are the mistakes made up front, particularly during analysis. Communication with users is of the utmost importance. The other point already made is that we want to avoid selecting a language during the analysis phase (particularly a language that is fading in importance in regard to new development).

The example problem.   Our problem is to specify a base-level process: calculate shipping price. That is what was written in the base-level bubble. In order to do this calculation we need to apply a discount to the sales price according to a simple formula that takes into account *amount of sale* and *creditworthiness*. After we apply this discount, we add on the handling charges to deliver the product, giving the shipping price. We will forget about sales tax.

Figure 4-6 shows two tables we may need in order to do this calculation and present it clearly to the user. The alias table in Figure 4-6(a) relates the common user name to the COBOL name, which identifies the data in this shop. The COBOL

ALIAS TABLE	
User name	COBOL name
SALES-PRICE	LOCAL-INP-SALES-PRICE
CREDIT	LOCAL-TAB-CRED-LOOP
DISCOUNT-PRICE	LOCAL-WAI2-DISPRIC
HANDLING-CHARGE	LOCAL-LIT-DD-HAND-CHG
SHIPPING-PRICE	LOCAL-WAI2-SHIP-PRIC
25.00	GLOBAL-LIT-DD2500-2
25	GLOBAL-LIT-DD25
50	GLOBAL-LIT-DD50
1.00	GLOBAL-LIT-DD100-2
1.50	GLOBAL-LIT-DD150-2
2.00	GLOBAL-LIT-DD200-2

(a)

CREDIT TABLE	
Score	Status
0–25	Bad
26–35	Marginal
36–50	Average
51–100	Excellent

(b)

Figure 4-6   (a) Alias Table, (b) Credit Table.

name, as is usual in structured systems, expresses not only the user meaning but also the data design, which attempts to get rid of literals such as "25" and also attempts to share data on a program library through global and local constants, such as "25". The user name is invariably the alias.

The credit table in Figure 4-6(b) shows the creditworthiness score. This table, besides demonstrating the importance of tables to this specification, also figures importantly in all solutions shown.

Figure 4-7 shows the process specified by the structured-English methodology as taught by the educators in the structured method. (Assume that in place of English any other language will do, whether or not the language supports COBOL.) Reading through this specification, it is clear that a satisfactory solution to the problem has been achieved. What is not clear is whether the product sales managers will want to be able to understand the syntax, which is strange to them. The nested IF within IF becomes really difficult to follow in a more realistic problem. Make a note for future reference also, that this method results in seven IFs and six operations.

Figure 4-8 shows another structured-English solution of the same problem. In this case the alias is not used, implying that the primary communication in this analysis phase is thought of as being between the analyst and the designer-implementer. In order to read this, please refer to the alias table for a meaningful

```
CALCULATE SHIPPING-PRICES.
   CALCULATE DISCOUNT-PRICE.

      IF CREDIT IS EQUAL TO OR LESS THAN 25
                        MOVE SALES-PRICE TO DISCOUNT-PRICE
      IF SALES-PRICE IS LESS THAN 25.00
                        IF CREDIT IS EQUAL TO OR LESS THAN 50
                           COMPUTE DISCOUNT-PRICE EQUAL TO
SALES-PRICE − 1.00
                        IF CREDIT IS GREATER THAN 50
                           COMPUTE DISCOUNT-PRICE EQUAL TO
SALES-PRICE − 1.50
      ELSE
      IF SALES-PRICE IS EQUAL TO OR GREATER THAN 25.00
                        IF CREDIT IS EQUAL TO OR LESS THAN 50
                           COMPUTE DISCOUNT-PRICE EQUAL TO
SALES-PRICE − 1.50
                        IF CREDIT IS GREATER THAN 50
                           COMPUTE DISCOUNT-PRICE EQUAL TO
SALES-PRICE − 2.00

   CALCULATE SHIPPING-PRICE

   COMPUTE SHIPPING-PRICE EQUAL TO DISCOUNT-PRICE +
                        HANDLING-CHARGE.
```

Figure 4-7   A Structured-English Solution.

```
CALCULATE-SHIPPING-PRICES.
  CALCULATE-DISCOUNT-PRICE.
    IF LOCAL-TAB-CRED-LOOP IS EQUAL TO OR LESS THAN GLOBAL-LIT-DD25
    MOVE LOCAL-INP-SALES-PRICE TO LOCAL-WA12-DISPRIC
    IF LOCAL-INP-SALES-PRICE IS LESS THAN GLOBAL-LIT-DD2500-2
    IF LOCAL-TAB-CRED-LOOP IS EQUAL TO OR LESS THAN
                                    GLOBAL-LIT-DD50
    COMPUTE LOCAL-WA12-DISPROC EQUAL TO LOCAL-INP-SALES-PRICE -
                                    GLOBAL-LIT DD100-2
    IF LOCAL-TAB-CRED-LOOP IS GREATER THAN GLOBAL-LIT-DD50
    COMPUTE LOCAL-WA12-DISPRIC EQUAL TO LOCAL-INP-SALES-PRICE -
                                    GLOBAL-LIT-DD150-2
    ELSE
    IF LOCAL-INP-SALES-PRICE IS EQUAL TO OR GREATER THAN
                                    GLOBAL-LIT-DD2500-2
    IF LOCAL-TAB-CRED-LOOP IS EQUAL TO OR LESS THAN
                                    GLOBAL-LIT-DD50
    COMPUTE LOCAL-WA12-DISPRIC EQUAL TO LOCAL-INP-SALES-PRICE -
                                    GLOBAL-LIT-DD150-2
    IF LOCAL-TAB-CRED-LOOP IS GREATER THAN GLOBAL-LIT-DD50
    COMPUTE LOCAL-WA12-DISPRIC EQUAL TO LOCAL-INP-SALES-PRICE -
                                    GLOBAL-LIT-DD200-2
  CALCULATE-SHIPPING-PRICE.
    COMPUTE LOCAL-WA12-SHIP-PRIC EQUAL TO LOCAL-WA12-DISPRIC +
                                    LOCAL-LIT-DD-HANDCHG.
```

Figure 4-8   A COBOL Friendly Structured-English Solution.

translation of the data fields or elements. In defense of this method, it is very close to being readable input into a COBOL compiler, which would create a machine-ready load module. This type of solution is not at all a rare occurrence in organizations that practice structured techniques for system development.

Figure 4-9 shows the decent-outline solution. It is extremely user-friendly (shall we say instead *user-genial* for those readers who have come to hate the words "user-friendly"?), but not at all congenial to the designer-implementers, who must figure out the details of the process flow for themselves. This does not bother the author, who likes this method as one alternative. Tight English is also a decent outline but more complicated, since it is based on the syntax of structured English, which is then rewritten. The results are quite worthwhile, but one hopes that in time the practitioner of tight English will go straight to a decent outline.

Regarding the designer-implementers, in the long run it is less expensive and more job-enriching to let them in on some of the more interesting development. To the extent that they have some control, they will bring their own expertise to the problem solution. Meanwhile the analyst has represented the reality of the system in completely unambiguous terms.

Before leaving the decent-outline solution we should bring up some old history which is of some interest in connection with tables as solutions. As far back as the sixties language researchers were looking into using decision tables as a direct

A. Calculate shipping price

  1. To discount order, figure out the discount using table.

DISCOUNT TABLE		
Order value	Credit score	Discount
Less than or equal to 25.00	0–25	0.00
	26–35	1.00
	36–50	1.00
	51–100	1.50
Greater than 25.00	0–25	0.00
	26–35	1.50
	36–50	1.50
	51–100	2.00

  2. Figure discount-price by subtracting discount from sales-price.

  3. Figure shipping-price by adding handling-charge to discount-price.

  4. Save discount-price.

  5. Save shipping-price.

Figure 4-9   A Decent-Outline Solution.

source language which could be input to a compiler. The General Electric Corporation carried this approach to the point where it began doing and advocating system development through decision tables. Their decision-table language was called DETAB-X. When GE decided to bow out of the computer mainframe business, their method seemed to die with their exit. The author has drawn a horizontal dashed line near the bottom of the table in Figure 4-9. If the reader were to move up steps 2–5 under that dashed line in the box and call them actions and make some easy changes to the decisioning above the dotted line, what would result would be a decision-table source language suitable for machining into a DETAB-X compiler.

Finally, Figure 4-10 shows the flow chart solution as developed in the last chapter as a possible solution to base-level specification within the structured analysis method of data flow diagrams. We show the input, output, and data resources with the flow chart. Also available as a data resource are the credit and alias tables. The flow chart is clearly easy for any concerned person to read. Flow charts are old hat to users, particularly those who have been to business school.

Note that the process is defined with four simple decisions and five operations. Remember that when structured English was used, seven decisions and six operations were required.

This flow chart solution is, of course, also available as part of the neoclassical methodology, in which case the entire specification through all levels and partitions is accomplished with the one charting technique.

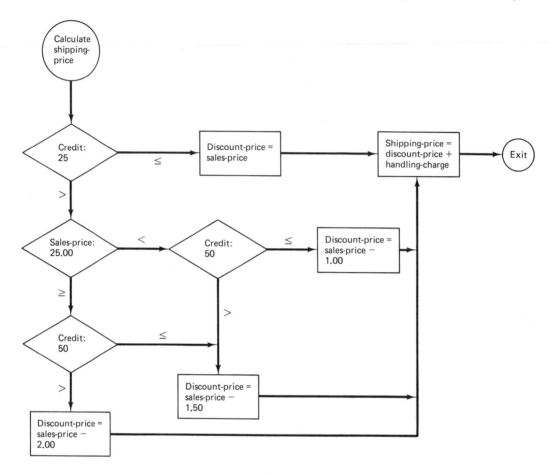

Input: Sales-price $\lfloor N\ N\ N.N\ N \rfloor$

Output: Discount-price $\lfloor N\ N\ N.N\ N \rfloor$

      Shipping-price $\lfloor N\ N\ N.N\ N \rfloor$

Data resources: Credit table

                Alias table

Triggers: Continuous

Other resources: 1 Level-2 clerk

                  1 IBM PC

Figure 4-10   A Flow Chart Specification Solution.

ADVANTAGES OF THE STRUCTURED METHOD OF TRACKING DATA THROUGH THE USE OF DATA FLOW DIAGRAMS

1.   The data flow diagram is the best bridge between the user and the people who will design and make any new system. If done skillfully, it is supremely readable. It makes the details of the system, including the automated internals, available to

the user with a minimum of training in reading the diagrams. It sets a clear course for the designer, who now understands the task ahead. It reveals specification errors. It serves as documentation better than any serial language (such as English, French, German). A good picture is worth a thousand words. The problem is not the Victorian novel as specification, it is that some data flows are so complex that they cannot be expressed well enough except with comprehensive pictures that exist on two levels: the one-thing level and the whole-chunk level. Staring at the elm tree tells us little about the forest topology. Grasping the parts relationship of the forest tells us little about that one Elm. DFD's do both.

2.   Data flow diagrams are more flexible than flow charts for showing nonserial flow. Any number of data flows can be shown going in and out of partition bubbles. If a data flow is too complex to be understood in terms of one or two words on a line arrow it can be divided into smaller data flow parts. The outputs from a bubble can go off in many different directions. This models exactly what is happening in the data flow through an organization.

3.   Data flow diagrams can show the flow of data through manual operations just as easily as through automated operations. No other method has any advantage in this respect.

4.   Tracking data is better than tracking process when the analyst works for an organization that has, say, thirty transactions going through four hundred processes.

5.   Data flow diagrams use structured English, a data dictionary, and decision charts and trees. The essential methodology does not rise or fall on any strict interpretation of the law. If the user wishes to see tight English rather than structured English as the published documentation, that is quite all right.

CONCLUSIONS ON CHARTING METHODOLOGY

The author sees no need to create a duality regarding process and data flow, since they are two sides of the same coin, and so will leave it to the readers to find their perfect place regarding the use of these methodologies. Remember that in either case what we are about here is defining base-level processes. And it is not an either-or situation. Sometimes we may want to mix and match or use one or the other.

These methodologies are the core specification that describe the system processes. The analyst will also be responsible for the cost-benefit analysis and in general any financial indexing that is useful to describe the existing system. The analyst is also responsible for visioning any new physical implementations, such as batch versus on-line solutions. The analyst is also responsible for the full range of information transmission within the system, including voice, video, and facsimile, and must be able to represent the user in discussions with in-house, consultant, and vendor communication experts. Telecommunications exists because of system requirements, and the analyst cannot begin to do such things as cost-benefit analysis without having some knowledge about the alternatives and potential in this area.

Therefore we have a number of other charting methodologies to learn in future chapters concerning data, physical systems, and terminal dialogues before we are able to complete the entire system specification and derive realistic cost and benefit forecasts.

A NOTE ON APPENDIX A

For further drill in the use of data flow diagrams and other particulars of the system specification in a Fortune 500 corporation setting, the reader is advised to turn to Appendix A at the back of the book.

5

REQUIREMENTS ANALYSIS

Mentioned in previous chapters was the development of a logical current system specification. It was suggested that the conversion to logical from physical system could be done either as one step (directly to logical) or as two steps (physical then logical). This conversion removes particular references to workplaces and individual workers. This is important, because when we create our new system specification we are not necessarily going to replicate the current system jurisdictions. What we will try to do is develop the most logical new system possible according to the terms of the new system requirements.

Very early on in the system development life cycle we began to lay the groundwork for the requirements analysis. The very reason that analysts were called in to look into an existing system was most probably because there was a perception on the part of the users that new requirements existed. The executive letter of authorization that started the ball rolling touched on the new requirements.

The very first document produced by the systems analyst was the overview current system and requirements study. This is shown on Figure 1-3 as the first operations box on the life-cycle flow diagram. Figure 2-1 shows the way new goals are narrowed down to specific goals and tied to requirements in such a document.

At the point in the analysis when the logical current system has been charted and otherwise specified to the satisfaction of all principals involved, it is time to apply the general statement of new requirements already documented and come up with a complete statement of requirements for the new system.

It doesn't matter which methodology, structured or neoclassical, has been used; the requirements solution is the same. The analyst working closely with the users must go through all the base-level processing logic along with the associated data definitions and cover the detail requirements upon which a new system will be specified. Covered in the requirements analysis are the process (also know as functional) requirement and the data requirement for each base level (whether bubble or subroutine rectangle) that is in the domain of change—that is, each base level that must be newly specified to meet the new general requirements. In addition, there may be new requirements for which no processes or data exist at all in the current system.

It is customary to show the new requirements as both those that need to be included in the new specification and those that do not. In other words, all possible requirements are listed by bubble/box and then a Yes or No is placed next to that requirement possibility. Also rate and volume requirements are often included in such an analysis. For instance: Must be able to bake 1000 loaves of white bread an hour. Yes.

Having the list of requirements for each bubble/box in the domain of change with a Yes-No indicator serves two purposes:

- In case of a hassle over whether some requirement was left out or wrongly included, this document serves to settle the issue. One suspects that dentists take so many before-and-after x-rays for the very same reason.
- The creative effort by which the analyst comes up with the new system specification is enhanced, because all the ingredients of a solution that can come out of a methodology are present: there is a comprehensive logical specification for the current system and a detailed set of requirements for the new system.

The analyst, operating now in the most creative area of systems analysis, will come up with version upon version of a new system in the form of rough sketches to show the user, who is participating actively. In this process new requirements and obsolete requirements keep emerging as the system is reformulated. At the end of this exercise, when the new system specification is finalized, we also have a net requirements analysis document consisting of process requirements and data requirements.

In some organizations a requirements analysis document is not maintained and published as a formal document, since there is redundancy here. After all, the new system specification along with its associated data definitions is a requirements document showing the Yes requirements only. In organizations where there is a high trust relationship between analysts and users and where the users are good enough at their business to spot a missing requirement, a separate requirements document is not wanted. What is important is that the analyst do the requirements analysis.

Figure 5-1 shows an example of functional and data requirements. Refer to Figure 4-4 to see the base levels from which this requirements list was prepared. The process detail for these base-level modules is not shown. This detail would include all the Yes answers shown in Figure 5-1.

A well-done requirements analysis should cover the inputs a base level requires, the outputs that are necessary, the data stores that will be needed, the operations that must be performed, and the resources that will be required. Re-

FUNCTIONAL REQUIREMENTS

2. TURN IN MONEY, STALE, ROUTE ORDER

Y   2.1   Make school orders.

Y   2.1.1 Maintain phone list for schools.

Y   2.1.2 Maintain name of contact person at schools.

N   2.1.3 Maintain back-up contact name at schools.

N   2.1.4 Call schools daily for order at 3:00 P.M.

Y   2.1.5 Call schools when open for order at 3:00 P.M.

N   2.1.6 Take special orders for administrative staff of school.

Y   2.1.7 Record phone order on school route form.

Y   2.2   Check route order.

Y   2.2.1 Super check form for accuracy of product.

Y   2.2.2 Super check form for accuracy of quantity.

N   2.2.3 Contact customer if unavailable.

N   2.2.4 Make substitutions if necessary.

Y   2.2.5 Notify route salesperson of route order problem.

AND SO ON

DATA REQUIREMENTS

2. TURN IN MONEY, STALE, ROUTE ORDER

Y   2.1   Make school orders.

Y   2.1.1 Phone list.

N   2.1.2 Contact name list.

Y   2.1.3 School route order form.

Y   2.1.4 List of closing days

AND SO ON

Figure 5-1 Requirements Example.

garding resource requirements the neoclassical approach is probably more sound, since this method integrates resource requirements into its method, as we have seen. Input and output workload statistics are also very useful in determining new system specifications.

TWO KINDS OF YES PROCESS REQUIREMENTS

Taking a narrow view of new system specification (which we will not do often in this book), we list a new requirement as a Yes only if it is to be implemented with an automated computer solution. Remember, we have to bound those processes that are to be automated, at least in regard to their information support needs, apart from the nonautomated parts of the system. In this narrow view it is necessary to give requirements only if automation is involved. A broader view would be that opportunities exist for improvement, and we will improve toward excellence using computer information automation as an implementation method along with other methods for which we as analysts are responsible. In this view there are two kinds of Yes requirements. Fortunately, only one Yes answer is needed for any given requirement, since a process will either be automated or manual, not both.

Requirements analysis does not need to be done for that part of the system that is not involved in new system specification. In other words, processes for which no change is needed need not be involved.

6 AN INTRODUCTION TO DATA ANALYSIS

Earlier chapters made considerable mention of the relationship between data and the specification of systems. Data-in-motion was described as the inputs and outputs of a system at whatever level, whether at the whole systems level or the base level. Data-at-rest was described as a major system resource, existing in the form of data stores.

The specification of the current or new system was said to consist of the decomposition of the whole system into levels and partitions within levels, using either the neoclassical or structured charting methodology, until we had defined all the base-level chunks of processing. These base-level chunks were then identified in complete processing detail, using one of several methods. This technique was said to be related to data all the way down the line. Up to now we have not gone deeply into how we describe the two kinds of data in order to produce a complete specification (and in order to understand the data in our organization). The next several chapters will be concerned with understanding data in all those aspects which are of concern to the systems analyst.

Understanding data and the structure of data can hardly be underestimated in terms of its importance. The opposite of well-controlled data is systems disorganization to the point of chaos: this is the consensus opinion of scientists working in the area of information theory. There is no more important (or more interesting) area in systems analysis study than the analysis of systems data.

This introductory chapter will present some of the theory and principles of data analysis and outline the material to be covered, so that the following chapters on different data topics will relate to the whole subject.

Our first concern in this introduction is to understand the relationships among data, information, knowledge, and communication.

DATUM, DATA, INFORMATION

Consider the following:

31

31? What is a 31? It does have a code and therefore is not gibberish. It is a two-digit number unrelated at present to the rest of the universe. It is a datum. 45 is a similar number with the same apparent code system. 31 and 45 are data.

(With the kind permission of the reader the author would now like to dispense with the singular *datum* and use the word *data* to represent both singular and plural, so that we can say 31 is data, 45 is data, and 31 and 45 is data. This is the common American usage of the word data. It should be clear from the discussion when the singular is meant.)

Well, then, what is meant by the data named 31? A data is a *fact*. Data are facts or at least potential facts. That is the dictionary meaning of data.[1] We accept 31 as some kind of a fact which has a code and attributes but which is of no use to us at the moment.

If it had use to us at the moment it would be *information*. What does 31 need in order to be information? It needs:

- relationship
- apprehension
- surprise

If 31 is a subject, which it is, it needs relationship to another subject to be information. Let us say that the relationship is IS. Let us say that the other subject is Fahrenheit-Temperature. Putting this together, we have

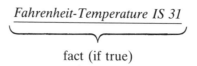

Fahrenheit-Temperature IS 31

fact (if true)

If there is a thermometer showing a readout of 31 on a tree near an abandoned house on a deserted property, we do not have information, because there is no

---

[1] Funk and Wagnalls, *Standard College Dictionary* (Pleasantville, N.Y.: Reader's Digest Books, Inc., 1968).

one to apprehend the readout. If one of us gets up on a rainy morning and goes to the window and reads the number 31 on the outside thermometer, we have information. Our data 31 is now information. We can make inferences about this information which can lead to useful action. The information that 32 is the freeze point, that there is probably a freezing rain falling since the temperature is 31, which may make the roads hazardous for driving (three inferences) is certainly the type of information that is meaningful.

Information can involve subjects in a string of relationships from which we can make complicated inferences. "Today's low will be 31 and the high will be 45." We can infer from this that the roads although perhaps hazardous at this moment, will not remain that way all day.

Relationships can be implicit as well as explicit. Somebody with a gloomy look can utter dolefully the word "rain . . . " and we understand that *Person DISLIKES Rain* is the information being presented for our apprehension.

Information always comes as a surprise. Each instance of information is unpredictable. If the thermometer were broken at 31, the readout would be meaningless, even though the code system was complete. We would learn nothing and could infer nothing from that 31. That would not be information.

Let us then say that data are facts and:

Information is unpredictable data with a predictable code. It has attributes and relationship to other data and may be apprehended.

Some sources including dictionaries call this information entity including relationships DATA. This is not precise enough a definition for the working analyst, and for this reason we have gone to some pains here to define data as distinct from information.

Regarding the information that "the temperature is 31" we can make the following decomposition:

- Temperature is a data fact.
- 31 is a data fact.
- IS is the relationship between these two data facts
- The code of 31 is: two markers in the decimal number system that conform to all the rules of decimal numbers.
- The attributes of 31 are: a range with a lower limit of -40 degrees Fahrenheit and an upper limit of 120 degrees Fahrenheit on this thermometer.

—and the following general inference: It is cold but not exceedingly cold outside, but it is raining and 32 is the freeze point, so there is a problem, since it is 31. This inference combines information with knowledge.

We should make two points here about relationship between subjects. First, the subjects in this example—now that we know what 31 is—are Climate Measuring

Devices and Observations, of which 31 is one of many. The second point is that relationships are not always one-to-one and can become quite complex—to the point where language is not a proper code structure to carry the information. In this case we might go to mathematical notation in order to maintain the integrity of the relationship and be able to manipulate it.

We will find out when we go to build data structures such as transactions and databases and communications systems to transmit the moving data that the more precisely we can define our data, the more chance we have to build intelligent computer systems or any systems at all.

Particularly important is the code understructure. Unless we understand the code system, the data is meaningless and the information is garbage. It is what the information theorists call *entropic*—from Newton's law concerning entropy, where (in this case) data starts as highly organized and coherent and ends up in a state of inert, meaningless disorganization. Since it is not hard to prove that biologic and social systems are held together and made meaningful because of information (e.g., the DNA double-helix code or the formation of antibodies by the immune system tailored to the precise information specifics of the invader or the mailing list of the mail-order business), we can see that what we are dealing with here in terms of data code analysis and the general precision of our data is profoundly significant, because by definition it is *antientropic* and leads toward cosmos rather than chaos.

To use a mundane example, the poignancy of this choice is made clear to any hapless person who has been marked wrongfully as a poor credit risk by a computer system using berserk data code.

Precise information allows systems to survive. Lose the code and we lose the system. Fully grasp the data and the code of the data and the attributes of the data and the information relationships between the data and we maintain and enhance the system. This is antientropic.

One person did more than anyone else to establish a useful theory of information as antientropic and code-based. That person is Claude E. Shannon, who did his most interesting analysis of information while working for Bell Labs[2] in the late 1940s. His papers on information theory are one basis on which the new science of information has been developed.

INFORMATION AND KNOWLEDGE

Information is data-in-motion, available data, organized according to the definition given already in this chapter, which when perceived (apprehended) by a recipient capable of *comprehension* of the information may be stored by the recipient as

---

[2] Claude E. Shannon, "A Mathematical Theory of Information," *Bell System Technical Journal*, 27 (1948). Also, about Shannon and other pioneers in information theory read: Jeremy Campbell, *Grammatical Man* (New York: Simon and Schuster, 1982).

data-at-rest (an available data store). Then information becomes knowledge. From this information other information can be derived and stored.

This information, as it becomes knowledge, is perceived, understood, retained, elaborated upon, evaluated, formed into principles, and made retrievable as problem solutions. Then information becomes useful knowledge. Then we have a knowledge base, which can be used to evaluate new information relevant to this knowledge base. Take the following scene as an example:

A man walks into an office, where sit two colleagues. "Regina Summerhill called and wants one of you to call back right away," he says. The man is a new manager, recently transferred from the London office.

His message meets all the code requirements and other indications that the information is valid. But the knowledge base of the other colleagues instantly rejects the information as invalid and produces the correct information. In less than two seconds one recipient replies, "John, I think you mean that Ruth Somerwhile called. Regina Peters works for her; that is where you got the name Regina. Summerhill is or was a famous progressive school in England." The other person nods in agreement with this information reformulation.

"Oh," says John, "right you are. I'll catch onto all these new names soon enough."

We will return again, more than once, to the task of understanding how incoming information is validated, both by inherent structure and by comparison to a relevant database. Meanwhile, ponder on how difficult it will be to program the type of inference engine necessary for a computer to be used to give John a comparable interpretation of his original message.

COMMUNICATION

Communication is successful information transmission which is based on successful data transmission which is apprehended and comprehended. In the preceding example we have both successful data transmission via John and successful information transmission thanks to the recipient's ability to relate the message to the information already in the knowledge base.

FURTHER COMMENTS ON DATA EVOLUTION

What we are attempting to understand in our data studies is

- the accurate and economic transmission of data-in-motion, which we define broadly to include all communication of data
- the making of information out of received data
- the integration of information into the knowledge database

- the accumulation of knowledge into culture, knowing that all that separates modern humans from stone-age "savages" is this cultural accumulation

The nature of the problem before us is to study and improve systems that are based on the evolutionary development of coded data into knowledge, wherein the binary-coded hexadecimal data in eight-bit chunks becomes the two digital characters *31* becomes 31 degrees F. becomes knowledge about weather becomes archived, accessible knowledge.

To support this system we will build and transmit data-in-motion and store and hold for retrieval data-at-rest. We will combine the two types of data through an input-to-output processing system and arrive at knowledge which will allow us to solve problems.

In order to manage our information we will be concerned with the following subjects in this part of the book:

- data-in-motion in the form of messages or transactions in and reports out
- data-in-motion transmission
- data-at-rest records
- the administration of data in an organization
- the documentation and control of data in the form of the data dictionary

DIFFERENT VIEWS OF DATA ANALYSIS

We will be looking at data flow and storage from three different perspectives (using the words data and information to mean the same thing in this discussion):

1. data ownership
2. data development alternatives
3. the overall computer solution to data automation

Data Ownership.

Four levels of data ownership usually come to mind.

1. *Private data* (also known as *workstation data*). This might be data stored on microcomputer disks such as memo files or spreadsheet files. It might be data in desk drawers or filing cabinets. It might be files resident on mainframe disks which are reserved. It would include all access to external databases such as newspaper and newswire electronic files, legal and medical and stock-quote databases, and the like.

2. *Shared data* (also known as *workplace data*). This would include data used by

a group of people in the organization which is not of interest to the rest of the organization except in a summarized form. This data would include vehicle, machine, and facilities maintenance schedules, short-term project reporting, workload scheduling within a department, and so on.

3.   *Common data on limited databases.* Limited databases contain data that is of interest to many but is owned at the subsystem level, such as a materials management database of interest to the sales and financial people. In this case a communications link or interface has to be established between the limited database and the primary organization database. As a further complication these limited databases are often vendor-oriented in that they are part of vendor application packages or part of a fourth-generation vendor database management system and language.

4.   *Common or integrated organization data* (also known as *corporate data*). This is the set of subject-oriented databases developed as an organizationwide resource following a data resource plan for the whole system along with a transaction system to service the integrated database. Shared and private data still exist and can be related to this primary data, including the downloading of systemwide data for local use.

Data Development Alternatives.

As this book is being written, and surely long after it has been published, this issue will be with us as one of the major decisions an organization must make regarding information management. Do we develop corporate data, or do we take the applications development approach which is our inheritance from the past, or is there some middle-ground alternative? It is very expensive to develop corporate data, and a lot of time is lost in up-front research which is not productive in terms of deliverables. On the other hand the confusion caused by redundant and inaccurate data not independent of applications is so costly for the large organization that information systems tend to become useless as they accumulate these problems. Large organizations therefore tend to be more motivated to go for integrated organizational databases than smaller organizations that can't afford the up-front writeoffs and that feel they can still control the data at the applications level. There are three types of alternatives available.

1. Applications systems.   When there is a specific application in mind, a database and a set of transactions is defined for that application. If product description or employee number is needed for this application, it is defined and implemented as owned data, no matter how many other applications are also using the same data. Every effort is made to provide information links between this application and other systems, such as accounting systems, that need this information. Some attempt is made to insure that product description, for instance, is the same in all applications, but it is not usually a top priority.

The advantages of application developed data systems is significant. There are many productivity tools to help the application get done quickly. These include packages and fourth-generation languages. There are far fewer people to consult regarding data definition. The projects can show payoffs almost immediately. Mistakes, when made, are more circumscribed and less costly. Prototyping is easier to accomplish. Political considerations are reduced or eliminated, since usually only one user jurisdiction is involved.

2. Integrated organization (corporate) systems.   Once you bite on and digest the bullet of slow start-up and group hassling over data meaning, the advantages of integrated systems are even more significant. The number of data entities that need to be defined and set up for are less than one might assume. The number of application views of this same data tend to be more than one might assume. It is not unusual to find 100 data entities or data relations supporting 2000 or 3000 data views. The usual profile is surprisingly simple data relations supporting surprisingly complex data application permutations.

If the applications have been well selected in terms of order of implementation, it is not unusual to find that, after doing a certain core of applications using this method, the rest of the applications get done very quickly. This is because the data has been fully defined, including the transaction set and the validation rules. Also there is a lack of back-end hassling due to the fact that the data has been clearly defined up front and maintained in the data dictionary. It is a lot easier to maintain data in a data dictionary if the system data has been integrated.

Data modeling of new data becomes possible with integrated data, since the new data views are added to the growing snowball of all the data views and a new set of relations with high affinity in subject databases can be constructed quickly with the use of modeling software. This data evolution can be accomplished without the need to rebuild all the application programs that use the data. This, by itself, is no small accomplishment.

It cannot be denied, nevertheless, that doing an integrated database for a large organization is an ambitious, costly undertaking that has led to many disappointments and failures. Senior organization management, in particular, tends to get impatient with the delays and false starts that have accompanied the grand scheme of integrated subject databases.

3. i−1 data integration.   This approach has been turned to, often in desperation, by refugees from both the applications method and the corporate integration method. There are advocates of this method with very impressive credentials.[3] In this plan integration is attempted one level down from the whole-system level, hence the use of the label "$i-1$." There is some interesting general systems

---

[3] William P. Grafton, "IMS: Past, Present, Future," *Datamation*, November 1983. Grafton implemented the first IMS production system and managed it for three years while working at Rockwell.

theory concerning the importance of the $i-1$ level on a broader social canvas than the single organization.[4]

Under this scheme, a large corporation with nine divisions, would integrate data at the division level and upload corporate data from each divisional system according to a corporate plan. Many of the advantages of corporate data administration could still be retained, such as a corporate data dictionary.

In the case of an organization with such distinct $i-1$ groups as sales, manufacturing, finance, distribution, and engineering R&D, the highest integration would be at that level with an uploading scheme in place. The same would be true of a government executive branch or a university of many colleges and graduate schools.

This scheme has either the best or the worst of the two opposite philosophies. Agreement and data management are possible with good organization planning and executive control, but the possibility of data confusion is still there. Think of all the resources that are shared by different schools in the university environment, such as admissions, registrar, plant facilities, faculty files, student files, and so on. Think of all the cross-school training, such as when a student in the graduate school of business needs to cross-train in the graduate school of information sciences (not a bad idea, by the way).

Which answer is the best for a given organization? The tradeoffs are too complex for us to construct a table of choices. There are too many psychological and political factors involved. What can be said, most emphatically, is that a master organization plan documented in a comprehensive data dictionary is a necessity for survival in the data jungles. Ask yourselves (for the rest of your careers): how many applications problems are really data problems in disguise?

The Overall Computer Solution to Information Automation.

We might also call this *the generic information-processing solution*. An automated solution using computers is not essential to this formulation.

There appears to be just such a generic solution concerning the evolution of bits of data into knowledge systems. Putting aside systems whose only purpose is to control data flow, the logical purpose for data-in-motion to enter a system is to become knowledge from which data-in-motion outputs can be produced by triggers. An example is the order transaction that generates a shipment-of-product label output. First the order creates an update or an add to the database, and then a millisecond later the shipment label is generated from the database because (trigger) an order shipment is known to be pending.

Figure 6-1 shows this model for information processing. This model will be

---

[4] John Platt, "Hierarchical Restructuring," in *General Systems*, Yearbook of the Society of General Systems Research, Vol. XV, 1970, pp. 49–53.

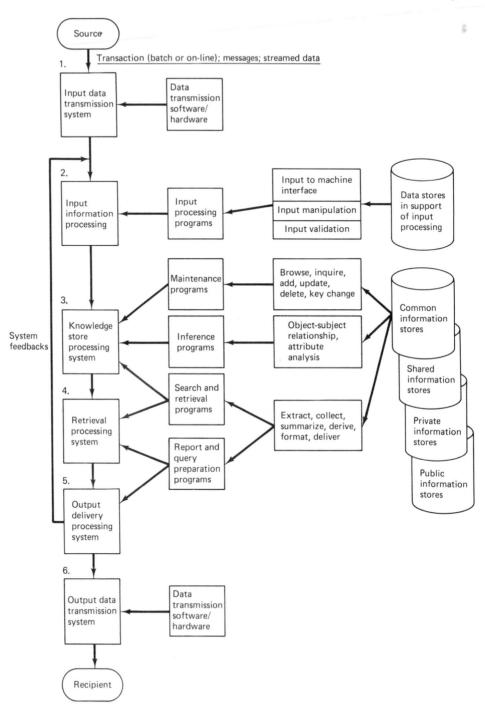

Figure 6-1   The Overall Computer Solution to Information Automation, Batch or On-line.

discussed as an overview here and will serve as a guide to the rest of the discussion in this part of the book.

Figure 6-1 shows data-in-motion coming from environmental sources via a data transmission system, entering the system, going through a refining process, becoming part of the stored knowledge base, generating outputs to recipients, generating system feedbacks, and refining and delivering outputs to recipients. This is all in good agreement with our definition of a system, where inputs are processed into outputs by triggers using resources including data resources. It is now time to be more specific about this data flow from input to output using data stores and processes.

1. The data transmission system.   This system uses programs and devices to collect and separate data and move data along the transmission system from collection point to collection point. It is interested in efficient and safe data throughput. Data is accompanied by code information to indicate start and end of data message and enough code information to secure the data reliability and allow the data to be deblocked and reblocked during transmission. Transmission may be in plant, in plant complex, local area, or remote area. Data is taken to mean the full range of analog and digital transmission involving message (as teletype), voice, video, digital, and combinations of these, such as partially digitized video.

Transmissions involve both terminal and computer end points in any combination. Data to and from terminals is immediate and people-oriented. Data to and from computers is deferred for eventual use by either people or computers.

Data transmission systems and programs are not much interested in data as information. Relationships between data subjects, the heart of information understanding, is outside the scope of data transmission systems.

2. The input information-processing system.   The data is now understood to be three kinds of information: (a) *Transactions*, whose each-instance contents are unknown but whose field definitions are completely known and prearranged. (b) *Messages*, consisting of text or voice or visual information, where very little is known about fields or field attributes and where almost no implicit relationships are known in advance, except that there is a sender, a group of recipients, a time stamp, and perhaps a keyword or keywords and some length or duration parameters. This information provides a kind of knowledge but is not subject to manipulation by inference-generating programs, at least at this stage in our development of intelligent systems. (c) *Streams* which consist of information where only the transmission facility has been planned for and whose content is unpredictable and where the stream may have begun before we "tuned in" and may continue after we "tune out." Included in this category are public media and database services and (in a more circumscribed form) such communication links as video teleconferencing.

Concentrating for the moment on the very important transaction form, we have a set of associated processing programs to do the following:

- provide a good interface between humans and machines in order to enter information from terminals or otherwise prepare input for machining
- combine, split, or enhance transactions with the use of data stores
- align transaction key to relate to database key system
- validate the transaction input fields with data store support
- in the case of batched transactions, sort and merge the transaction data sets to the collating order of the database

3. The knowledge-processing system.   This system stores information flows in the form of transactions and combines programs and program tables to create knowledge. The programs can do cognitive processing such as computation and decisioning and do inferencing from data relationships. As we said earlier in this chapter, the input information is perceived, understood, elaborated upon (incorporated), evaluated, formed into principles, and made retrievable as problem solutions. This can be done under program control with human oversight and input, which is why we call these data stores together with these programs a knowledge store or base. The following types of programs support the knowledge store:

- maintenance programs that allow browsing, updating, adding, deleting, and inquiring
- cognitive and inferencing programs that create output flows to solve problems

4. The retrieval processing system.   This system consists of programs that extract information from the data base. These programs are refineries that

- extract, collect, derive, summarize, format, and export answers

5. The output delivery processing system.   These programs or program functions combine processing logic with triggers to produce two kinds of information:

- recycle information in the form of transaction feedbacks, because the nature of the transaction is that it needs either repeated passes through the database to fully perform its function or a feedback to a human being waiting at a terminal requesting additional information in order to make the necessary inferences to answer questions
- output information in response to requests or preprogrammed triggers in the form of queries and reports

6. The output data transmission system.   This system is similar to the information transmission system from the source. In this instance the data goes to the recipient, where it will be interpreted as transactions, messages, and streams.

We will now focus on this flow, store, and flow of data in sufficient detail to fully understand the analysis of systems.

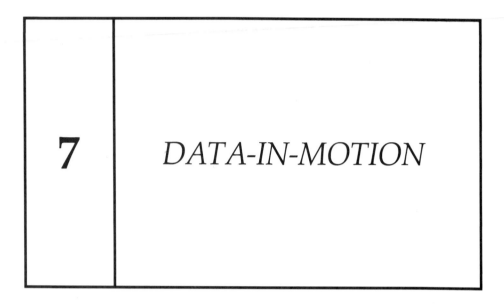

7   *DATA-IN-MOTION*

THE CONCEPT OF A TRANSACTION

In this chapter we will be discussing data in the form of input transactions, messages, and information streams and in the form of output reports/queries, messages, and information streams.

A *transaction* is an exchange of information between two systems. This definition is correct but not detailed enough for our analysis. Here is a further definition:

A transaction is a data transmission of an information record consisting of precise fields (with attributes capable of a high degree of validation) and simple implicit relationships, or such precise fields along with textual fields which have an imprecise set of attributes (and therefore difficult to validate) and relationships both implicit and explicit.

Here is an outline of that sentence:

- a transmitted information record
- precise fields

- with attributes capable of a high degree of validation
- simple implicit relationships

or

- precise fields
  - with attributes capable of a high degree of validation

and

- textual fields
  - with a less precise set of attributes (and therefore difficult to validate)
  - both implicit and explicit relationships

The author agrees that, even with outlining, this sentence is capable of causing indigestion unless swallowed a little bit at a time. We will return for a detailed examination of this sentence, but first let us look at Figure 7-1, which shows an example of a transaction consisting of more precise and less precise fields. Please note that fields and data elements are one and the same, and the words may be used interchangeably.

Figure 7-1 can be seen to be a payment transaction consisting of nine fields. A sampling of these fields shows the following codes, attributes, and relationships:

Field 1.   The code consists of three alphanumerics in the ASCII code set and furthermore consists of printable characters for all available printers. The attributes consist of two numerics in the range 01–20 and one alpha, which is either an I, a U, or a Z. If the first digit is a zero, a Z cannot be present. This transaction code can only mean an update of an existing record in a data store. The implicit relationship is Transaction-Code Defines Field-Set. This is a precise field, suitable for routine computer validation.

Field 2.   This field, the customer identification number, has the same code as field 1. In addition it has a check digit in the tenth position of a numeric field, which is calculated from an algorithm using the other nine digits. The check digit is arrived at by the following formula:

1. Double all the odd digits and add up the sum of all digits as shown in the example. Use the first nine digits only.

   $$213251731 = 4 + 1 + 6 + 2 + 1 + 1 + 1 + 4 + 3 + 2 = 25$$

2. Take the complement of the low order digit of the resulting sum. Since the sum low-order digit is 5, the complement is 5. Therefore 5 is the check digit in the tenth or low-order position.

The attributes of field 2 are that it is a ten digit number with all numbers lower than 8*nnnnnnnnn*.

Field #	1	2	3	4	5	6	7	8	9
Field name	Transaction-Code	Customer-Identification-Number	Product-Code	Product-quantity	Unit-Price	Payment-Amount	Customer-Name	Customer-Address	Payment-Description
Field type*	3 AN	10 N	6 AN	4 N	N N . N N	N N N . N N	24 AN	60 AN	144 AN
Field example	12 I	2132517315	36 MT12	2	4.50	9.00	F. Ferguson	61 Baltic Court Box 106 Altamxgoola CT 06333-0001	Customer paid with check #10570 on 11-12-85. Check had a cross-out in the name but was good and rung-up with supervisor's OK.

Figure 7-1 A Transaction Made Up of Fields.

*N   = NUMERIC
AN   = ALPHANUMERIC
N N . N = NUMERIC (IMPLIED DECIMAL)
E.G.: 3AN = THREE ALPHANUMERIC CHARACTERS

The implicit relationship is Customer-Identification-Number Identifies Customer. F. Ferguson may get married and change last names, F. Ferguson may move and change addresses; the identification number remains the same.

Field 8.   This field is enough to give a Data Administrator migraines. The code is all right. It has more redundant a's than redundant x's. A human might question the x in the city name, but that is not a code problem. In terms of analyzing the attributes the field cries out to be decomposed into street, auxiliary information, city, state, zip, and carrier route. If these were the fields, they would be much more precise. We could then look up Altamxgoola against a good city data store by zip and either reject the transaction from the input stream or, if we wish to be a class act, substitute the right city name. Who needs the city name for a payment, anyway? The only information really needed to process this transaction is contained in fields 1–4 and 6. A systems analyst could really go to work and improve this transaction, even if the user insists on all nine fields.

The implied relationship for field 8 is Customer-Address Is-Residence-Location-Of Customer-Name. The reverse relationship is Customer-Name Resides-At Customer-Address. (The author has a long-range reason for giving this much drill on subject relationships. This will become clear in the chapter on data dictionary.)

Field 9.   This field is the only example in this transaction of less-than-precise text information. Even here we should separate out the check number and the check date as two separate precise fields. We could even make a field out of supervisor OK (yes or no or blank). To the extent that this field is free-form text, it is hardly worthwhile to define attributes to be used for validation checking. Notice that as we go to text, we go away from implicit relationships and go into explicit relationships, which make up so much of the content of text. Examples of the explicit relationships here are Check Had-A Cross-out and Rung-up With Supervisor's-OK. Evidently there are a series of reasons for checks to be rejected because of their appearance and a series of reasons for accepting an unacceptable check. A systems analyst working in this area would recognize these relationships and try to document these rules, in which case this entire transaction could be converted into precise fields with no free-form text.

Now let's go back and reexamine our sentence defining a transaction.

A transaction is a data transmission . . . because transactions are data-in-motion both logically and physically. Logically, because in this case the transaction is the result of a payment entering from outside the system into the system. Physically, because the payment was data-machined or scanned and/or transmitted in some way to the environment where the knowledge base resides. Transactions are stand-alone information, but they have no consummation in terms of the system until they react with the knowledge base. This payment will update F. Ferguson's master record whether it is in a filing cabinet or a database on an IBM 3380 disk drive. It will be applied as a credit against this account and the accounts receivable records, and F. Ferguson will remain a promotable customer and will not receive a past due notice and the product 36MT12 will show a better profit-and-loss earnings record.

. . . *of an information record* . . . This means that the transaction makes sense by itself to the extent predetermined by the designers. Perhaps it is a payment by someone for something.

During remote data transmission this transaction could be arranged into packets that include code and delivery information, and these packets could be separated in transmission—so that, say part of the transaction goes by way of Chicago, part goes by way of St. Louis, and part is shot into outer space—and then arrive in New York from San Francisco at different times and in the wrong order. Have no fear: payment transactions do not suffer from bandwidth lag, and the code will put the transaction back together again for delivery to the user at the end node. Again we see the importance of a good code to save our transaction from disorganization and entropy during the rigors of travel.

. . . *consisting of precise fields (with attributes capable of a high degree of validation) and simple implicit relationships* . . . This could be the basic definition for the successful system. Social organizations are determined to a large extent by what they do with their inputs as they process them into the outputs by which they are known. This definition can be taken as a normative value against which we might examine all data-in-motion of the transaction type. To the extent that the fields or subjects are precisely defined and have clearly defined attributes and simple relationships that are built into the structure (as in the case of the payment) the whole system will be tightly organized and efficient and antientropic. One of the main tasks of the analyst in specifying new or revised systems is to carve such clean transactions out of a messy landscape full of noise.

. . . *or such precise fields along with textual fields which have a less precise set of attributes (and are therefore difficult to validate) and relationships both implicit and explicit.* Some applications have data transmissions which consist largely of text. These data transmissions are not what we would want to consider as transactions, although they are certainly information. Consider, for instance, the ability to have as input the contents of remote libraries throughout the land without leaving our office or den through the use of a microcomputer with a modem. Certainly, this information was entered into a data store at one time as pure text and was not subject to the same precise field validation that we have been discussing. There were other forms of validation, such as eyeball validation. When the text was machined, it was done a page at a time through an optical scanner.

THE MOTION PATH OF A TRANSACTION

Figure 7-2 shows the flow of a transaction into a system to the point where it loses its identity as a transaction by merging with the database. What follows are checkpoints along the flow showing choices to be made.

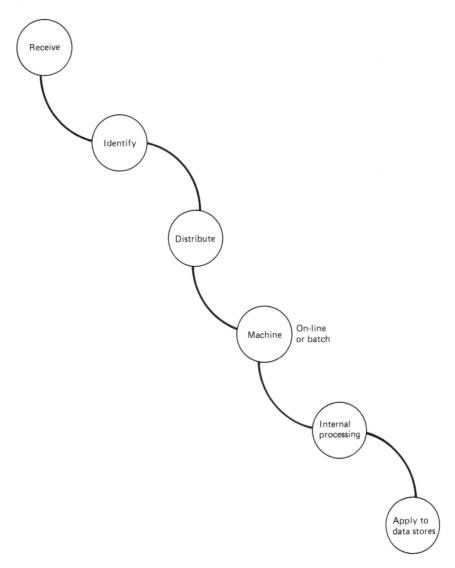

Figure 7-2   Transaction Processing in System.

Receive

An input origination source outside the system sends information to the system where it is received at that function in the system responsible for incoming information. This function may be an airline or hotel reservation desk, a bank teller station, a salesperson at a remote location, an incoming mail receiving department, a radar tracking device, a sonar tracking device, a telephone answering machine, or any one of many other possibilities. The received transactions are logged and counted unless they are ignored.

Identify and Distribute to Next Workstation

Since the system usually handles a number of transactions, the transaction must be identified and distributed. It may be that payments are handled in the incoming payments department and orders are handled in the incoming-order department. Even if the same person does both, it is necessary for that person to switch hats.

Prepare for Machining (assuming an automated system)

First we must strip the transaction to the essential input. If it is a bank deposit along with ten five-dollar bills, we will count the money and verify the bank deposit-slip amount. Then we may add information to the transaction. If the bank deposit was accompanied by a check, we may want to write down the check number on the deposit slip. At any rate, the money goes one way and the deposit slip the other. The deposit slip is the transaction.

Batch or On-line Machining

This is a major choice that the analyst must make when specifying a new system. Do not assume that on-line entry is the only option available. If we are dealing in volume and if we are dealing in turnaround documents that are scannable and if it is not essential to the organization to know immediately about the transaction input, batch processing should be considered. The costs of on-line are several times higher than batch.

Consider the case of our payment that was a turnaround bill. The information that you already knew about the customer could be put in a "keyline" somewhere on the document where it could be scanned on an optical scanner. Say that 98% of the bills that are sent back as payments are straight payments in full. Say that the products sell for one of ten different prices. Then, we prepare in advance something called a "header tub file." This "tub file" contains a deck of scannable documents called *headers*, one deck for each of the ten different prices.

When the payments come in, the straight payments are batched several hundred at a time in back of a common price header. These batches with a header are read through a high-speed optical scanner, which is programmed to make an output record on tape or disk for each customer payment, redundantly adding the one price from the header into each customer payment record. The reader should be able to think of many other applications where this type of transaction processing is possible. Short of videotex, it is the ultimate in high-speed, inexpensive transaction processing.

Videotex is an interesting example of transaction processing where the customer is identified, the physical carrier of the information is electronic rather than wood pulp, and multiple turnarounds are possible. A customer whose terminal, let us say, can be identified, is shown a promotion (transaction 1). The customer places an interactive order (transaction 2). The customer's bank account is made available to the promoter to the extent of the order. The promoter arranges for a prepayment from the customer's bank (transaction 3). Immediately, a ship order goes out from the promoter to a distribution point within 15 miles of the customer's residence, and the goods ordered are delivered to the customer within one hour following the promotion. The customer inspects the goods and, on acceptance, signs and takes possession (transaction 4).

Videotex—or whatever else promoters may dream up—doesn't change the essential dance for the systems analyst. A transaction is still information is still a transaction. It is important in a time of great tactical change to see the relatively unchanging strategies and goals in systems interactions. This is the nice part of doing systems analysis. Unlike systems design, nothing really changes except our understanding of what was really there to know all the time. Basic systems structure is as close to an absolute as we will find in our perceptions of how the world works.

The trick for the analyst is to apply this general understanding to the particular problem at hand. To do this the analyst must understand the underlying code system. (For example, how do bits become bytes become data become information become, for instance, transactions? And how do transactions compare to other forms of data/information-in-motion?[1] And what does it take to make knowledge out of data-in-motion? Hopefully, the underlying pattern is beginning to become palpable to the earnest reader.)

On-line Entry

On-line entry of transactions into an automated system usually involves dealing with a predefined menu-panel system, where the menus guide us to the action panels (screens) where we interact with the information format presented. These are fill-in panels, and we can do five kinds of transactions:

- browse
- inquire
- add
- update
- delete

These are really transactions about transactions, or metatransactions. We

---

[1] Information-in-motion is the better usage, but data-in-motion is in such common use that the two expressions are employed in this book to mean information-in-motion. An exception occurs in any discussion about data transmission: transmission systems are not interested in information content.

may be entering an order transaction to a customer's account, and to do this we may browse the accounts in a certain range and target in on one account to do an update, or if it is a new order we may add a new customer record from the incoming order. If we make a mistake, we may delete the record. Later on, to be sure the update worked, we may do an inquiry.

Keys

Information chunks, in order to be useful, must be addressable. We will be discussing this subject and defining what is a "usable" information chunk in the chapter on data-at-rest. However, in order to understand data-in-motion we must at least know that this information must be addressable once it has a temporary store. When we machine a transaction, it is resident on a disk somewhere in an enormously complex neighborhood. The *key* is the address for all the relationships on the transaction. The key for our payment is the customer identification number. The key for our temperature reading of 31 is Measuring-Device-Code + Date/Time-of-Reading (this is called a *concatenated* key because more than one key element is needed to address all the data elements "owned" by the concatenated key).

Internal Transaction Processing

Now that we have machined the transaction through either a batch or on-line method (there are other forms of on-line techniques that were not mentioned, such as prompting or voice recognition, ending up with the same results), we proceed with the data-in-motion through the internal journey.

Whether we have entered the transactions one at a time on-line or in batches behind headers, the transactions receive the same internal treatment. This means that each transaction runs the gauntlet of the transaction information-processing programs one at a time and is accepted or rejected one at a time. Five types of processing may be called upon, all of them optional. It is quite possible to build a system which allows transactions to be entered, perhaps through a screen system, without any further processing before file update. This is not recommended except for certain types of prototyping. The rule is: process every field in a transaction in every way possible whenever possible.

Validation, for instance, is not always possible. Consider personal computing, where database management systems are available as packaged software. Often these packages cannot be modified to add validation, and none is provided. In this case we must simply use careful eyeball validation.

THE SPECIFICS OF TRANSACTION INPUT PROCESSING

In order to really understand the specifics of transaction processing we will have to dig into a realistic case problem and show the data evolving to the point where the transaction can be applied to the database. This problem we will call A Shop

Workload Transaction. By the time we finish examining the transaction input processing in this case, the reader will have a good hunch about what is on the database and what the output reports might be about. This problem is part of a like group of problems in systems analysis where users want to track a set of quantitative fields by a set of indicative fields. In other words, the user might say: "Show me x and y by A by B by C" (which means C within B within A, where A, B, C are the indicative fields and x, y are the quantitative fields). "And, by the way," says the user, "show me x and y by any permutation or combination of A, B, C which is possible."

The reader who grasps the commonality in this set of problems will find it quite easy to apply a generalized solution to all like problems which are encountered doing systems analysis.

Figure 7-3 shows three interface methods by which our basic transaction can be entered from the external human-oriented system to the internal system of computer programs to do input processing to prepare the inputs for database maintenance and subsequent output reporting.

Do you remember, back in Chapter 2, that we discussed presenting the user with alternative new system specifications, ranging in development and expense from minimum to maximum recommended? Figure 7-3 shows such a case. Often the interface between the person and the database is the area where the main cost increments occur.

In this example the user is running a manufacturing operation. Over 25% of the shop staff is clerical or supervisory. The manufacturing operation is large-scale and does not require a high level of operator skill, except in the area of quality control. The job consists mainly of keeping the machines loaded with input materials, removing the refined outputs, making a careful inspection under microscopes, and keeping a log of information about what is happening. This log is the source of all reporting and querying needed by manufacturing, accounting, plans and controls, and sales regarding the operations and allocations from this processing. The current system is completely manual and doesn't work at all. The management is seriously considering jobbing out the manufacturing operation to outside vendors, who can do the shop work for less money than the current in-house operation. This they are reluctant to do, because the quality of finished goods is better in-house and the organization is proud of its reputation for quality products.

There is one operator to each machine and a group leader for every four or five machines. The log for all machine operations is shown in Figure 7-3, part 2(c). We can see that there is not all that much data to enter. We need the operator's initials, the job code, the machine identification, the start date and time, the end date and time, and the volume of work done. It doesn't sound like much information, does it? The management turn to the systems analyst for help in improving this operation. They hope that automation will help but are vague about details. The operations management have looked into completely automating the machine operation and have concluded that it is not feasible.

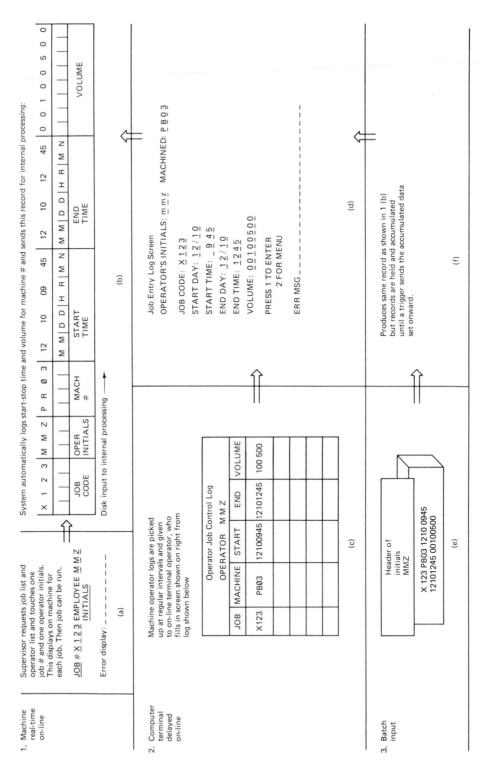

Figure 7-3 Shop Workload Transaction: Three Interface Methods for Entry.

Not too long after beginning the necessary research the analyst makes an initial hypothesis as to where the gold is buried. The reporting from the system is just no good and costs too much. The vendors don't allocate 25% of their staff to clerical work. The machine automation must be taken a step at a time. This situation is very similar to another project the analyst worked on, where good information eventually made machine automation feasible.

The situation is as bad as the management thinks, because no one knows clearly how much anything costs. The sales department cannot do a good P&L, the accounting department cannot do costing, and so on. The analyst begins to work—first working with the user to define the output reporting requirement, then establishing the need for application data stores, and finally dealing with the transactions needed to accomplish the output reporting plus future possible needs. A rough nonparametric cost-benefit analysis convinces the analyst that improvements are possible that would· make the operation competitive compared to vendor operations.

Three input interfaces are proposed to receive the machine logs.

- machine real time on-line
- computer terminal delayed on-line
- batch input

In Figure 7-3 parts 1(a) and 1(b) show the machine real-time solution. All machines would have simple terminals wired to a master terminal on-line to the main computer system. Plant supervisors would assign operators to machines and enter these assignments through the master terminal. This would involve entering job code and operator initials, which could display on the machine also, as shown in 1(a). The machine identification is known to the system. When the machine is started, the clock is started, as is the volume counter on the machine. When the machine is stopped, the start date and time, the end date and time, and the accumulated volume are transmitted from the machine buffer to the master terminal to the computer system along with the other logged information that makes up the complete log shown in 1(b). The machine is turned on and off only from the master terminal.

This solution is expensive—many times more expensive than the other solutions—but the analyst feels, after seeing like systems work in other shops, that the data will be much better and that operator–group leader clerical time will be eliminated. Having sixty machine operators enter data inaccurately onto a log sheet by each machine is also expensive.

This system creates a log record on a disk file ready for internal processing. Anything entered wrong, and caught internally, will create an error display.

The big advantage of this solution is an intangible. Given this much automation of information in place and working successfully in production, a good operations manager will take a new, fresh look at automating the delivery of parts to the machine stations and the handling of output quality control and next-station transfer. After all, the machines are already connected via the master terminal.

A second solution is shown in Figure 7-3 parts 2(c) and 2(d). This is the computer terminal delayed on-line solution. Logs are still done at the machine in the old way, by the operators, and the logs are still checked by the group leaders. The group leader brings the logs to an on-line terminal where they are stacked for a computer terminal operator to enter into a temporary database. Entries that are rejected by the validation program are periodically listed out for correction and reentry. The operator works at night on the third shift, and by 7:00 A.M. the previous day's input has been entered and updated to the database and the hot morning work-in-process reports are ready to be run. The good output to disk is exactly the same record produced in 1(b).

A third solution is the batch-input solution shown in Figure 7-3, parts 3(e) and 3(f). Sometime during the slow third shift the accumulated machine logs are delivered to the central data entry department. The logs are entered key to disk at this location, using the header-detail technique, whereby the initials are entered only once per operator but are incorporated in all the output records. The key entry is done sometime during the next morning and the accepted records are applied to the database. The rejected records are listed but, unfortunately, not included when the reports are run. Corrections are made during the next day. This is the least expensive solution, since the entry work is done outside the department by people who specialize in entry.

SHOP WORKLOAD TRANSACTION SPECIFICS—INTERNAL PROCESSING

Whichever interface solution the user decides on, we end up storing the same transaction. This transaction is now presented to the system of programs that convert this raw transaction into the finished transaction (or transactions) presented for update to the databases.

Whether the user has selected an on-line solution or a batch solution, the program logic is mostly the same. The on-line logic has the program decomposed into modules, and these modules must be *reenterent*, which means that two like transactions can use the same program module at the same time. The batch-transaction data store must be sorted and merged to the sequence of the databases for updating.

Figure 7-4 shows the detail of the internal processing. Shown here are the types of processing that make up the sum and substance of postinterface internal processing before database update. These processes are:

- acceptance
- validation

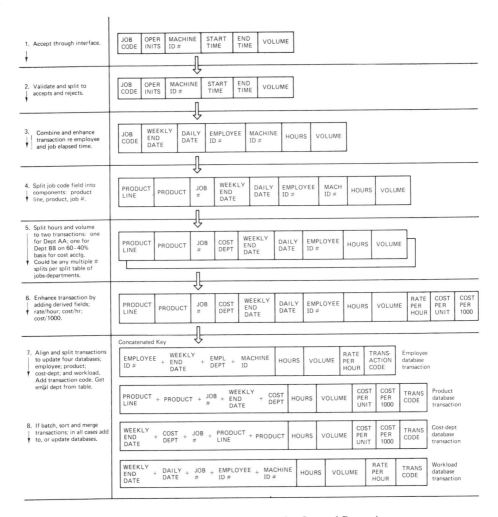

Figure 7-4   Shop Workload Transaction Internal Processing.

- combination
- enhancement
- derivation
- decomposition (splitting)
- alignment
- presentation

By going through the processing shown in Figure 7-4, we will see an example of each of these processes.

We start with an input, which is the output of Figure 7-3. An example of presentation and acceptance is shown in Figure 7-3, part 2(d), where an on-line job entry log screen is shown. The data is filled in on this screen and presented to the program module which validates this data. If validated, the transaction will be accepted. When the transactions are finally refined, they are presented to the database for add or update.

Going to Figure 7-4, item 2, we see that the first internal process is validation. Nothing can happen until the presented transaction has been validated. We have already discussed the need for precise fields in order to do validation. Knowing how to do validation of raw input insures the security of the entire system. Ninety percent of all validation (as a guess) involves lookups to tables and data bases. The other 10% involves string checks, range checks, code tests, and cross-reference checks. *String tests* check the input against a set of acceptable values. *Range tests* check for lower and upper acceptable limits. *Code tests* test for numeric, or alpha, or no special characters, etc. *Cross-reference checks* check transactions for field correlations. If field-*a* is so, then field-*b* is such. Let us look at the transaction in item 2, Figure 7-4, and pick out the validation. The iron rule is: if you *can* validate it you *must* validate it.

1. *Job code* is validated against a table of current job codes maintained by the sales department. We will see that job code is expanded into several sales-oriented codes.

2. *Operator's initials* are validated against a table of operators or against a commonly held employee file. Probably a local shared file is used to look up initials, get a hit, and pick out employee number. Then employee number is used to get a hit against the common employee file. If a hit is made on this file, the initials are good.

3. *Machine identification number* is validated against a local shared machine table or against a common facilities table. A direct hit is required.

4. *Start time* and *end time* are validated in the same way. A cross-reference is possible here, since end time must occur later than start time to be logical; otherwise we would have zero or minus elapsed hours. These times are in the form *MMDDHHmm* where *MM* is month between 01 and 12, *DD* is day between 01 and 28, 29, 30, or 31, *HH* is between 00 and 23, and *mm* is between 00 and 59. We see here a combination of range and string tests.

5. *Volume* is validated by being in numeric decimal code only and in series of ranges depending on elapsed time and job code. Certain rates per hour are believable and certain are not, and we would calculate rate per hour here and compare it to a table of reasonable rate-per-hour ranges.

If our transaction passes all these validation tests, it is passed on to the next stage in input processing. We can see how much validation we were able to get out of precise fields.

Item 3 in Figure 7-4 shows an enhanced transaction with derived and added fields. The daily date field has been derived from the end date. The weekly end date (used for many statistical and accounting reports) has been looked up to a date table using daily date. Start time and end time are gone, and in their place we have hours (elapsed hours to two decimal places). Operator's initials are gone, and in their place is employee identification number (alias payroll number).

Item 4 in Figure 7-4 shows decomposing or splitting at the field level. Job code was a device for efficient and accurate entry of several pieces of information: product line, product, and job number. We now have some valuable sales information and can supply costs of manufacture by product and product line.

Item 5 in Figure 7-4 is our first example of decomposing or splitting at the transaction level. We now have a new field: cost-department. In this case we mean the department, usually a sales department, to be charged for manufacturing costs. This is a debit against the product-line and product P&L and vital to a proper calculation of profit or loss. There is an agreement between sales departments in this operation that costs are to be split between two departments 60%–40%. This was found by looking up splits on a job number table. This table, like all others mentioned, must be maintained up-to-date by responsible persons in charge of a valuable organization resource. Now we have two transactions instead of one. The difference is a different department number indicative field and a 60–40 split of hours and volume.

Now, in item 6 of Figure 7-4, we have some derived fields to add to the transaction. We have made a decision to add this data at input time rather than update or output time. Without going into the details of the calculations, except to say they are not difficult, we have taken a standard labor and fringe rate, a standard materials rate for this product, and a standard machine rate, all of which are changed weekly, got them by table lookup to a weekly rate table, and combined them with hours and volume to produce three new fields. These fields are rate per hour, cost per unit, and cost per thousand. Now we have the fully refined transaction with quite a bit more information than the few fields provided from the machine operator or supervisor.

One final step is needed to produce transactions which can be presented to the databases. That step is shown in item 7 of Figure 7-4. (Item 8 shows the sort and merge operations for batch processing, which we have discussed.) It is here that we take the completed transaction and decompose it to suit the physical structure of the databases. It is here that we produce the all-important KEY which will allow us to apply transactions to normalized relations in the form of a relational or hierarchical database (to be discussed in the chapter on data-at-rest). We see in item 7 that there are four databases to be updated. These databases (let us say) have been constructed to optimize access. In this system there probably are five retrievals for every update. In most systems the ratio is greater—20 to 1 and beyond. Retrievals in an on-line environment must have appropriate response time. If the user is involved interactively with the terminal in an intellectual pursuit where thought process must not be disturbed, then the response time cannot be greater than two seconds. This need for quick response time is what all the database discussion is about.

The four databases involved are employee, product, cost-dept, and workload. To update these four databases we need four transactions in different sequences of indicative fields. Notice that the KEY is a concatenation of those fields necessary to deliver a normalized quantitative field, so that an add transaction can build that moveable quantum of information we will be calling a *relation* or a *segment*.

Now the transactions are ready for presentation. The complexity shown here is not unusual for input processing. This is one reason why input processing is still very much an in-house handmade set of programs done in low-level languages such as Command Level COBOL. We will see the same is not true for output retrievals, where higher-level languages are in common and successful use. Several vendors are starting to produce a good high level language for input processing, and success in this area is inevitable.

THE OTHER INPUTS: MESSAGES AND STREAMS

Messages

Except that it is prefixed by certain obligatory fields, the *message* is different from the transaction in that it is not made up of fields or data elements but rather of information following the normal rules of grammar.

This information is not public information like a newspaper. It is sent from somebody to one or many recipients. It can be either private or shared information.

The usual fields associated with a message may include:

- identification of sender
- addressable location of sender
- identification of recipient
- addressable location of recipient
- message keyword
- subject or reason of message
- request for action such as reply or forwarding
- length of message
- body of message

Examples of Messages and Nonmessages.   A message is not a fully interactive communication; that would be a stream. A telephone call to a recipient (which may end in seconds or go on for hours) is a streaming-type communication.

Voice Mailbox.   A transmission to a remote voice mailbox is a message. In this system the sender has another terminal alongside the regular telephone terminal which allows identification and addressing of both sender and recipient, both of whom have private data stores on a common database. The analog voice message is digitized and stored in digital form on the database at the recipient's address. The recipient calls in for pending messages on a similar telephone setup and receives a reasonable analog approximation of the message. In most systems like this the sender's voice is recognizable. We get what we pay for, because the voice transmission can be sampled on input to the data store. The more sampling, the larger the temporary data store. We pay for storage space. The recipient has the ability to delete or save the message.

A system like voice mailbox would be useful, for instance, if a field service agent called in to get the next assignment from the home office. We have all experienced the "I'm here you're not there" syndrome. In this situation communication is established but the recipient is not at his or her usual location and later gets a written or oral message from whoever answered the phone call. When the recipient returns the call, the sender is not there, and so on throughout the day.

One possible future problem with voice storage is that snoopers can also store the message and then try to use computer programs to flag keywords for selective human eavesdropping. Imagine the saving to a totalitarian government not to have to assign two or three human eavesdroppers to listen to all of the chatty "useless" conversation that goes on during surveillance. Even the most dangerous of anti-government characters have a penchant for idle telephone chatter and call-out food orders. In a more serious vein, what we are confronted with here is the right to privacy.

This last paragraph opens up the whole issue of text recognition. Someday soon we will develop the ability to really recognize subjects and subject relationships in message or textual-type information. The ability to do this will revolutionize information processing and will change our basic ideas about what is and is not precise information to be validated and used to maintain databases. For the present, that day has not arrived; when it comes, it may (or may not!) come slowly in terms of meaningful applications.

Other messages.   Other messages, both input and output, that come to mind include:

- electronic mailbox messages sent to systems for data storage available for monitor display or printing. Very similar to voice mailbox and more common. Voice mailbox cannot be displayed or printed at present.
- textual addenda to standard transactions.
- computer output messages (to be discussed in this chapter under data-in-motion outputs)
- all kinds of verbal and informal written communication that can be only generally defined as either messages or information streams

Streams

To describe the three types of information streaming perhaps we can borrow some terms from the world of data transmission systems. These expressions are *simplex* transmission, *half-duplex* transmission, and *full-duplex* transmission. The use of these words in this context may upset the data communication experts, so let us say that their common meaning is transmission in one direction only (simplex), in either direction (but not both simultaneously) (half-duplex), and in both directions simultaneously (full-duplex).

We will call such activities as turning on the radio or the tape recorder and possibly calling up the radio station later to say we liked the program *half-duplex streaming*. Videotex, which we have already mentioned, is half-duplex streaming.

A face-to-face discussion, a telephone conversation, and a video teleconference are all examples of *full-duplex information streaming*, all being highly interactive. Among some civilized people, of course, the notion of conversations with many people talking simultaneously is offensive. But surely the type of complex interaction that can occur between three or four or more remote locations in a video teleconferencing session, with more than one participant at each location, can hardly be adequately described as "two-way interactive in the half-duplex mode".

The reason we are discussing such activities as telephone calls and video teleconferencing in a book on systems analysis is that these functions are all part of the information usage of the social organization. There is no whole-system understanding without them. Also, analog and digital information now share the same physical transmission facilities. Not only that, but data increasingly is being passed back and forth between analog and digital code forms, as we have seen with the voice mailbox. We also sense that the future of information processing lies in the direction of trying to master information messaging and streaming as we have transaction processing.

DATA-IN-MOTION SYSTEM OUTPUTS: MESSAGES, STREAMS, TRANSACTIONS, REPORTS, AND QUERIES

Messages and Streams

The character of the messages and streams we have just discussed as input remains the same on the output side. Half-duplex information streams characterize the retrieval operations of public databases. It is possible to do keyword retrievals from these databases. An example is a request to a common database that carries newswire reporting to download to a personal computer all stories which contain the keyword "Lebanon" for a given date. This request results in the delivery of an output stream from the common database to a particular computer.

Messages are a vital output of computer based systems, particularly on-line systems. Look again at Figure 7-3. Note in part 2(d) on the job entry log screen the bottom line with the words ERR MSG followed by a response line. This stands for ERROR MESSAGE. If the operator enters a JOB CODE of 4123 instead of X123, the message NO SUCH JOB CODE ON FILE will appear on this line, and the operator will have to correct this mistake to continue. This response is certainly an output from the system and is not a transaction. It is a message. Good computer systems are full of messages to users, telling them when inventories have fallen dangerously low, when pressure is dropping in a cabin, when mistakes have been entered into the system, and so on.

Another vital type of output systems message is the request for additional information about subjects and the attributes of subjects and the relationships between subjects that is an output of the artificial intelligence (AI) system. One of the key attributes of an AI system is the ability to produce such messages and then to receive the response transaction and incorporate it in the knowledge base during interactive processing.

Transaction Outputs

These transactions generated from a system are either a response to an incomplete transaction update or the result of some trigger going off. Sometimes a transaction needs to thread through several computer systems, updating several databases according to variable conditions, in order to be a complete transaction. The system generated transaction that is output from one system and becomes input to another system can be said to be such an incomplete transaction.

Another example of this kind of transaction output is the incomplete feedback transaction. The transaction lacks enough information to be applied the first time and must be recycled. An example is an order to a customer file that is arranged in geographical order. When the order is applied, we must ship the goods to the right address. If the customer has just moved from Connecticut to New York and the order tries to apply to the Connecticut address (in a batch system in Zip order), the first time through the new address is added to the transaction and the transaction is recycled to apply at next running at the correct New York address.

There are many examples of transaction feedbacks. Perhaps something in the database indicates that certain transactions are to be deferred for a given time period. This transaction would then recycle output back to the input backlog.

Reports and Queries

We will deal with reports and queries together for good reason. There is no good definition of one of these outputs which would not also suffice to define some examples of the other output. What we can do is define the attributes of reports and queries which show a tendency to form two affinity clusters.

It is worth giving over some time to this, because the main reason for the type of system called the *decision support system* (*DSS*) is to produce reports and queries. The other general type of system we find is the *operational system*, where the information is dynamically driving the processes of the organization. Even in the case of the operational system the reports and queries are vital in that, if interpreted properly by humans, they prevent system chaos. When in a later chapter we discuss data-at-rest and consequent database tradeoffs we will see that problems involving databases correlate closely to requirements concerning querying and reporting, particularly in regard to query response time.

Report attributes.   Reports tend to

- be printed rather than displayed
- answer several questions rather than just one
- cover a time range (such as quarterly or year-to-date)
- be predefined rather than a "what-if . . . " response
- have a regular distribution list and time
- have less security concerning information
- have better cosmetics (pleasing appearance, packaging)
- be oriented toward linear information
- have more official value (a payroll check is a report, not a query)
- have more derived fields
- have more support text
- have more elaborate summarization, totaling, and cross-footing
- be more likely to be archived (saved)
- be more likely to involve remote transmissions either in collection or distribution of the information
- be more likely to need sorting, merging, and centralized processing and distribution
- go through more rigorous validation
- have more relationship to other reports

Query attributes.   Queries tend to

- be displayed rather than printed
- be responsive to and support interactively the human thought process
- answer one question at a time
- look at things as they are now more than how they were
- be spontaneous requests rather than predefined
- be personal rather than shared information
- not be too concerned about appearance
- not have too much relationship to other queries or reports

- be more likely to incorporate nonlinear information such as color graphics
- have few derived fields, not much support text, simple use of summarization, totaling, and cross-footing, and generally have less complex information, simply presented (which implies that for best querying the data store should have already incorporated this complexity!)
- be less likely to be archived
- be less likely to involve remote transmissions, at least for output
- be more likely to be retrieved from a derived rather than a primary database (owing to the need for easy access to complexity and fast response time)

Note. The primary database is the database that is updated with the system input transactions; the derived database is a condensed (probably) download from the primary database. The primary database is truly current in a real time environment; the derived database is never current but rather is refreshed periodically.

- not need sorting, merging, and centralized services
- not go through a rigorous validation by comparison to other outputs that should balance or correlate
- be used in-line to handle emergency operational situations

Comparing querying to reporting.   It is not enough to say that reports have their place and queries have their place. It is legitimate to raise such serious questions as: Do we really need to go to the enormous expense of new database development and instant response time just to make a report into a set of queries? And, from the other point of view: What good are reports on wood-pulp physical carriers anyway? Don't we always need one answer to one problem at a time? Why don't we go paperless and make our stored information report-fragment genial and what-if query genial and make our terminal displays capable of information splitting so that we can see several data views at one time? As far as having official value, as in the case of the payroll-check report, why not output a transaction to electronically update the employee's banking account and make the confirmation a message in his or her electronic mailbox?

Now that we have asked those questions—some of the big issues in current information analysis—let us say that reports have their place and queries have their place. It is not likely that a respected, useful report will be replaced by a set of queries. However, there is definitely a tendency in the direction of getting the specific answer needed to solve a specific problem and getting it quickly.

From database to final use: the internal evolution of the report or the query.   Whether a report or a query is entailed, the internal computer system processing (and of course any manual system which does the same processing) remains the same. Here are the steps by which information in data stores is refined into reports and queries:

- syntax checking
- collection
- reduction
- extension
- ordering
- accumulation
- formatting
- presentation
- validation

1. Syntax checking.   A request for data from a database goes through several important pieces of software. A query goes through a terminal-computer interface system that manages on-line transactions. An example is IBM's CICS software. The query must conform to the expectations of CICS. Next, the transaction query goes through the database management software, such as IBM's DL/1 interface, to the IMS database. Next is the system operating software, which handles all data-in-motion transfers for the system and manages multiple programs running simultaneously. The query must be acceptable to all these layers of software. Before going through this system software the query must first be acceptable to the syntax rules of the language it is written in. Most queries are now written in fourth-generation languages, and if the query is acceptable to this first syntax examination the prospects are bright for making it all the way through. This is more an input than an output discussion, but in a query session good syntax is the starting point. Reports have the same kind of problems regarding syntax, but the problems have been resolved at an earlier point in time through debugging.

2. Collection.   By this is meant the extraction from the database of all the data needed to satisfy the retrieval. This is the most expensive and troublesome operation in information retrieval for reports and queries. Everything else is really quite straightforward and inexpensive.

For optimal collection the trick is to look at as few relations or segments as possible. The relation or segment (both having the same content in a normalized database—the smallest moveable chunk of data, consisting of a key and attribute fields) is examined to see if it contains data needed for the retrieval; if it does, the data is taken and added to the universe of data needed for the report or query. The way to look at as few chunks of data as possible is:

- arrange the database in such a clever way that it is possible to minimize the number of chunks of data that need be retrieved and examined in the collection process. (This could mean well-thought-out physical database structures, secondary indexing, derived databases, or relational databases—all subjects for a later chapter.)

- *qualify* the request in such a way that only a small part of the database need be collected in order to supply the data, or, if a good amount of data must be collected, arrange to do it as a batch job in the wee orders of the morning when co-workers are not contending for retrieval resources. Qualification means to ask only for what you need, eliminating all else.

Figure 7-5 gives an example of the effects of qualification on collection. It shows an organization that stores different kinds of apples (e.g., Macintosh and Delicious) in different kinds of crates (e.g., wood and cardboard) in different kinds of trucks (e.g., diesel and gasoline) at different locations (e.g., St. Johnsbury, Vermont, and Boca Raton, Florida). The boss, mulling over some opportunity, removes his cigar from his mouth and says, "Hey, kid, how many Macs we got over three inches in diameter?" Our first thought is to hire at least one division of the Yugoslav army to make this count. It is clear from the query that we have to look into every crate in every truck at every location. This kind of a question,

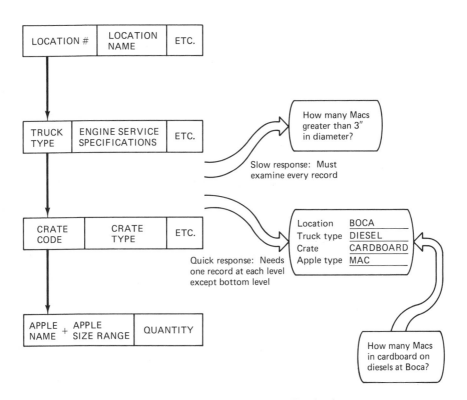

Figure 7-5   Qualification Effect on Retrievals.

which apple magnates like to ask, puts a tremendous strain on information resources.

Fortunately we have an information system to answer this query, but unfortunately our system has a simple VSAM file or a hierarchical database without secondary indexes and therefore we must do what the Yugoslav army might have had to do: go through and collect every segment in the database in order to do the next step in retrieval processing.

The other example in Figure 7-5 shows the type of query that does not bring down the system and kill response time. The boss, still puffing on his cigar, says, "Hey, kid, how many Macs we got in cardboard in the diesels at Boca?" Right off we can eliminate all locations but Boca. At Boca we are interested only in the diesel trucks. On these trucks we care only about the cardboard boxes—hardly a counting job needing a full division of the Yugoslav army. This is the type of request suitable for an on-line environment. (But don't forget—we have other tricks to play in construction of the database in order to handle more complex requests.)

The important consideration here is for the analyst to realize the forces at play and the tradeoffs involved in collecting data from data stores for retrieval servicing.

3. Reduction. This is the process whereby we take those data elements we need from those moveable quantums of data we have collected and discard the rest. One question to ponder here is: are we collecting data for a string of reports or for one report or query at a time? If we are collecting for multiple reports (which is very common), we will want to go to a derived temporary database or databases and make our reports from there—either that or laboriously construct our report programming in low-level languages such as COBOL.

4. Extension. If we have hours and volume on our database and need to show rate per hour, we will divide hours into volume and store the result in a new derived field something like RPH. This extends our data retrieved to include all possible derivations.

5. Ordering. Most data-at-rest has some key order. Usually this is child key within parent key and algebraic sequence of keys at each level. The logic of the report or query may require a reordering of the data store sequence. This is usually done by sorting, and, except for very small volumes of collected data, is not suitable except for batch report processing—which is to say that if you need immediate response to a retrieval request, the request order must match the database order.

6. Accumulation. This is the process whereby we go through our collected and reduced data and accumulate those subtotals, totals, and summaries which are required to meet the retrieval request specifics.

7. Formatting. This is the process wherein the final report in one big stream from top to bottom is given the final refinement and packaging needed for a good presentation to satisfy the input requirements. Title lines, time stamps, text, column headers, footers, and similar packaging are applied to make the completed format.

8. Presentation.   This is the process by which the top-to-bottom retrieval is framed to fit the terminal device to which it is presented.

9. Validation.   In a situation where a group of reports are produced and delivered to a user, the user is responsible for cross-checking the validity of the reports. The analyst is responsible for writing the procedure by which reports can be checked. It is not only banks that need to balance reports at the close of the business day. Too many shops check their input but not their output validity. More work needs to be done in most shops to develop automated cross-report-checking procedures.

A final word on reports and queries concerns the use of fourth-generation languages to produce required outputs. A later chapter will deal with languages, but let us observe here that most reports and queries have a routine programming solution. Languages have been written which use a simple nonserial syntax set that can fully define all but the most complex reports.

Report Specification Requirements

The analyst must build systems that allow users to make their own spontaneous queries with a simple language syntax. Also, for more formal and lasting reports, the analyst must see that there is a design specification which shows a picture of what the report will look like. This picture should be accompanied by a brief report specification page, showing field attributes and other specifics of the report, such as when triggered for printing. Like all other design specifications, this could be done in detail at system specification time. There is no industrywide rigid protocol for this. At the very least, the systems analyst must indicate what information is in the report in a new system specification.

An example of such a specification is shown in Figure 7-6. This shows a graphic picture of the report along with report attributes.

1. REPORT EXAMPLE

Location # name		Truck type	Crate type	Apple name	Apple size (greater than)	Quantity
1	Boca Raton	Gas	Wood	Macintosh	2	1,200
					3	1,300
				Delicious	2	900
					3	1,400
			Cardboard	Macintosh	2	600
					3	700
				Delicious	2	1,200
					3	1,600
		Diesel	Wood	Delicious	3	2,200
						11,100*
2	St. Johnsbury	Gas	Cardboard	Macintosh	2	2,300
					3	2,700
						5,000*
	Total					16,100**

2. REPORT DEFINITIONS:

Field	Type	Description	NOTES
Location #	1 N	Code for location	Either Boca Raton or St. Johnsbury
Location name	15 AN	Name of town or city	Either cardboard or wood
Crate type	10 AN	Box material	Macintosh or Delicious only
Apple name	15 AN	Name of apple	
Apple size	1 N	Inches in diameter	
Quantity	6 N	Quantity of apples	

Note subtotals on location and grand total at end

3. REPORT CHARACTERISTICS:

This report gives current apple inventory stored in refrigerated trucks
Purpose: Fill orders from two locations
Program: APTRK 002
Frequency: Weekly on Monday
Copies: 3

Requirements: Phone for status
Friday 4 P.M.
Telex report Monday 9 A.M.

Sequence:	Field	Algebraic
	Name	Ascending
	Truck type	Descending
	Crate type	Descending
	Apple name	Descending
	Apple size	Ascending

Report-type: RAMIS II

Users: HQ, location managers

Figure 7-6   Report Specification Example.

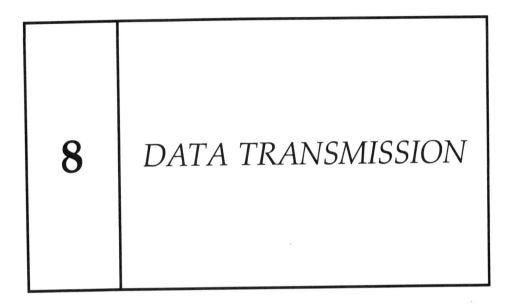

8 | *DATA TRANSMISSION*

Part of the new system specification is the first conceptual view of the new physical system. Without this imaging of the new system no cost-benefit analysis can be done. The application computer central processors and their storage devices are one part of the physical system. All the rest of the physical system is concerned with data-in-motion. This part includes the terminals and the transmission links and the support hardware such as front ends, modems, cluster controllers, multiplexors, concentrators, PBXs, and other components that make up a data communication system. It also includes the support software—the standards and the protocols by which the system is run. It is in this context that we will deal with data transmission.

The data transmission system, then, is the relationship between the host computer's running applications, the terminals that connect the users to the applications, and the transmission linkage that ties this complex together—all with different capacities for data transfer.

Terminals are defined as including any end point for data use which interfaces with humans or with devices under data control. Included are display screens, printers, machine controllers, telephones, video, facsimile, teletype, and so on. All social organizations have manual and automated communication systems (mail service, messenger service, planned meetings, chance meetings, lunch meetings, and beepers on belts are all elements of manual communication systems). Even the smallest organizations now have at least a microcomputer with a display monitor, printer, and keyboard. Many such stand-alone micros also have a modem connection to a voice-grade outside line. At the other end of the spectrum there are organizations with thousands of terminals and many computers connected by complex transmission networks which cross national boundaries and span oceans.

The two basic data-oriented questions to be answered by the physical system specification are: where do we want to store data, and: where do we want to present data for external use? Add to data the need for voice, message, and video communication. These considerations are the necessities which predetermine the range of our solution options. The problem is, given the tradeoffs, what is the most cost-effective system which meets the organization's information requirements? When it comes to data transmission, the options available for complex systems are now quite impressive—and so are the start-up and operating costs!

OPERATORS AND TERMINALS

Terminals interface with people and devices. Putting aside for now the computer–terminal/device connection, we have the terminal–user/operator connection. Also setting aside the ubiquitous telephone, by far the most common relationship involving terminals is that between the human operator and display screens. This relationship can be divided into the following work functions:

- on-line to primary databases
- on-line to personal information resources
- off-line batching to temporary stores
- typing facilities

Examples are the large data entry department converting source documents into batched electronic transactions, or the department clerk entering transactions on-line, one at a time, or the typist converting source material into memos and letters, or the engineer using graphics for computer-assisted designing.

Information-Retrieval and Transaction-Entry On-line Systems

These systems exist in two forms:

- readout-only
- two-way transactions

Examples of *readout-only* are customer inquiry systems responding to questions about bank, insurance, or retail account status or an inventory availability system.

Two-way transactions might involve calling customers to see if they wish to renew a classified ad in the phone book from a readout and then updating the account record according to the response—or an airline reservation system where flight-seat availability is first checked and then the seat is removed from the available inventory of seats on that flight.

Personal Information Resources On-line

This service is directed to the managers and professional staff and their assistants in an organization, as well as to students. Typical available resources would include

- word processing
- private databases
- graphics
- spreadsheets
- access to remote databases via modem and voice-grade line
- application software, such as financial analysis programs
- word and text analysis, such as spellers
- terminal-to-terminal direct communication
- terminal-to-terminal time-delayed mailbox communication

Information-Entry Machining Systems in the Batch Off-line Environment

This group of systems (which have replaced card keypunching) should be serviced by as many dedicated minicomputers as is needed to keep response time under two seconds. Some data-entry departments involve hundreds of screen terminals. In large departments service personnel should be on hand at all times to keep the computers running. The output of these front-end computers goes to disk or tape files for entry into the main update system on the mainframe computers. Direct input to the main frame is risky for large-volume operations.

The main problems found with this type of terminal use are

- operator comfort—good working conditions
- operator productivity, measured in stroke rate per hour
- operator quality, measured by verifying entire transaction batches or samples by having a second operator duplicate the entry of the transaction previously entered by the first operator

Most good dedicated batch-entry systems have associated software which measures operator productivity and quality and distributes entry costs by cost center.

Good working conditions involve the usual ergonomics (hygienes) of sufficient and well-placed lighting, adequate workspace, low noise levels, and wise placement of keyboard and display screen. In addition, rest time is very important. There should be a reasonable rest period in the morning and afternoon and a proper lunch period for this group of workers (30 minutes is not enough!). This type of work is particularly exhausting, and operator turnover, drops in productivity in the afternoon, and poor-quality system input are directly proportional to fatigue level. And yes—this is a proper concern for the systems analyst.

It has been possible to accumulate good statistics regarding the work patterns of data-entry operators. Since they are on-line to a computer, the system can know when stroke production is interrupted for longer than, say, ten seconds. The average operator works about two-thirds of the time he or she is logged on to the terminal.[1] The interesting fact is that operators with below-average ratios of stroke time to logon time often are more productive. These operators are the accurate binge workers who turn out more and better work than the slow plodders who are stroking all the time. The point is that analysts of these stressful data processing departments should recommend experimenting with increased planned rest and lunch time, using whatever statistics are available to monitor the experiment.

Information Machining Systems On-line to Primary Databases

This is the same kind of difficult work just described, with the added burden of correcting mistakes in entry on the spot in response to error messages. Another vexing problem in this mode of working concerns response time. So many other group needs are impacting the primary database that poor response time is common. This is not only costly to the organization but a cause of stress to the individual worker. Also common to these systems is downtime. It is easier to crash an integrated organization database and the supporting communications software than it is to bring down a personal computer or a dedicated minicomputer that has transaction machining as its only function.

Typing Facilities

This involves information machining at the department or personal level, usually handled by the same kind of equipment and software—which, if the work is computerized, is a microcomputer network sharing a printer. Many organizations operate in two modes, having a central typing department as well as secretary typists in department workstation clusters. Each operator has a loadable word-processing program, a storage medium (such as 8-inch floppies), and a terminal. The letter-quality printer often is shared.

Secretaries, besides machining, have the responsibility for editing documents for spelling, grammar, and often substantive accuracy and felicity of expression.

---

[1] Statistic from the author's personal experience.

To this extent they are not machining but also participating in the work process as assistants. Secretaries have the same ergonomic problems as the other machine operators and the additional problem of constant interruptions to handle their other functions, such as answering the telephone and answering all sorts of questions. Often they are located in high-traffic areas outside of offices in an atmosphere not conducive to concentration. On the other hand they are often justifiably proud of the variety of their tasks, even though the atmosphere is stressful. Also the secretary or typist is required to master the word-processing software.

Information-retrieval and Transaction-entry Systems

These are a variant on the on-line systems where the primary database is involved. This environment usually involves a clerk dealing with a customer at a remote location, such an an airline reservations agent processing a flight request on the phone. Of course the customer could also be present in front of the clerk. There are two kinds of work functions: (1) a request for information, which involves a readout-only transaction, followed by (2) some entry transaction which alters the database.

Personal Information Support

We have already mentioned typical resources available to the professionals and managers who make use of personal information support systems. It is important to add to this list the access to the organization databases through whatever type of terminal the user has available. Making this information available to this group, along with the personal computing already mentioned, is a major challenge for the systems analyst. The benefits for the organization are enormous, sometimes more important than mass machining and the operation of the on-line systems by the clerical staff.

Needs and Costs

All these different groups of people performing these different organization functions have certain needs in common which must be addressed by the systems analyst. These needs govern the eventual configuration of terminals, computers, and transmission links. The only other relevant need is that of the organization not to go broke providing an optimum information-flow facility which is not cost-justified. At this point communications specialists and systems analysts have to confront the often baffling task of the CREATIVE COMPROMISE. Here we enter the highly technical area of TRADEOFFS, which is such an important aspect of data transmission analysis and design.

Needs of Operators Using Terminals

Putting tradeoffs aside for the moment, here are the needs of operators using terminals:

- feeling comfortable with the terminal
- feeling comfortable with the software
- being unaware of the transmission circuits

Feeling comfortable with a terminal means not having eyestrain, neckaches, backaches, cancer from radioactive pollution, or unnecessary fatigue. For the lucky individuals doing creative work it means enjoying your terminal as a personal familiar or uncomplaining friend, much as we enjoy that most intimate of all terminals, the telephone. If one doesn't have a personal friend but must share a terminal with others, it means not having to wander the halls, as a supplicant, looking for a terminal not in use. It is amazing how many such terminals have no user but are logged on as if the user had just left for a minute. To feel comfortable with a terminal means having one in the first place.

To feel comfortable with the software means not to be left somewhere in the system without knowing how to get somewhere else. It means good menus and action panels which clearly describe all the possible alternatives. It means a two-track access system which is reasonable to use as a beginner and easy to shortcut as an expert. It means no bugs. It means no messages that require access to documents and manuals to decipher. It means error checking to stop mistakes as soon as entered. It means maximum versatility for the expert, combined with a more modest subset of functions quickly available to the beginner. It means integration with other software—so that, for instance, the user can move easily from information access to various types of information retrieval to database update. It means good training facilities and available wise heads for consultation.

To be unaware of the transmission circuits means to be able to direct-dial a person in Stuttgart, West Germany, almost as easily and quickly as making an intraoffice call. Why should the user have to be concerned with in-house wiring, the PBX, the trunk line to the local common carrier, the link to the long-lines common carrier, the various communication networks involved, and so forth? The same should hold for accessing a remote computer or a remote terminal such as a facsimile copier or a screen display. It should not seem like a world-changing effort (even though it is). The quality of the transmission should be what we bought and paid for.

It should only be the organization treasurer or whoever is responsible for signing checks from the accounts payable database who is painfully aware of the impressive costs of the transmission links, not the terminal user who is using the system according to the system specifications.

THE HOST COMPUTERS AND THE DATA STORES

The term *host computers* is used to differentiate between those computers dedicated to the organization applications as opposed to those dedicated to data transmission. Two problems associated with computer hardware usage affects and are affected by the data transmission problem.

- the problem of centralized computing versus distributed computing
- the problem of computer costs versus network costs

The quality, reliability, and cost factor of a data transmission system will go a long way toward determining whether an organization with a need for distributed processing is able to act on that need. There are some compelling reasons to decentralize.

Reasons to Decentralize Computer Resources

There is a worldwide pull in the direction of decentralization on all organizational levels, including the highest political levels. At the current state of the world grasp of information systems there is considerable chaos in all densely hierarchical systems which have centralized controls. Information flows up from the base level of the hierarchy. This is where the processes work. This information, not having a proper code and delivery system, is distorted significantly by the time it reaches the centralized leadership. Serious mistakes are then made. Conversely, decisions arrived at by the leadership do not result in action at the base levels. The information system has failed, and chaos slowly takes over. This is felt across the world in an existential sense and creates an urge for local control in every consciousness.

If that thesis is too much for the reader to accept, then let us just talk about the dissatisfaction the regional offices have with the headquarters operation. HQ has the super computer and the corporate database. Regional can't get their information on time, correct, the way they want it, or at all. HQ information services are gridlocked and focused elsewhere. Regional acknowledges that HQ has a need for regional's summarized information and that some information should be kept at headquarters and downloaded to the region. The answer from the regional point of view is a distributed computer system supported by a network communications system. Then the salespeople in the field would be able to get the information they need to support the local sales effort better than in any other region.

And what about the new banking merger, combining into one organization fifty banks that until recently were three banking organizations with their own computational facilities and methods of organizing information? Do we propose now to do this work all over in a few months and create one centralized information entity? It is not likely to happen.

These are just two examples of the urge toward decentralized computer resources.

There are also compelling reasons not to decentralize the organization computational facility. Improved data transmission facilities also act to make a case for remote terminals linked, through satellite computers or directly, to a central computer center.

It is certainly easier to conceptualize the relationship between one host computer and all the organization terminals. This does not rule out personal computing. A microcomputer could double up as a terminal on-line to the databases.

This brings up the concern which is the most compelling reason for centralized processing, once the question of transmission quality is resolved. The common data of the organization, the integrated corporate primary database, really must be resident at one location. This database cannot exist as multiple updatable copies at multiple locations. Also the technical staff needed to administer this database and put out fires in the communications system really must be located in one place. It is hard enough to staff one location with competent systems specialists. Multiple-location staffing is not feasible.

Mainframe load sharing between computers at remote locations is advanced as a reason for a distributed system. As long as we understand that the common database is not distributed, we can work our way into using this approach. Now it becomes a matter of cost tradeoffs whether it is better for the organization to keep multiple mainframes at a central location or to distribute them. In either case there is load sharing and back-up. If there are shared and personal applications at the distributed locations that require independent access to mainframes, then this factor will greatly influence the decision.

Figure 8-1 shows several examples of terminal–computer configuration, each satisfying a different requirement. None of the examples takes into consideration link attributes such as distance. The host computers in example (d) might be located in the same physical room or across a continent; the logical relationships remain the same. These examples by no means exhaust the possible configurations.

Figure 8-1(a) shows a single host computer supporting a network of terminals. Many organizations have several autonomous systems like this with a minimum of cross-communication—perhaps only telephone, facsimile, and personal visitation, perhaps minimal communication of summary data across public dial-up lines.

Figure 8-1(b) shows a data communication system where a host computer is connected to three satellite computers, all with their own shared databases and terminals. The host computer has the common database. Communication between satellite systems is only through the host system. Figure 8-1(c) shows the same network with the difference that the satellite systems can speak to each other and share local databases on an any-computer-to-any-computer basis.

Figure 8-1(d) shows the load-sharing system based on three host computers. Note that the common database is resident at only one of the hosts. It is possible in certain organizations to have a fresh copy of the common database at each location and somehow update in a batch mode to the one common database, but this is the exception rather than the rule.

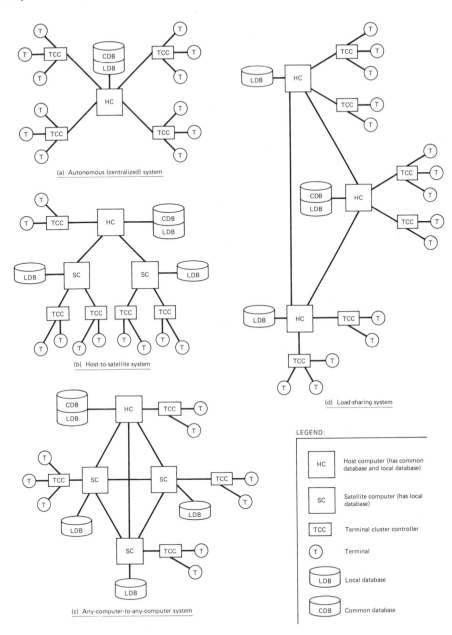

Figure 8-1   Terminal-Computer Configurations.

TRANSMISSION-LINK SYSTEMS

The third element in our computer–linkage–terminal physical system is the transmission function itself. This is a most complex subject, which we have tried to circumscribe by first discussing terminal and computer requirements. The role of data transmission is to enable communication between terminal and computer, terminal and terminal, and computer and computer.

The rest of this chapter will give the systems analyst as generalist enough of an overview to be able to deal intelligently with the communications design specialist. The reader is advised also to read certain key books that deal only in this subject[2] and to pursue further formal training. For the working systems analyst there are excellent short courses available.[3] It has already become impossible for the systems analyst to function in a large organization without a working familiarity with data transmission analysis and design.

It should be understood that any mention of *terminals* encompasses all terminals, including such communication devices as voice, facsimile, and video terminals.

Geographical Considerations In Transmission

From the point of view of an organization not in the communications business, but in need of organizationwide communication services, there are four geographical linkage problems:

- links within a workplace
- links between locally clustered workplaces
- links between remote workplaces
- links to the outside environment (such as customers)

These categories are not mutually exclusive. All organizations require links within a workplace and within workstations in a workplace. All require links to the outside environment. Most of these organizations require one or both of the other types of links.

Types of Links Available

The communication services available all boil down to combinations of the following basic choices:

---

[2] (a) James Martin, *Systems Analysis for Data Transmission* (Englewood Cliffs, N.J.: Prentice-Hall, Inc., 1972); (b) IBM, Synchronous Data Link Control—General Information, GA27-3093; (c) International Standards Organization, Data Communications, ISO 3309; (d) IBM, Systems Network Architecture—General Information, GA27-3102; (e) Mischa Schwartz, *Computer–Communication Network Design Analysis* (Englewood Cliffs, N.J.: Prentice-Hall, Inc., 1977), pp. 1–57, 340–347.

[3] Memorex Training Course On Data Communications Networking. Recommended.

- making your own private links
- leasing a private linking system
- leasing a group-shared linking system
- dialing up on a public system

Networks

Spelled out in a little more detail, the choices are:

- permanently connected private-line networks in workplace
- permanently connected private-line networks in local clusters
- permanently connected global private-"line" networks
- permanently connected exclusive leased-"line'" networks
- shared leased-line networks
- dial-ups to leased-line networks
- dial-ups to public common-carrier networks

Note that instead of talking about links we are now referring to networks of links. A *network* consists of all the connecting links that allow communication between computers and terminals resident in multiple locations. Networks also include the individual link switch points where data is collected and routed through to the next link, creating a total-path transmission from start point to destination.

In the case where the network is not a private system but is partly leased or dial-up, the organization network is mapped out onto a shared or public network. The same basic network structures exist in either case. The organization network is usually put together by using mixes of all the combinations shown above. Taking a long view and seeing the local workplace or workplace cluster as one point on a global network, Figure 8-2 shows some examples of different types of networks (different topological configurations), each serving to optimize some aspect of transmission tailored to the organization.

In the STAR network shown in Figure 8-2(a) a single host computer controls all transmission traffic. By host computer we also mean a special-purpose front-end computer that relieves the host of this job. None of the points communicate with each other except through the host. The transmission is either the distance between the host and each destination point or the distance from entry point to host to destination point.

The ANY-POINT-TO-ANY-POINT network shown in Figure 8-2(d) offers the shortest links and the most communication possible at a high level of complexity.

The RING and the TREE networks are both examples of multidrop networks, with each location a potential drop point. The message is so routed that it will be accepted only by the destination point, passing by other points without stopping.

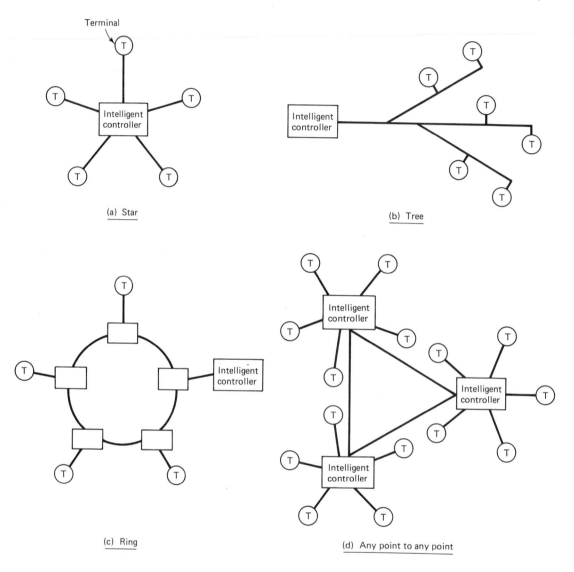

Figure 8-2 Network Topology Examples.

Systems Analysis and Design Considerations and Tradeoffs

We are trying to optimize a number of tradeoff considerations regarding data transmission, so that the organization may get the best possible results given the cost limitations. Of course, as pointed out earlier in this book, we will provide

alternative physical-system solutions at different cost levels. The design considerations (which relate to cost tradeoffs) are

- how much data
- how far
- at what delivery rate
- through what complexity (line capacity, traffic flow control, number of connections, routing algorithms, etc.)
- at what quality
- at what confidence level
- with how much security protection

These considerations are part of the job of the systems analyst. The communications specialist should be able to receive a well-thought-out answer to the
question: *how much data*? The volume of data by data type and delivery-rate
requirements or the number and length and distance of telephone calls will translate
to such considerations by the specialist as bandwidth requirements (where bandwidth is the quantity of data passed on a medium per unit of time). This will
translate to certain physical requirements for "lines."

How far? will translate to type of network and organization network topology.
At what quality? will translate into type of conditioned line (each "condition" having
a certain guaranteed level of performance and certain restrictions such as number
of drop points) and the acceptable error rate which a vendor must guarantee. *How
much security*? will also affect vendor selection.

At what confidence? will translate to the assurance the system has built in
that a sent message becomes a delivered message. *Through what complexity*? will
concern network topology and be a major cost factor.

Transmission Characteristics

When trying to define the different characteristics of data transmission systems we
enter into an area of imprecision. This difficulty arises because of the speed at
which new developments in this field are occurring and because vendors, in competition with each other, are defining their own product to stay competitive with
products using other technologies to accomplish the same results. Add to this vendor
scrambling a lack of industry standards.

Having given this caveat, we will now proceed to define the subjects with
which the systems analyst must have some familiarity.

1. Line speed.   "Line" is defined here to include both wired and wireless
physical transmission links. Lines fall into three speed groups: slow-speed lines,
medium-speed lines, and high-speed lines measured in bits per second (BPS).

Slow. Any BPS rate below 2400 is considered to be in the slow range. Personal-computer modems usually transmit at 300 or 1200 BPS. Message systems such as telex or teletype perform in the 150-BPS range.

Medium. BPS rates between 2400 and 9600 are considered to be in the medium range. This range consists mostly of voice-grade line transmission, such as the telephone lines. However, as this chapter is being written, AT&T has announced a high-speed data capability for voice-grade lines which will achieve speeds of 56K BPS (56,000 bits per second) and accommodate voice, data, high-speed facsimile, and other graphics on existing telephone lines.

High. High-speed transmission is considered to be any rate over 48K BPS. Lines operating at high speeds are called *wideband*.

Bandwidth. Data communications specialists use *bandwidth* to mean the potential of a line to carry a given BPS rate.

Line conditioning. Conditioning improves transmission quality by limiting drop points and compensating for line distortions such as the signal-to-noise ratio. Conditioned lines have grade codes such as C1, C2, D1, D2, and lease charges vary according to code type.

2. Lines and links. The physical carrier of information is perhaps better called the link rather than the line. A *link* is the physical connection between adjacent nodes and eventually between start points and destination points. The logical connection of the transmission is called the *path*. The path is invoked during a *session*, which is the connection of physical links to form a path. A direct line from start point to end point is a case where the link is always the same as the path.

The most used mediums for links are:

- *in-plant cable*—under the floor; above the ceiling; in the wall
- *telephone-type twisted wire*—both underground and on poles. The current transmission rate is 9600 BPS, but this is expected to be increased significantly by such public carriers as AT&T, so that the goal of public integrated service digital networks (ISDN) can be attained.
- *coaxial cable wire*—very high rates of transmission, exceeding one million BPS
- *fiber optics*—using either laser or light-emitting diode (LED) as the carrier. A fiber optics rod is made of glass or a glasslike substance and is only about 25 microns in width, which is just about visible to the eye; these rods are packaged together to form many lines through which the light passes, carrying data—truly an amazing and fully realized technology. The quality of the line

is superior to that of existing wire lines. Fiber optics lines are either strung aboveground on telephone poles or underground through leased conduits. Extremely high speed transmission

- *microwave*—limited to line of sight; 2–20 million BPS
- *satellite*—expensive, but 9600 to 50 million BPS and covers large areas

It may take several different kinds of links, used together, to put together an organization's communications network.

3. Modes for data transmission.   There are three modes:

simplex.   Data can move along the line in only one direction

half duplex.   Data can move along the line in both directions but in only one direction at a time

full duplex.   Data can move along the line in both directions simultaneously

4. Transmission units.   Data is transmitted either asynchronously or synchronously. In *asynchronous transmission* the bit pattern for each character, preceded by a zero start bit and followed by a one stop bit, is transmitted independently of other characters in the stream. The sending terminal unloads its buffer this way and the receiving terminal loads the buffer in the same way. Asynchronous is usually associated with low-speed transmission.

Synchronous transmission is block transmission and is used with data transfers at all speeds. In block transmission the data is prefixed and suffixed by coded data describing the block and the session. Based on the information in these blocks it is possible to break up the block into *packets* of data and transmit through a *packet switching network* which can optimize packet movement from node to node, varying the packet flow according to line conditions at the time of transmission so that one packet of a block goes along one path and another packet of the same block follows a different path. Whenever the packets arrive at the destination node and in whatever order, they can be reformed to become the exact original block.

5. Multiplexing data flow.   Multiplexing involves line sharing. A *multiplexor* gathers several transmissions at a lower BPS into one transmission at a higher BPS for transmission on a faster line. Each message retains its own data uniqueness. An inverse multiplexor at the destination point separates the messages to individual lines. Figure 8-3 shows an example of this process. A host computer at the start point sends eight 1200-BPS transmissions to a multiplexor, which gathers these messages to one 9600-BPS voice-grade line for transmission to a remote point. Since we are sending data over a voice-grade line which accepts only analog transmissions, we have a *modem* to modulate the data stream into the analog stream. At

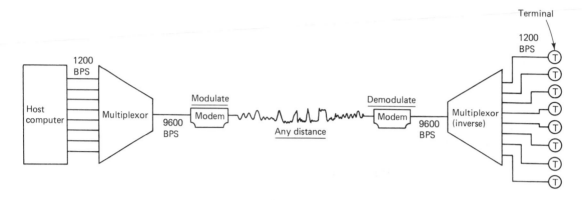

Figure 8-3 Multiplexors and Modems.

the other end of the line we have a modem to demodulate the analog back to a data stream and then an inverse multiplexing operation to separate the one line to eight lines, all going to terminals or computers on different 1200-BPS lines. (At this speed the end point would be either a terminal or a microcomputer.)

Three kinds of multiplexors are in use: TDMs, FDMs and statistical multiplexors. *Time-division multiplexors* (*TDM*) allocate a time phase to each line that is connected and manage the incoming traffic from the terminals through this time division. *Frequency-division multiplexors* (*FDM*) divide the available bandwidth up into separate channels, each assigned to a connected terminal. Both TDM and FDM multiplexors represent a technology that is being replaced for the more complex terminal connections by the statistical multiplexor.

The *statistical multiplexor* is a computer that can perform quantitative analysis of a physical system and program algorithms to optimize channel and line usage. It can also use *polling* to query terminals for available messages and improve message routing time.

Some multiplexors now have the built-in ability to do switching.

6. Front ends.   The computer that stands in front of the host computer doing multiplexing, being a programmed machine, can also do other functions such as block formatting. It is often advantageous to offload to a front-end computer some of the transmission operations that could be done on a mainframe. This frees up the mainframe resources for other work. Meanwhile, the host computer has such operating-system data communication access methods as IBM's VTAM (virtual transmission access method) to connect the application logic to the physical transmission facility.

7. Node operations.   Links in a packet-switching network system are connected by *nodes* which collect, check, and route data blocks and packets. This work at nodes is done by computers.

8. The private branch exchange (PBX).   The PBX serves as the connecting device between the internal organization lines and the external lines going out of the building in order to provide the best possible line service, so this is essentially a switching device. The switch is an important consideration in network planning and is considered to be the dividing line between the domain of the mainframe and the domain of the external communication system.

9. Local-area networks (LAN).   It is not easy to exactly define a local-area network. It is hard to get two communications experts to agree on just what is the geography of the local area. We will approach this much-discussed topic the same way we approached the definition of report versus query in the last chapter—by identifying what it is most likely to be.

First, a local-area network must be the other choice in networks besides global-area network. Therefore, we might begin by saying that the local-area network is geographically limited. An example might be a local bank with a headquarters and some branches. But some would say it would usually be within 5000 feet.

Second, a local-area network usually has as an important ingredient a high-speed computer-to-computer data transmission, although this is not the only transmission service offered.

Third, there is usually a perceived opportunity to significantly cut costs per device connection over using a common-carrier service.

Fourth, a local-area network may be part of a global network scheme, where a group of local-area networks are linked by a global network over large geographic distance.

Finally, the local-area network is usually private rather than shared, or at least has a major private component. In this case the organization, having a good communications group, feels secure enough to act as a sort of general contractor, putting together private lines, leased lines, and public lines (and all the supporting equipment and software) into the local-area network.

Maintenance of the network facilities turns out to be a very significant part of the total setup and operating cost. The organization must be prepared to maintain some of the equipment, arrange service contracts for the rest, and, in general, deal with a number of vendors. The alternative, which could make heavy inroads into the local-area-network business, would be for some government-regulated public carrier such as AT&T to offer such an attractive LAN package at a competitive price with such good service that any alternative to that solution would be foolish. (And, of course, AT&T and others have such a package in the works as this chapter is written.)

10. Value-added networks (VAN).   A value-added network system adds something of value to the basic transmission service of links and nodes. This is best explained by an example. In Figure 8-3 we have a transmission of data at 9600 BPS from point-to-point, but nothing is said about whether the data is in synchronous or asynchronous format. Perhaps we are sending a synchronous transmission but the receiving point receives only in the asynchronous mode. A value-added system perhaps might offer as an added service the automatic conversion of synch to asynch.

STANDARDS, PROTOCOLS, AND ARCHITECTURES

A good start has been made in establishing much-needed communication standards, protocols, and architectures, which we will now attempt to cover.

The ISO Seven-Level Standard

Earlier in this book we dwelled for many pages on the technique of leveling, through which we can understand any natural system. This leveling technique has been applied to data transmission to establish a standard. Figure 8-4 shows a seven-level standard defined by ISO (International Standards Organization). The standard defined by ANSI (American National Standards Institute) is virtually the same on this high level.

What we see in Figure 8-4 is the ISO seven-layer cake. The bottom level is the *physical*. This is the processor-to-multiplexor-to-modem physical description.

Level two is the *link* level, which is a functional description—a set of rules for information exchange between terminals. These functions include initialization, framing (block prefix and suffix data), link management, error control, flow control, and recovery.

Level three is the *network* function, such as addressing nodes and routing packets.

Level four, *end-to-end transport*, covers transport services such as link and node operations and confirmation of message (a standard whereby the sending point receives positive confirmation from the receiving point that the message did indeed arrive).

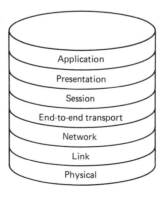

Figure 8-4 International Standards Organization (ISO) Data Transmission Standard.

Level five is the *session*. Recall that a session converts all relevant links into one optimal path. This level defines the physical-link path-control software.

Level six is *presentation*. Blocks of data are transformed at the end point so that formats are understandable and available to the application programs. This interfaces with communication software, such as IBM's CICS.

Level seven is the level of the *application* program that displays the data for an operator or processes the data in user terms, such as order and payment processing or electronic mail.

Architecture

In an *architecture* the protocols and relationships of data transmission are stated in detail and based on a set of standards. Figure 8-5 shows a very high-up overview of one such architecture: the IBM Systems Network Architecture (SNA).

The figure shows data flow from application area A (the sending application) to application area B (the recipient). Application areas involve either computers or terminals at the end points. For instance, in a hotel reservation system application area A might be a terminal requesting reservation information from a remote computer at application area B. A second SNA transaction would be the transmission from the remote computer (now area A) to the same terminal (now the application area B receiving terminal). SNA is not software or hardware; it is rules and regulations about the software and hardware.

The *application* is not a part of the network, whether it be a reservation system, electronic mail, a voice mailbox, or whatever. The system is architected in such a way that each level is independent of the other, which is to say that the application level, for instance, can be changed without changing the other levels. The network is transparent to the application if the network is operating successfully within design limits.

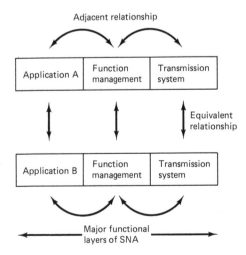

Figure 8-5 IBM Systems Network Architecture (SNA).

The *transmission system* is responsible for connecting, routing, moving and delivering data from A to B. It is therefore responsible for session operations where physical links are made into transmission paths (path control).

Function management is responsible for the presentation of the data to both the adjacent and the equivalent layers. This involves control and manipulation of the prefix and suffix data into which the application data is packaged. This data is used all through the transmission system for such purposes as packet switching, error control, drop addressing, and so on.

IBM's SNA involves a major effort which is only touched on here. SNA is much more involved than a set of protocols. It is truly an architectural specification, which leaves the builders of transmission systems very little room to dream up their own interfacing. This type of rule-setting within the official national and international standards should go a long way toward clearing up the confusion in data communications.

THE SYSTEMS ANALYST'S ROLE REGARDING COMPLEX PHYSICAL SYSTEMS THAT INCLUDE TRANSMISSION SYSTEMS

The brief catalog and description of data transmission elements given in the last half of this chapter should enable the general systems analyst to block out the type of physical system which will correspond to the user's wish list of requirements. This blocking out work must be done with the collaboration of the data communications specialist.

Let us say the user's wish list included a telephone on every desk, a display terminal at the workstation of every user, electronic mailbox, voice mailbox, at least one video teleconferencing room at every office location in the network, computer load sharing at remote location with private and shared data by location and central common data, maximum two-second response time at display terminals, no exchange busy signals on telephones, and museum-quality facsimile transmission. This is not to mention fast throughput of the systems development and application backlog that is to be specified and built. The analyst must be able to put a price tag on all elements of this wish list. In terms of transmission systems the cost to build and maintain must be compared to the cost to lease and the cost of dial-up to a common carrier. This analysis will be a joint effort, undertaken with the support of the communications specialist.

The analyst must also know how to translate a transmission system into comparative benefits. Only then can the user choose between possibilities. There is nothing unique about doing cost-benefit for a data transmission system. There are tangible benefits and indirect benefits that are hard to quantify. There are development costs to put the system on the air and there are operating costs subject to inflation. Operating costs can drag an organization down much faster than one-time development costs that do not involve major borrowing and have the effect of lowering operating costs.

What is different about doing transmission system costing is the high stake required to get into the game ($500K for one video teleconferencing room?) and the disastrous nature of obsolescence. An organization can spend a fortune wiring a LAN with fiber optic links and then have the local telephone company change the rate structure and the service offerings so favorably to the user that the original LAN is cost-obsolete before it is ever used. On the other hand, how long can an organization wait for the ultimate system before taking some action?

As a final word, let it be said that data transmission is one of the more exciting areas in which systems analysts must get involved. If the analysts find that they are becoming experts in this field, they will start running around the organization with green and red wires sticking out of their pockets and exotic metal boxes with interesting dials cluttering up their offices. A pencil and a yellow pad were never visibly very exciting tools anyway. Also, senior management is usually more enthusiastic about spending money for data transmission systems than for other data processing work. The software comes, all done, with the system, and the system goes up quickly as compared to an in-house system that needs design and programming. The benefits are immediate and attractive.

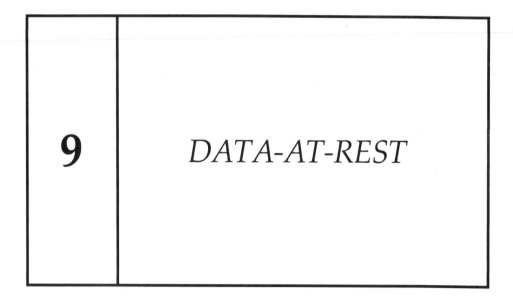

9 | DATA-AT-REST

HOW DATA-AT-REST RELATES TO DATA-IN-MOTION

The motion data in a system always determines the resting data.

We do not store data that never leaves the system in some output form unless it is some kind of control data or a trigger that fires off an output. We might make an investment in a data item that has no immediate output use, but we do so on the assumption that it very likely will have such a use in the future.

All the data in the data stores gets there by adds and updates resulting from transactions moving into the system.

In purely decision support systems the need for retrieval determines the data stores. The sum total of retrieval views is extrapolated back to become the minimally redundant stored data, which is then extrapolated back to become the necessary transaction data to produce the desired data stores.

In an operational system the transactions assume a separate importance. The validation of "raw" input transactions requires unique data stores, such as a table of acceptable data-item values or a table of replacement data-item values (see the discussion in Chapter 7). Perhaps 90% (author's guess) of raw-transaction validation and expansion involves lookups to data stores.

Also these operational transactions have a significance of their own, apart from the activity of providing retrieval information. Deposits or withdrawals to bank accounts have a meaning apart from such outputs as the monthly statement of activity on the account or the total money available for lending. As a matter of fact, many source input documents that translate into machined transactions are physically stored for long periods.

Therefore, to arrive at data stores (for an organization system, a subsystem of that organization, or a specific application in support of that organization) we perform the following three activities:

1. Establish the data used to satisfy the output requirements, which consist of:
   - reports
   - queries
   - output transactions (for multiple updates or feedbacks as discussed in Chapter 7)
2. Establish the data requirements of the input transaction set.
3. Establish the logical control information if any (such as imbedded switches) and the triggers. These triggers define *when* events will take place: events such as an output report or a purge of aged records.

Note. While it is necessary to consider controls and triggers relative to data stores, it is not advisable to imbed these functions in the data store if it can be avoided. An example of a control is a counter in a parent segment of a tree database that keeps track of how many child segment twins there are. Another control example is a trailer counter in an applications file that keeps the trailer count for programming loop control. A *trigger* is usually a point in time or a condition that fires off an output.

The user and the information department management should be suspicious of any proposed systems analysis and design relative to data storage that binds such information into the fabric of the data storage. Let us also include derived fields such as subtotals and totals and cross-foots in this discussion. We should particularly be skeptical of storing derived information that requires us to cross tables or files and cross databases. These approaches to data storage organization, taking a charitable stance, usually signify a point of view that is code-intensive in third-generation languages (such as COBOL) and cut the application development off from the best use of fourth-generation languages, which are nonprocedural (we tell the language what our specifications are; the language controls the step-by-step coding).

We exempt from consideration here device-oriented elegant code, such as CPU microcode, which moves to software functions that used to be in the hardware. This exemption can be expanded to include all tight coding involved in the software engineering connected with application support systems of all kinds. COBOL is seldom used for such systems, where speed and fit into limited space are critical.

There are many other examples where data storage organization must still support solutions to problems done in lower level computer languages owing to inherent time or space constraints or operational complexity. The point is, we should always question the wisdom of such a solution and expect a good business case justification in reply.

Once we have bought the premise that the system must be navigated internally by programmers, we have automatically multiplied our development costs by some factor. Regarding triggers such as dates, they should always be in tables peripheral to the main data stores, independent of program code.

Overwhelmingly, the problem of defining the data storage is that of synthesizing the inputs and outputs.

Of course, it is a long distance from deciding on the necessary data for storage to the proper arrangement of that stored data for optimal updating and retrieval—but more on that later in this and the next chapter. First we need to firmly establish this connection between the moving data and the stored data and also the connection between this whole evolutionary business of data definition and the systems analytic process.

Figure 9-1 reviews this evolution of the data definition from motion data to resting data and then extends the relationships further to summarize all the relationships we have discussed so far. Parts (a) through (e) summarize graphically the preceding discussion whereby data stores evolve from data-in-motion so that

GIVEN:

- a view of the output reports, queries, and feedback transactions as first retrieved as data groups from the data store and therefore in a raw form before all the internal manipulation discussed in Chapter 7
- a view of the above output in refined form ready for transmission to a terminal (such as a printer or screen)
- a view of the input transaction set as first presented in the raw state before all the internal manipulation discussed in Chapter 7
- a view of the above input transaction set as it is now represented in refined form ready for application to the data stores in the form of adds, updates, and delete functions

WE CAN NOW, THEREFORE: define the data stores needed to support the system.

These views we have just discussed (or at least the views of the raw input and the refined output) are known by the proponents of data modeling (a method to arrive at common, optimized data base storage) as *user views*. Therefore, using this methodology, we would substitute "user views" for all views of data and say that "Knowing the user views, we can construct the necessary data stores."

This is by way of being a historic statement, not to be passed over lightly. To construct a data storage system satisfactory to all the applications that need to

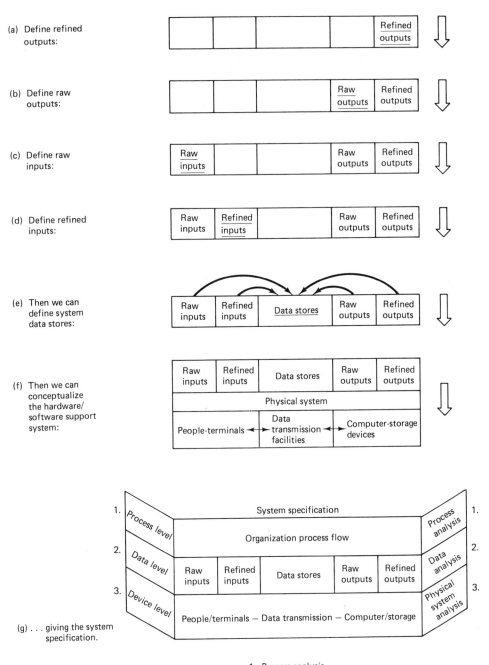

Figure 9-1   Data Analysis—Systems Analysis.

share the organization common data, each from different views of the data store, is to realize the search for the very best of on-line information processing. At present, using old, instinctive methods of building data stores by building systems at the application level, we have been rushing into some serious problems of data storage disorganization. The main problem is that data-at-rest is really not at rest at all but constantly changing to meet new organizational demands.

Figure 9-1(f) shows the relationship between systems data and the physical system that we discussed in Chapter 8. The purpose of the physical system is to store, compute, transmit, and present the data.

Figure 9-1(g) adds the final component in systems analysis. Data is the prime resource by which the organization performs the processes which define the meaning of the organization.

The perceptive reader may detect a certain redundancy between the data analysis we are now discussing in this chapter and the data flow diagramming discussed in Chapter 3. That diagramming involved a method by which we could arrive at and specify process flow by tracking moving data on its journey from transform to transform, intersecting with data stores. The present discussion of data is complementary to that discussion. We will define the moving data only once during the specification process. What is new here is the discussion of data store *organization*. Our concern is whether we should employ the "user-view" approach to first capture the data. We have not discussed this yet, so let us just say that if we plan to use vendor application software to assist us in optimizing data organization, we may find ourselves incorporating the user-view approach in the specification methodologies discussed in Chapters 3 and 4.

Regarding system specification, the reader will remember that throughout Chapters 1 through 5 we defined specification as necessary for both the current system study and the new system study, one preceding the other. Regarding current and new systems, the systems analyst, wearing a data analyst hat, will find the following situations where data storage organization is a factor:

1. In the case of the current system study the analyst will collect the motion data requirements, form a judgment on necessary data stores, and compare that judgment to the existing data stores to see how accurately and efficiently the data stores support the data-in-motion. Questions will occur to the analyst, such as: Are people manually working over output reports? Are terminal users who require two-second response time getting twenty-second response time?

2. In the case of the current system study where there are no new requirements for data-in-motion but where the data stores are reaching toward entropy (that state of disorganization discussed in Chapter 6), the systems analyst is faced with a need for data reorganization either local or global in nature. The usual symptoms of data store disorganization result from implausible access, redundancy, imprecision in data-item names (multiple data items same name, a data item with multiple names), data stores buried in program logic, and data–application program dependency, wherein a reorganization of data compels rewriting of programs—a

horrible thought when dealing with production programs that are finally thought to be debugged after five years of bug removal.

The resulting symptoms include degraded response time on-line, the inability to reorganize data, difficulty in programming new processes that need new views of the data, constant checking to see that two or more redundancies in a data item do not differ in values, and other such alarming results of poor data store organization. Without being alarmist, it is fair to say that after thirty years of applications development our present worldwide data store organization is rusting out at an accelerating pace. We should bear this in mind when we discuss the expensive, incomplete, and often frustrating alternatives which are now available and will be presented in the next chapter.

What must be done to a current system in this dilemma, even though the process requirements have not changed, is to do a new system specification. This will be a specification to reconstruct the stored data and migrate the data values to the new structure. This type of operation can happen only once in a generation and should always be the result of a long-range strategic plan for data upgrading as well as any process upgrading—such as going from batch to on-line systems— that can be worked into the plan. Also usually involved is an upgrade of the data access languages used for building input handling systems and output report and query specification. These involve user-genial access languages.

Perhaps the data upgrading will involve a transfer from department applications data to data accessible as common data either from primary data stores or extracted data stores, according to need. When we discuss *extracted data* we will mean data maintained on a primary data store and offloaded at periodic intervals to an extracted data store for outputs.

3. In the case where a current system exists that is working within acceptable performance boundaries but there are new requirements due to the changing nature of the operations, it is, as we have pointed out, first necessary to redefine the motion-data requirements (new or changed transactions, query subjects, and reports) and then redefine the data stores. To the extent that the data is independent of the programs that already exist, this may or may not precipitate a much larger project than that delineated by the new requirements.

4. In the case of a completely new automated system we will first define the motion-data requirements as shown in Figure 9-1. We had better start now to ask some basic questions to get this system off to a good start:

- How can we organize the data so that different accesses can have the best possible performance levels?
- How can we separate the evolution of the data stores from the evolution of the application programs so that both may evolve independently?
- What is the best balance between primary and derived data?
- What is the balance required of the system between real time, on-line, and batched data access?

- What can we do to stored data to increase flexibility and adaptability to change, so that we can move quickly in response to the organization's need to be adaptive?

A goodly portion of the remainder of this discussion will be concerned with methods that have evolved over the last ten years in answer to these questions. There is, however, no one method of optimizing data store organization which is best for all organizations. The use of integrated common data stores arrived at through data modeling techniques is not a universal solution for all. Consequently, we will make a best possible net case for three different approaches to data store management and organization suitable for different classes of users. We will call these three approaches

- the application driven approach
- the $i - 1$ approach
- the integrated common system approach

Before examining these methodologies, we need some more general discussion of information storage methods and a rigorous examination of a frankly absurd redundancy in terminology that confuses what is already a complex discussion. We will select a minimal terminology and provide *aka* (*also-known-as*) tables for the reader.

SEVERAL WAYS OF REGARDING DATA STORES

Data Store Ownership

In terms of ownership, data stores can be classified as common, shared, public, and private.[1] It is important to discriminate among these possibilities. We will proceed to design data structures and even select different application development languages based on what type of ownership is involved in the systems analysis problem.

1. Common ownership.   Commonly owned stores are also known as the integrated corporate database. This set of data stores holds information needed by many organization functions. Figure 9-2 is a simplified illustration of common data used by various business subsystems in a manufacturing corporation. It shows the different views of the data that each business subsystem requires and the fact that,

---

[1] Daniel S. Appleton, "Law of the Data Jungle," *Datamation*, October 1983, p. 229. Dan Appleton uses the terms: private, shared, common.

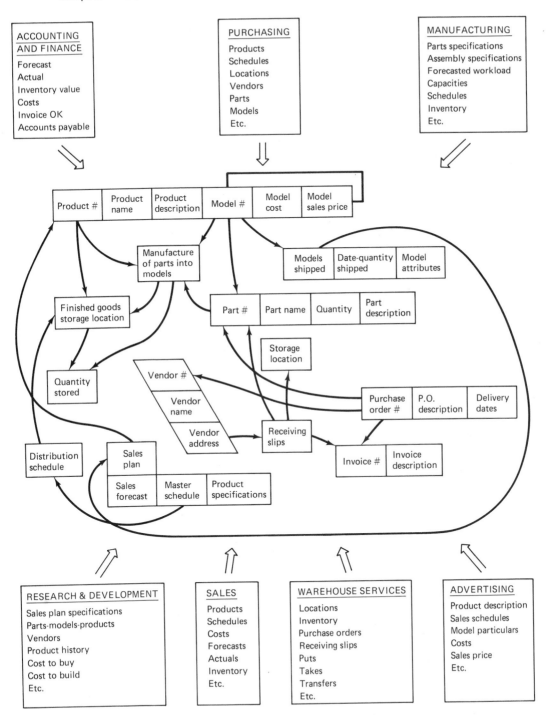

Figure 9-2 Common Data—Different User Views.

however the data is stored, there is a redundant need, which is to say that each user may have a different view of the same data.

For instance, in Figure 9-2, the purchasing department is interested in looking at the data from the point of view of the purchase order. For each purchase order they need one vendor, many locations, and many sets of part information. One parts order may be used for several different products and several different sales plans.

From the manufacturing point of view the product is most important. For each product a parts description is needed. For each part an inventory location and quantity available are required. The master schedule, which is related to a single product in some views, is needed for the sum of all products in this view.

Accounting needs to reconcile the sales forecast to the actual shipments of product, needs to know the cost of parts, the cost of manufacture, and the product and part inventory status. They also need to compare purchase orders to invoices for invoice approval . . . and so on.

Without going through all the possible user views we can see that the data is common and it cries out to be organized in such a way that it is available to all. We can also see that some of the data must be available for immediate access and some not.

For instance, if a truck cannot be unloaded until the purchase order and the location assignment are available, then immediate access in real-time is called for. On the other hand, the information needed by the advertising department may or may not require immediate access, and definitely there is no realistic real time requirement (*real time* meaning the instant availability of data as soon as it is updated).

The need for immediate access to real-time information is a vital determinant of whether an organization needs common primary data storage (*primary storage* implies retrieval from the data store that is updated without regard to timing factors). The ability to access data, either real-time or slightly aged, on-line through a terminal rather than batched and delivered two hours later from the printer in the computer center is one of the two factors involved in making this determination. The other factor—which is even more important—is whether the information must be real-time. Do we have a few minutes to sort an output retrieval to a different user view? Is immediate access to an extracted (downloaded, uploaded) data store which is refreshed twice a day satisfactory to the organization? Answer "Yes" to either of these two questions and advise going slow on giving everyone a terminal, and one stands the chance of becoming a dangerous, destabilizing individual capable of creating an economic recession. It may be that one's organization stands to save a bundle by concentrating on excellent computational facilities delivered without benefit of immediate access to on-line information. This is not in the best interest of the many vendors who are making projections on a certain potential market for terminals, transmission facilities, mainframes, and software of all kinds. Even the schools are assuming that on-line in real time is the only method to be used and taught.

What is the responsibility of the systems analyst? Which comes first, the common good of the culture or the good of the organization? A tough question. Sometimes there is no one answer satisfactory to all concerned. The author's view, for what it is worth, is to

- stay simple and appropriate
- be nothing more than an interested observer of new technologies until they are proven
- do not unwittingly become a beta test site for new software
- do not try to ring a doorbell by firing at it with a 20-mm cannon
- know the alternatives
- don't get stampeded
- be prepared to move decisively when something new proves credible and cost-justified, based on good testing

All this is by way of saying that the common-data-store approach should be a necessity rather than a luxury. There is too much to lose, mostly in delay of application development and high costs. At the present level of technology in high-performance, fourth-generation database management systems, local data stores run by local applications get express treatment regarding development throughput time and performance reliability.

2. Shared ownership.   Almost all large organizations have a need for a common data store. The question for them is: what is common data and what can be called shared subsystem or department data? For a smaller organization the need for common data will be obvious. Does the organization deal with customers over the telephone, where customer records needed cannot be known in advance of the phone call? To respond to the customer's questions does the terminal operator need to know what products are in stock? At the same time, do the purchasing agent and the warehouse worker have a need to see or change the same information? Did the refrigerator the customer wants just get added to stock two seconds earlier? A "Yes" answer to these questions implies that good use can be made of a common data store, no matter what the size of the organization.

Shared information is information owned at the subsystem level of an organizational system, shared by workers in that subsystem, and available to one degree or another, at less-than-optimal performance levels, to other members of the whole organization.

The shared data can be either on-line to display terminals and printers or off-line to batch processing or a mixture of both. What really matters is who owns the data. Whoever owns the data maintains the data through updates, adds, and deletes or delegates an agent to do this for them (a central data entry). Usually subsystem data stores are controlled by applications that spin off transactions to other application centers or to the common data system.

Figure 9-3 shows a typical operation where an application in a department owns a data store. In this case, in order to control the manufacturing operation the shop managers need to know work-in-process and work-done by volume and elapsed time by date, by product, and by job number (they also need to know work-to-come and parts availability, but this system does not provide that). With the information from this data store, which includes machine lost and idle time, the plant managers can run their shop.

But other departments also must have access to this information on a less-than-real-time basis. Accounting can not distribute labor and machine costs to product and cost centers unless this manufacturing system spins off a transaction to that system. The same is true for employee-performance data needed by personnel, product-cost information needed by sales and accounting, and machine-performance information needed by facilities management.

So, a transaction is spun off, from data store to data store. The transaction is guaranteed by the sender. There will be redundancy; there may not be harmful redundancy. The transactions are usually done in batch, so delay in response time is predictable at the receiving end. If the application changes at the sending end, the constancy of the transaction is guaranteed not to change. It is usually not unreasonable to change a department application system and maintain the integrity of outgoing transactions.

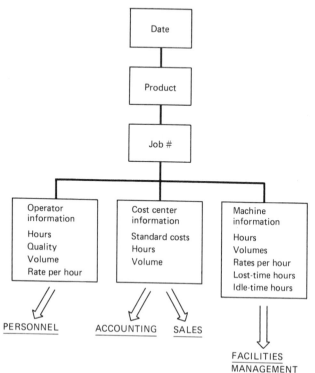

Figure 9-3 Manufacturing Data Store Showing Output Spin-off Transactions.

If the recipient of the transaction, such as the accounting department, goes to a new type of accounting system, or if sales wants different information and if a couple of dozen departments are passing similar transactions from a couple of dozen different application systems, we have quite a different story. This could be trouble in a shared-only system.

3. The privately owned data store.   There is no question but that the private data store must be on-line, real-time, and have the fastest possible response times associated with it. This is very meaningful to the author, who is pounding this information into a personal computer with two on-line disk drives controlled by word-processing software, right now, in real time. (The publisher would not mind having a common database, or at least an exact copy meaningful to their production department and editors.) Common database or not, the personal data store supported by good applications software is a tremendous step forward in information processing.

It is interesting that the big debate in the software-for-personal-computer market as this book is being made into a data store is whether to have integrated software capability, where different packages by different vendors work together using common data store standards, or whether to have multipurpose software packages using owned data stores not usable by other packages.

The present situation is that software cannot talk to software using common data store standards, and the author is hard-put to find a speller and an index generator that can read the word-processing data store for this book.

There are two choices in privately owned data stores. Either the data store is resident on floppy or hard disks, with the owner being responsible for back-up and sanitary work conditions (no Coca-Cola spills on the disks), or the primary data store is resident on the computer center disk storage but privately owned.

4. Public data stores.   One of the most interesting aspects of a privately owned computer is the access this provides to public data stores. All the data stores of the world will soon be available to the private computing facility that has a modem connection to a public dial-up facility. This is not the chapter in which to discuss in any detail the revolutionary culture-changing potential of this technology. Suffice it to say, from the author's viewpoint, that it would be tremendous to be able to do research at a library in Ann Arbor, Michigan, without packing a suitcase and buying a plane ticket to Michigan from Connecticut.

Data Storage Mediums

Data stores occur in the following carriers in organizations: in a single consciousness, in a group consciousness, in paper stores, in programs, in serial tape storage, and in immediate-access disk or drum storage or computer memory. Card and paper-tape storage is now just about gone.

1. Consciousness as a data store.   A surprising amount of stored information exists nowhere but in the heads of people. Sometimes one person can provide an answer to a problem; sometimes the answer exists only in a group consciousness and a meeting must be held to gather people who have a piece of the answer into a group that can extract and collect the necessary solution from the common "data storage."

Sometimes a crisis arises which is similar to one that arose several years earlier but was not documented at the time. The crisis managers gather as a group, knowing that an answer must be forthcoming from the meeting. There is intense group concentration, and invariably pieces of information from the past are put together with new insight and the problem begins to become resolved. There is nothing that could or should be done to eliminate this kind of data store. However, it can be recorded and documented.

2. Paper stores.   Paper stores are all over the place in an organization, ranging from a typewritten list of items on a piece of paper kept in a glassine folder to an elaborate Kardex file, perhaps of a parts inventory. Paper files are private, convenient, and on-line. That glassine folder can be pulled out of a desk drawer in less time than it can be retrieved on a terminal screen. But the dreadful truth about paper stores is that they are notoriously inaccurate. Thirty copies of a sales plan are distributed to key people. This sales plan is revised five times. Further revisions are made by memo or phone calls, and each of the thirty copies gets pencil notes in the margins. By the time the plan is work-in-process, thirty key people are each working from different revisions and the margin notes are all different. The information content of each paper file has diverged from the other copies, and disorganization results—enter the crisis managers, who keep the correct copy of the sales plan in their head.

3. Data stores in programs.   Most computer programs are, to one extent or another, driven by tables. A very considerable amount of programmer maintenance involves changing time-dependent "tables" that should really be disk files maintained by the users. The problem is that the tables are not recognizable as tables but instead look like program code. Here is an example of a data store burnt into the procedural code of a COBOL program:

```
IF TEST-NUMBER EQUAL TO 02 OR 05 OR 07 OR 12 OR 32 OR 54 PERFORM
        PROMOTION-01 THROUGH PROMOTION-01X
                GO TO END-R1.
IF TEST-NUMBER EQUAL TO 01 OR 03 OR 04 OR O6 OR 11 OR 28 PERFORM
        PROMOTION-02 THROUGH PROMOTION-02X
                GO TO END-R1.
  IF TEST-NUMBER EQUAL TO CON13 OR CON15 OR CON26 PERFORM
        PROMOTION-03 THROUGH PROMOTION-03X
                GO TO END-R1.
```

Next time these recurring promotions are run, the programmer will go into the program and change the code (that is, use a new set of literals and constants that really should be an external data store).

4. Serial storage on tape.   A reel of magnetic tape is about a half-mile long with data packed at densities of 6250 bytes per inch. Tape is becoming a less popular data store medium than it used to be because it is not randomly accessible. Typically, if we want information from a tape file, we will wait several days from the time of the request to the time when a printed record of the response is presented. If this is adequate for the purposes of the organization and exceedingly large files are involved, tape should certainly still be considered.

The tape drives are expensive and usually are connected to a mainframe computer. The tape reels themselves are inexpensive and reliable but are labor-intensive, since they must be delivered to the computer room from the tape library, mounted and demounted by an operator, and then returned to the tape library for logging and storage. If a data store on tape is to be updated, there will be an input tape from which the data store is read and changed, one record at a time, and then written out to an output tape. Three generations of such tapes are usually stored to protect against read errors.

If a tape cannot be read as input, it must be recreated, using the previous generation (the "father") in the process. Every so often a grandfather about to be "scratched" is stored off-site to protect against disaster situations at the data center. Many organizations would collapse if their data stores were lost. Disk data sets are usually unloaded to tape for clean-up, back-up, and archiving.

A logical record, which can be either fixed or variable in length, is stored on tape as blocked records. The operating system builds blocks out of logical records (such as a customer history record) until the next record logically "written" by the programmer exceeds the designated block size. At that point the operating system actually physically writes the block via the computer I/O channels to the tape. If the logical records are of fixed length, there is no control information written to tape, and the sum of the logical records as seen by the programmer is the block as physically stored. If the blocking factor is 100, and the fixed logical record length is 100 a physical record of 10,000 characters is recorded on the tape. If dusted with iron filings, this block of 100 logical records can be read. What the programmer and user think of as a record is internally (physically) recorded as part of a block between two interrecord gaps (IRG). The IRG is to allow deceleration and acceleration of the tape drive to read-write speed.

If the records are of variable length, the programmer must include a record-length indicator (RLI) at the front of each logical record, and the operating system computes a block-length indicator (BLI). This BLI is the only instance when the programmer's conceptual view of the data differs from what is physically stored. An example of logical records stored as blocks on tape is shown in Figure 9-4.

Considerable atttention has been given here to the almost one-to-one rela-

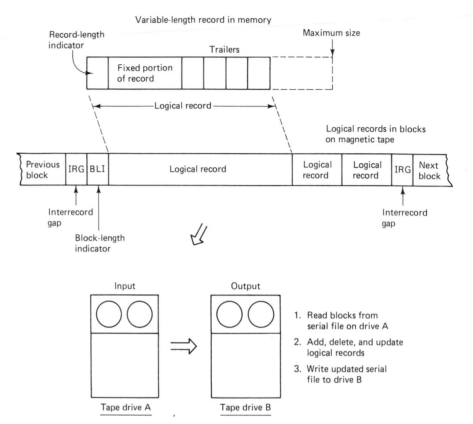

Figure 9-4   Tape Serial Storage.

tionship between internal and external data store views when using tape records. Please bear this in mind as a contrast to the same situation involving database record storage on a direct-access storage device.

5. Immediate-access (direct-access) storage.   A direct-access storage device (DASD) is the most used random, immediate-access device for data stores. Drums are another form of on-line storage available but are rarely used. Immediate-access disk storage is really a concept rather than a literal truth. Any on-line storage device such as a tape drive or a disk drive has as its Achilles' heel the electro-mechanical component. The central processor may be working at nanosecond speed, but the read-write head on a disk device is part of a mechanical unit that operates in the millisecond range—a million-to-one disparity.

As a matter of fact, it is this disparity that lets us multiprogram a central processor so that several programs may be run at the "same time." While a read of a record in one program is forcing the mechanical movement of the read-write device, the program can be interrupted by another program for central processing unit (CPU) work or other I/O work until the mechanical action is terminated.

This electromechanical action is also involved in the DASD-to-memory action involved in the paging concept, upon which virtual memory storage is predicated. If part of a large program or its working storage is assumed to be in memory (virtual memory) but as a point of fact is not, it must be paged in from the DASD.

The DASD is essentially a stack (a pack) of records (platters) spinning around. The electromechanical seek-read-write device is like a very large pocket comb poised beside the spinning platters. At the end of each tine of the comb is a read-write head which can read one track on a platter. Each read involves one block of logical records, just as with the tape device, but with the DASD there are more internal data overheads than the BLI on the tape block. There are indexes and pointers.

The electromechanical comb device dictates the basic physical data organization of the disk pack. Putting sequential records on one platter or disk doesn't make much sense, since the read-write head would constantly be moving from track to track. The point is to avoid moving the multiple read-write heads any more than is necessary. This brings us to the concept of the cylinder.

The *cylinder* is a logical tube within the physical disk pack, as shown in Figure 9-5, wherein the sequential data blocks are placed on tracks directly under each other so that they can be read or written one after the other without movement of the read-write device. On an IBM 3380 disk drive, for instance, this makes about 700,000 bytes available as actual immediate access.

Therefore the DASD is a disk pack consisting of vertical *cylinders* and horizontal *tracks* on which data stores are placed in such a way as to optimize the electromechanical component. The fact that the data records can be connected by

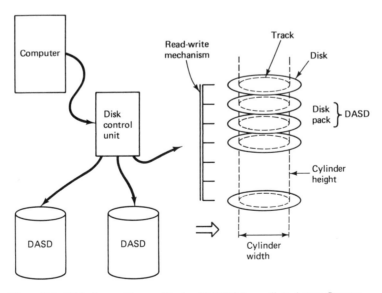

Figure 9-5 Disk Access Storage Device (DASD) Immediate-Access Storage.

pointer data and index data because of the random-access nature of the device is the basis upon which we can build a database. We have not yet defined a database, but we shall see that, whether we are referring to a *tree*, *plex* or *relational* database, the physical device defined above, giving the ability to establish relationships between chunks of moveable data groups, is the ground upon which database organization is built.

Vendor- or Organization-Developed Data Stores

We will only touch on this view of data stores here, since this subject is the concern of a later chapter.

Four kind of vendor systems provide data stores:

1. *The minimal database management system (DBMS)* that links the database or other data store to well-known second- and third-generation languages, such as BAL and COBOL.
2. *The congenial DBMS* that provides full-service decision support and application development services, including automatic screen painting, reporting, and querying, in addition to the basic services necessary to maintain any database or other data store.
3. *The vendor-developed application system* that does such an application as payroll or inventory and contains a data store to which the organization has only limited access.
4. *The combination of systems 2 and 3*, where the congenial DBMS is offered with fully developed application packages written in the DBMS language, which is also available to the user organization. Also available to the user is the data store itself, which has the same code system as any application the user might build within the context of the DBMS. Developmental work at interior levels of the DBMS can be carried out by the vendor almost transparent to the user, such as changes in the physical database organization.

We should also mention, in passing, the *data management system (DMS)*, which is sold by the vendor without a vendor data store (subtracting the B from the DBMS). A DMS provides one or more of the following, for certain well-known databases or file-access methods:

- nonprocedural reporting (a nonserial specification language)
- nonprocedural querying
- input transaction development including screen painting
- applications such as audits
- procedural language interfaces (such as a link to COBOL)

The reader should be aware of the very significant difference in approach between the DMS and the congenial DBMS. With the DMS we remain independent as an organization from any given DBMS vendor. We depend on the ability to integrate into an organization's information system a variety of software offerings and orchestrate our own software configuration. We must be able to integrate multiple vendor software and in-house-developed software. We must develop expertise along many different vendor lines. This implies a large data processing department with a variety of application development specialists (and systems programming support for those specialists).

The alternative is to go into a virtual business partnership with a congenial DBMS vendor and hope the vendor is as trustworthy and enterprising as, of course, we are.

Such are the conflicting choices—to which there is no completely correct answer—that await the systems analyst with high-level responsibilities.

The Primary Data Store and Extracted Data Store

The *primary data store* keeps information in a nonsummarized, nonderived, updatable form. The information is decomposed to the lowest possible level so that various combinations are possible during extraction. Very seldom is this data derived from other primary data. For instance, we may store quantity and cost per unit but are not likely to store quantity cost on a primary data store. (This is not a rule but a proclivity.)

The primary data store is in a state of constant flux. The term data-at-rest used in this book describes only a whole-system view of stability, not a base-level view (in a human the personality may be stable so that we recognize our friends from day to day as the same persons, but while we engage each other in pleasant conversation, wholesale death and birth are taking place in each of us at the cell level).

In a primary data store on-line and batch updating, output transactions, queries, and reports occur at a furious rate, contending for the data store resources. Many different user views of the same data are involved. Users spanning across nations and oceans expect two-second response-time service not only from the transmission system component but also from the database access component.

As an antidote and relief to this type of primary data store we have the *extracted data store*, also-known-as *derived*, *uploaded*, and *downloaded*. Uploading implies reordering, customizing, summarizing, and adding derived data. An uploaded data store is used for decision support where real time information is not required. A download usually involves transfer of all or part of a data store without content change, such as browsing the *Wall Street Journal* column titles and deciding to request a data transfer to a personal computer of a particular article.

The extracted uploaded data store is usually "refreshed" at periodic intervals and is usually (hopefully) used only for information retrieval. In rare situations of distributed processing with multiple data stores at multiple locations, updating of extracted data does take place, but this dubious practice is for the sophisticated gamblers among us.

One of the basic questions a systems analyst must receive an answer to, as we have stated before, is whether an application needs real-time data storage facilities. It is not a major problem to refresh a data store two or three times a day (although it may be irritating to the warehouse worker who cannot store materials unloaded from a truck to be told all on-line access is tied up for twenty minutes due to batch extraction of data stores).

THE DATABASE

What kind of a data store is the database? We will go to the writings of James Martin for two different definitions:

> A data base is a collection of data which are shared and used for multiple purposes.[2]

or, in more detail:

> A data base may be defined as a collection of interrelated data stored together without harmful or unnecessary redundancy to serve multiple applications; the data are stored so that they are independent of programs which use the data; . . . . One system is said to contain a collection of data bases if they are entirely separate in structure.[3]

Accepting Martin's definition of a database imposes on the author and the readers the responsibility for defining those data stores which do not conform to it. The author has private data stores for use on his personal computer. Some application systems use data stores which are not stored to be conveniently shared and are not program-independent. We shall define these data stores as a *file* of logical records, wherein the physical storage concept is pretty much the same as the external logical view and the data is owned by one application.

There are also very stable files of simple construction which are available for sharing and are program-independent.

Two other points are worth making before we study progressively more complex databases:

1. Redundancy is more likely to occur between databases than in one database.
2. There are two kinds of program dependence upon data; one dependence is much more serious than the other. Less serious is program dependence where

---

[2] James Martin, *An End Users Guide to Data Base* (Lancashire, England: Savant Research Studies, 1980), p. 3.

[3] James Martin, *Computer Data-Base Organization*, Second Edition (Englewood Cliffs, N.J.: Prentice-Hall, Inc., 1977), p. 22.

only the input-output buffers are involved. This is the situation where the moveable chunk of stored data, whether it be called a relation or a segment or a file record, is changed so that the process of moving the buffer to the input work area or the output to the buffer area has to be rewritten but the rest of the process logic of the program remains the same. More serious is a change in the meaning of a data item or the relationship between the data item and the rest of the data. If a data item concerning a quantity field is changed from seven to ten numerics and now has two decimal places rather than being a whole number, the entire program has to be reviewed. Of course, this has less to do with database organization than with good data administration.

Without question, the data dependency and redundancy and poor access which are associated with almost all the databases in current use constitute a major problem involving many billions of dollars and a huge waste of resources. We will be dealing with the available antidotes to this problem in the next chapter.

Levels of Database Complexity

From simple to complex, here are the available physical database formats in common use:

Simple single-level database.   This is a very satisfactory database, provided that the inefficiency of the redundancy involved is acceptable to the organization. Figure 9-6(a) shows an example of this form. If data group A occurs uniquely 10 times, data group B occurs uniquely 100 times, and data group C occurs 1000 times, then the storage for the A group, instead of 10 times the byte size, is 1,000,000 times the byte size. D is a repeating data group occurring n times on the C level. In the example shown, $n = 3$.

Probably most databases around today are of this type. The simplicity of the data organization is such that applications development goes at express speed. However, when the data group contains data items that must be configured in many other different ways with other data items for other applications, then the harmful effects of data redundancy take the bloom off the rose of this approach to data storage. Data items tend to diverge in meaning and values. There are, however, methods by which this can be controlled—such as multiple updates from a single transaction, constant auditing of acceptable data-item redundancy, and data-dictionary control—so that if the accounts payable amount from view A is x then the accounts payable amount from view B is also x and not $x \pm 50$ cents.

The *relational database* is a variation of the simple single-level form. This calls to mind the story of the Zen Buddhist who as a young man stood and faced a view of a winding river, snow-capped mountains, and a blue sky. The young man had many a journey in life, did many great things, and grew in spiritual stature until he became a Zen master. In the full wisdom of his old age he returned to the

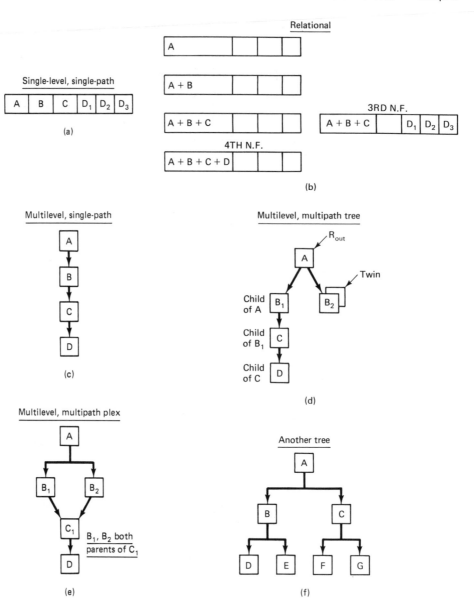

Figure 9-6 Levels of Database Complexity.

same place and stood looking out—and saw a winding river, snow-capped mountains, and a blue sky. This little anecdote points out the journey from simple to complex to simplicity. The relational database will require very complex software to achieve this simplicity. It is stored with redundant key data items as shown in Figure 9-6(b) but otherwise with mostly nonredundant key owned data items. The complexity needed by the external view of the application is generated at retrieval execute time by the DBMS software. The internal storage is not only one-level, one-path but is much simpler, in that it is fully normalized to either the third or fourth normal form (a little bit of jargon to be discussed in the next chapter).

Hierarchical databases: The multilevel, single path form.   This form, shown in Figure 9-6(c), is available in such DBMS packages as RAMIS II, a product of the Mathematica Corp. division of Martin-Marietta. If the hierarchy is simple enough and the levels are shallow enough, this form is excellent, because it simplifies the operations that the DBMS is required to perform in order to support automated entry and retrieval processing and database maintenance. For many users this level of complexity is well suited to their applications, particularly since different databases can be virtually linked so that one database is the host for the application and the other databases (files) are the "associates." For different views of the data the DBMS will sort the retrievals to the desired view. This can be done on-line or in batch.

The multilevel, multipath database.   This is the form most technical people now think of when the name database is invoked. Figure 9-6(d) or (f) shows the *tree*, which is a hierarchy consisting of a root which has children which in turn have children, and so on through perhaps fifteen levels, which is the case with IBM's well-known tree database, IMS. Each child has only one parent. Repeated instances of a child are known as twins, which would be the equivalent of rows on a table. These familial terms are known and used by all those connected with this type of database.

A variant of the tree database, which is less well known, is the *plex*. In the plex multilevel, multipath database a child may have two parents. Figure 9-6(e) shows an example of the plex. In this example the plex can be made into a tree by introduction of the following redundancy: A still the parent of $B1$ and $B2$; $B1$ the parent of $C1$, and $B2$ also the parent of a redundant $C1$.

The multilevel, multipath database, all these years after it first became available commercially, is still the state-of-the-art in large-database organization. In the fast-changing world of computer science, database is such a major investment that change is less rapid than in, say, hardware development. What has been changing is the amount of thought that goes into analyzing and organizing the data and data relationships on such a database.

The reader should be completely clear about the difference between a database, such as we have just been discussing, and a group of such databases which forms the organization's *database system* or *data bank*. An application view of the data may cross over two or more databases. Such a view is seen by the users and programmers as their database. This illusion can be made a reality by the custodians of the physical database system, the database administration group (DBA), which establishes the linkage through the language provided by the DBMS. In the case of IMS this language is called Data Language/1 (DL/1).

The next chapter will complete our study of database, concentrating on a full understanding of database levels, database design through data modeling and a comparison of the different methods of designing databases. Thousands of organizations rely on databases. Not all organizations have the same needs. No one approach will satisfy all cases. In fact, right now, when it comes to complex data requirements there is no approach which will fully satisfy any case.

<table>
<tr><td>**10**</td><td>*MODERN*
DATABASE
DESIGN</td></tr>
</table>

REVIEW

The systems analyst cannot begin to do a current systems study, determine new system requirements, or specify a new system for a large organization that has up-to-date computational facilities without being expert at database design. This does not mean possessing the particular skills of the database administrator concerning the physical implementation and monitoring of a database system. It does mean being adept at logical database design.

Before we commence this study, the reader may want to review certain illustrations in this book in order to grasp the evolutionary path we have taken so far regarding data stores in general and the database in particular.

Figure 1-1 at the very outset of the book shows the generalized systems model, where processes use resources to refine inputs into outputs when triggered. The vital information resource is what we are now dealing with: the data, the data store, the database.

Figure 1-2 shows the specification document as the main output of the systems analytical process. Figure 3-11(a) gives the baseball-team specification, where the data is shown with the process. At the lower left of this illustration the data store resources are displayed. What is shown there is very high level, perhaps the names of subject data bases. The details of the data store were left for development in future chapters, with just a reference to the fact that we would probably be using a data dictionary to record the detail. No matter how we show the detail design and contents of the data resource, it will surely become just as important a part of the specification as the process flow. In many information systems the data and data organization far outweigh the processing in importance.

Figure 6-1 shows the overall computer solution to information processing as input development, knowledge storing, and output delivery. The role of the data store in these three functions is graphically illustrated.

Figure 9-1, concerning steps in data analysis, carries the concept developed in Figure 6-1 forward to show that we design all data stores by first defining the raw and refined inputs and raw and refined outputs. Having defined the input-output of the system and knowing that the reason for the processing is to convert raw inputs to refined outputs, we are in a position to define a stored data resource to satisfy these input-output requirements. The efficient design of such a data storage system—increasingly, such a database system—is what we are now about.

DATABASE DESIGN TERMINOLOGY

The author has spent many thousands of hours over the years trying to communicate with colleagues—face-to-face, on the phone, in writing, and in groups—concerning rather simple concepts which we were all making very complex simply because of the many different words we used meaning the same thing. Or, conversely, we used the same words meaning different things. One can never be sure when some-one uses an also-known-as word that the person really is referring to the same thing we are assuming. There is a time for a rich and subtle variety of expression (after hours at the watering hole) and a time for precise nonredundant expression. The latter is the case with database design.

For this reason we will now review the most important terminology used in the design of logical databases and arbitrarily select one expression in each subject field for use in further discussion in this book.

Figure 10-1 is the also-known-as (aka) table. The selected term in column 1 is also-known-as the other terms in that row.

Bypassing *schema* for the moment, let us discuss *table* expressions, as shown in Figure 10-1.

Tables

A database is made up of moveable chunks of information that have relationship to each other. These chunks (just like systems) can be thought of as being in vertical levels, in that some chunks have a one-to-many relationship to other chunks (as one class has many students). Each chunk can be called a *table* if the data can be represented by columns and rows.

In a relational database the physical table relationships are not burned in

Selected Term	Also-Known-As (aka) Terms					
1	2	3	4	5	6	7
External schema[1]	Application view	Subschema[2]	Information group	External subschema	User view[3]	IMS PSB (program specification block)
Conceptual schema[1]	Schema[2]	Canonical schema	Organized sum of user views	Logical database	IMS logical DBD (database description)	
Internal schema[1]	Physical database	Physical data[2]	IMS physical DBD (database description)			
Table	File	Entity set	Data set	Relation		
Data group	Record	IMS segment	Tuple	Entity	Table row	Logical record
Data item	Data element	Field	Key or attribute	Group-level item (COBOL)	Elementary item (COBOL)	
Domain	One table column set	Field values set				
Attribute	Dependent item	Dependent attribute	Owned item	Owned element		

Expressions Regarding Keys — All Have Different Meanings

Key	Primary key	Secondary key	Candidate key	Concatenated key

1. ANSI X3 SPARC
2. CODASYL
3. User views also defined as input-output views

Figure 10-1   Also-Known-As (aka) Table.

(with predefined pointers) as hierarchical structures as they are in tree and plex databases but the hierarchy is burned into the key structure of each table so that *many* hierarchical applications views can be joined together upon request. This is the lure of the relational database—we don't have to "hardwire" the physical database into one hierarchy with a maze of pointers to allow the extraction of other logical hierarchies that are required.

We speak of normalization of tables as desirable. Normalization is used to reduce the table chunk to the smallest reasonable unit of information that has cohesiveness and is unlikely to need splitting in the future. It stands to reason that we can build more data relationships from minimal simple chunks than we can from large complex chunks. In the latter case the cohesiveness becomes a curse, since we will find ourselves breaking apart our basic chunks and having to rewrite parts of those programs that use them. It's like wanting to replace a 60-watt with a 100-watt bulb and finding that the bulb and the socket are one chunk, so in order to go to 100 watts we have to take the whole lamp apart. Sounds silly? Welcome to much current database design.

We will discuss normalization later in this chapter. For the present we can say that a table is a file is an entity set is a relation. All of them have a key by which the table is identified and a set of dependent or attribute data which enhance our understanding of the key. For instance:

TABLE: checkstub (<u>check #</u>, date, transaction description,
check amount, balance)

describes a table called checkstub which has a key called check # (note the underline). The key check # has dependent attributes which is the other data shown. The full listing of all the rows (all the check #s) defines the table.

Instead of showing the table as we just did, we could show it as one normally sees a table:

CHECK #	DATE	DESCRIPTION	AMOUNT	BALANCE
207	11-02-85	Jones Hardware	12.95	497.00
208	11-05-85	Bill Smith	17.00	480.00
209	11-08-85	Stop 'N Shop	20.00	460.00

. . . and so on.

Another way of showing a table is the user-view bubble chart, a first-rate charting technique which we will cover later.

Data group.   One row of a table is called a *Data group*, aka a record, segment, tuple, entity, table row, or logical record. Figure 10-2 shows the data group as a row in a table.

Data item.   The *data item*, aka data element, field, key, attribute, COBOL group-level item, and COBOL elementary-level item, is the smallest logical unit of data. A data group is a related collection of data items consisting of a primary key and a set of dependent attributes or items.

We will not dwell here on the data item which can be broken into smaller data items recursively. This is the case of the COBOL group-level and elementary-level data item such as:

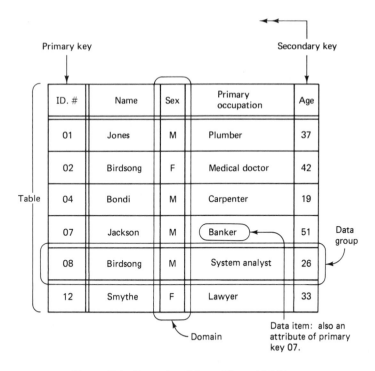

Figure 10-2   Examples of Some Figure 10-1 Terms.

```
05 date.
  10 month picture 99.
  10 day picture 99.
  10 year picture 99.
```

Here, date is a data item of six characters consisting of three data items of two digits each.

Domain.   Figure 10-2 shows an example of a *domain*, aka a table column set and a field values set.

Attribute.   The *attribute* data item is also known as dependent item, dependent attribute, owned item, and owned element. A primary key and a set of dependent attributes make up the data group. A set of data groups makes a table.

When discussing keys, we use the following expressions:

- *primary key*—the data group identifier
- *secondary key*—an attribute item which also acts as a key, pointing to a primary key in a one-to-many relationship—that is, pointing to the primary

key of its own or any other table. A *secondary index* is a table of secondary keys used to locate primary keys.

- *candidate key*—one of the keys that could be the primary key of a new table. In the case of the example in Figure 10-2, Id.# and name are candidate keys, but since Id.# is nonredundant and name has redundancy (Birdsong), we select Id.# as the primary key
- *concatenated keys*—are two or more keys that together are the primary key for the table

Therefore, at the table level or lower we will use the following database terminology:

- table
- data group
- data item
- domain
- attribute
- primary key
- secondary key
- candidate key
- concatenated key

The working professional who prefers a synonym should check Figure 10-1. It should be there.

Database Terminology Above the Table Level

Figure 10-3 shows the five levels of database system organization. We will deal first, briefly, with the two outer levels, which are usually taken for granted in discussions of database design.

As shown in Figure 10-3 on the left, the application program view is the physical makeup of the program working storage that calls the needed data from the database. In COBOL terms this is the linkage section, and it is the difference between a PCB mask as defined in the program and the PCB as defined in Data Language/1 when dealing with IBM's IMS database. In other words, it is the way the data looks to the application program language as opposed to the way the data looks to the database language.

On the far right-hand side of Figure 10-3 is the actual storage of data on the DASD, as opposed to the physical storage mapping which the database administrator presents to the DBMS that actually manages the database. The DBA defines the access methods and the actual byte sizes of table segments and the physical pointers that link the segments within and between physical databases (thus forming

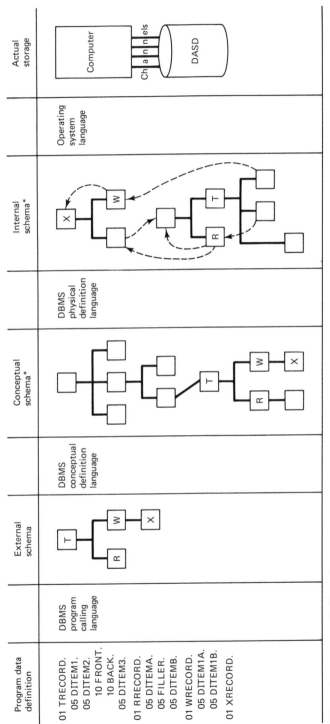

Figure 10-3 The Five Levels of Database System Organization.

*These schemas not exactly illustrative
of each other in this drawing.

the internal schema), conceiving of the data as a continuous flow. But in the actual DASD storage the data is much more "messy" and discontinuous than in the internal schema definition. There are overflows, additions stored out of line, and access overheads with which the DBA is not concerned except to see that the database is periodically dumped, reorganized, and reloaded for more efficient operations.

AMERICAN NATIONAL STANDARDS INSTITUTE DATABASE TERMINOLOGY

This leaves us with the three-layer database architecture as defined by ANSI/X3/SPARC. These are the three middle layers in Figure 10-3. This three-layer concept is the essential difference between the modern database and the data stores of the past. This structure allows for data independence and database control language independence between the three views of the data. This means that up to a certain limit the physical internal database can be changed without reorganizing the conceptual view and that the conceptual view can be reorganized without altering the external application view.

This also holds for changes to the database programming languages. A new access method might be introduced without disrupting the other levels. In terms of the database design, a new pointer relationship can be physically designed without invalidating the current relationships. A new table can be defined with only small changes to the middle level and no changes at the external level. Thus we achieve a reasonable level of data and program independence by inserting the all-important middle conceptual level between the application view and the physical database.

Let us now add to our database design terminology list three new expressions: *internal schema*, *conceptual schema*, and *external schema*. The author has selected these three from the other aka expressions shown in Figure 10-1 because the word "schema" can be factored out of the expression, since all three levels use this word the same way. A *schema* is an abstract representation of a process.[1] This leaves internal, conceptual, and external as our three database levels, which is a nice, clean definition, perhaps preferable to the other aka choices.

Internal Schema

This internal schema is also-known-as the physical database, the IMS physical DBD (database description), and the physical data. This, as we have said, includes the description to the DBMS by the DBA of the physical database, the access method, the physical DASD device, the pointers within and between databases, the relationships between the table loads or segments such as parent-child, and the physical

---

[1] *Schema*: "A diagrammatic presentation." Webster's Ninth *New Collegiate Dictionary* (Springfield, Mass.: Merriam-Webster, Inc., 1983).

byte assignments for each of these elements. It is on the one hand, the ground upon which the DASD updates are made, and on the other hand it provides the limitations and possibilities for the next higher level, the conceptual schema.

The internal schema is shown as the bottom level in Figure 10-4. An example of the way the DBA communicates this schema to the DBMS is shown in Figure 10-5(a), where via DL/1 (data language/1) the DBA defines the internal schema (physical DBD) to IBM's IMS database system.

Conceptual Schema

The conceptual schema is also-known-as the schema, the canonical[2] schema (minimalized and "pure"), the organized sum of user views (user views being input-output views, in whole or in part, needed by the user to solve an organizational problem—so let us add *user view* to the list of expressions with which we will be working), the IMS logical DBD, and the logical database.

As shown by the example of a conceptual schema in the middle section of Figure 10-4, there is almost no end to the logical relationships between tables (segments) which can be defined within the context of the physical pointers of the internal schema.

What we have here is a set of combinations and permutations on the physical database system which supports *all* the external schemas needed to use the organizational common, on-line information. As new needs develop for external schema information, the DBA reviews the conceptual schema to see if an add on to the conceptual schema will suffice or if the internal schema will also need a change. In any case the objective is database independence from the existing organization application programs so that many programs, do not have to be recompiled and tested because a new application view needs to come on-line. Of course, if the *meaning* of the data changes rather than the organization of the data, we are possibly in for a bit of trouble under any database organizational concept. That is one reason why the data administrator (DA) function is becoming so important.

An example of the DL/1 coding needed to define the conceptual schema to the IMS database management system is shown in Figure 10-5(b). One fact worth noting here is the absence of reference to bytes or devices or access methods such as we see in the internal schema shown in Figure 10-5(a).

External Schema

The external schema is the application's view of the data store resource needed for the process logic. The application user view, which is usually for one application, is the external schema shown in Figures 10-3 and 10-4. It is just that data in the form of moveable tables or segments needed to accomplish the application task.

---

[2] James Martin, *Computer Data-base Organization*, 2d ed. (Englewood Cliffs, N.J.: Prentice-Hall, Inc., 1977), p. 684.

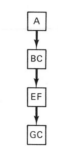

EXTERNAL SCHEMA (ONE APPLICATION VIEW)

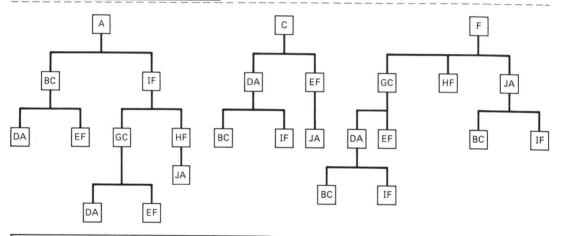

Note: The two letters combined show the pointer relationship.

CONCEPTUAL SCHEMA (LOGICAL DATABASE SYSTEM)

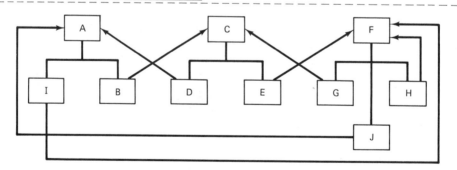

INTERNAL SCHEMA (PHYSICAL DATABASE SYSTEM)

Figure 10-4   The Three-Level ANSI X3 Database System Organization.

------------------- MARKET ANALYSIS DATABASE ---------------------

```
DBD   NAME=XMMMLN1,

           ACCESS=(HIDAM,VSAM)

DSG1    DATASET DD1=XMMMLN1,DEVICE=3380,SIZE=4096,FRSPC=(2,30)

        SEGM NAME=MARKANAL,PARENT=0,BYTES=200,FREQ=7000,

             PTR=TB,RULES=(PLP)

        FIELD NAME=(MARKLNO,SEQ,U),BYTES=10.SSSTART=1,TYPE=C

        LCHILD NAME=(MMLN1IX,XIMMLN1),PTR=INDX
```

(a) Physical DBD (internal schema)

```
DATASET    LOGICAL
    SEGM     NAME=SEG,PARENT=0,SOURCE=((SEG,DATA,DDSPSEG))
    SEGM     NAME=SEGDTE,PARENT=SEG,
             SOURCE=((SEGSEG,DATA,DDSPSEG),(DTE,KEY,DDSPDTE))
```

(b) Logical DBD (conceptual schema)

Figure 10-5   An Example of IMS Data Language/1 (IBM).

Important Note: This external schema, or some subset of this external schema or some group of external schemas, is the data store resource which is identified with the charting process explained in Chapters 3 and 4. This data is also available on the data dictionary. When we have described the data dictionary contents, we will have closed the circle on the specification which is the heart of the systems analysis methodology. The final charting methodology we will need to do a current or new system specification (before the design phase, where there is yet another charting methodology to learn) is the *user view bubble chart.*

THE PURPOSES OF USER-VIEW BUBBLE CHARTING

The purposes of user view bubble charting are to:

1. Show user views and tables in such a way that not only keys and attributes are clearly defined, but also all the relationships between key data items and attribute data items in all the possible combinations.
   - key to key
   - key to attribute
   - attribute to key
   - attribute to attribute (not so good, as we shall see)
2. Use the user-view bubble chart as one good way to define the inputs and outputs (data-in-motion) of a system.
3. Design our databases from our data-in-motion (input-output) requirements. Use the user-view bubble chart technique to set down the data-in-motion in such a way that it can be accumulated into a systems user view and then made into a best case possible conceptual schema. This will be done, if we are required to do it by the complexity of our data-at-rest, using a formal mathematical process which has become known as *normalization*.

This section of Chapter 10 will show user-view diagramming and the process just mentioned, whereby the conceptual schema (logical database system) is designed from the collective user-view diagrams. The internal schema (physical database) is modified or created from this conceptual schema model.

Here, again, are the steps we will take to achieve the best possible database to support our application requirements. This method is known as *data modelling*.

DATA MODELLING STEPS:

- make a user-view bubble chart for each motion data view or part thereof
- combine this user view with the other user views that reflect the common motion data
- simplify this accumulated user view down to the minimal combined user view by doing such things as eliminating redundancies, resolving many-to-many relationships, and establishing secondary keys
- decompose the accumulated user-view bubble chart down into tables
- normalize these tables to the third or fourth normal form, which means further decomposition to even smaller tables which are now unambiguous
- connect these tables to each other in terms of the known relationships which we have discovered from our user-view bubble charts (i.e., location table is known to contain many bins tables, which contain many parts tables for manufacturing, and also parts tables can be at many different locations)

- help the DBA build this common data model into the conceptual schema that fits the DBMS which will be used to contain the physical databases—pay particular attention to those pathways which support the most important application access requirements
- look over the DBA's shoulder as the DBA builds the internal schema (the physical database system) out of the conceptual schema and as the DBA provides application access paths as needed (for instance, program specification blocks in IMS) (if the DBMS is more modern than IMS, you won't be able to look over the DBA's shoulder, since this work could be a part of the DBMS internals)

Note: Doing these steps that involve adding the latest user view into the accumulated user view may involve only some or none of the other user views, so while the data modelling involved is formidable, it is not impossible. Several companies now market data modelling software packages that do much of the work we have just listed.[3] These packages work like this: after we have drawn the individual user view, it is entered into a computer run that also enters the current accumulated user views. The software combines the new user view (which has been coded in a data modelling language that is quite simple) with the cumulative user view and then reduces this new schema to a minimal (*canonical* is a word commonly used for minimal or basic, as in canon law) schema. Then the software decomposes this schema to tables in the third or fourth normal form and shows relationships between tables in a series of reports.

This means that the inputs to the software are individual user views and the outputs of the software are normalized tables ready to be built by the DBA into a conceptual schema (logical database system).

If data modelling as presented here is right for a particular organization, then it is advised that data modelling software and instruction in the use of the software be contracted for. But please do not think that this data modelling software will reduce the enormous job of designing a satisfactory database system to anything simple—or even reasonable for an organization with limited resources. We will discuss some of the pitfalls of data modelling for a common data bank at the end of this chapter.

NORMALIZATION

James Martin defines *normalization* as "The decomposition of more complex data structures into flat files (relations). This forms the basis of relational data bases."[4] As things have worked out, some organizations have used normalization to con-

---

[3] (a) Holland Systems Corporation, 3131 South State Street, S-303, Ann Arbor, Michigan 48104; (b) Database Design Inc., 2395 Huron Parkway, Ann Arbor, Michigan 48104.

[4] Martin, *Computer Data-base Organization*, p. 692.

struct other than relational database systems, as we wait for relational database implementation to evolve into a commercially acceptable "industrial chugger" the likes of an IMS or IDMS or TOTAL database system. The assumption is that normalization is good per se and organizes the data stores in a way that will make further changes easier.

The author believes that the very process of normalization, combined with other data administration functions such as establishing a strongly supported data dictionary, has a very strong prophylactic effect on large organizational information systems. There is probably not one heavy user of information systems around who is not in the unfortunate position of having to live with seriously disorganized and inefficiently stored data.

The organizations that devote the necessary resources to ordering their data will have a great advantage in the future. Making data stores independent of programs, eliminating redundancy, clarifying meanings, enhancing modern decision support, and understanding the relationships between data (as discussed in Chapter 6) so as to support artificial intelligence systems where the user views are much more complex and less predictable—all this is predicated on control of our current data stores. Normalization and the other data modelling techniques under discussion here are important steps in the right direction.

The Steps In Normalization

A table is un-normalized when a domain has repeating data items—that is, when a dependent data item has a many-to-one relationship to the table key that owns it. What we want is a *flat* table. This becomes very clear when we look at an example. Figure 10-6(a) shows a table giving information about employees. Any one of these employees could have any number of skills. Perhaps Carlson is a race-car driver, an investor in the stock market, a baseball player, and a speaker of Tagalog in addition to being a cabinetmaker and a supervisor of final assembly. Carlson's skills have an open-ended many-to-one relationship with Carlson's employee #. That doesn't make for a flat table.

By decomposing the table in Figure 10-6(a) to the two tables shown in 10-6(b) and (c), we get rid of the repeating skills information by making employee # + skills code a concatenated primary key. These tables are now normalized and are in at least the first normal form. It happens that they are fully normalized and are in the fourth normal form.

The difference between third and fourth normal form is one we may want to ignore, speaking just of third normal form. According to some experts a table is in fourth normal form when there are no closed, repeating data items. For instance, a data item for each of the oceans or continents involves a many-to-one dependency on a key, but we can be fairly sure that this is a stable, closed repeating data item, since it is unlikely a new continent will rise up. In third normal form this is allowed; in fourth normal form it is not. The author's recommendation is to not allow such

UN-NORMALIZED

Employee #	Name	Current job title	Date of birth	Date of hire	Skills code	Skills description	Skill level
0021	Jones	Group manager	12/1/26	4/7/64	L1	Spanish	Fluent
1262	Wabash	Machinist — first class	11/3/61	11/2/80	F2	Pilot	Beginner
8033	Carlson	Supervisor	3/4/70	1/12/72	M1	Cabinet maker	Expert

(a)

NORMALIZED

Employee #	Name	Current job title	Date of birth	Date of hire
0021	Jones	Group manager	12/1/26	4/7/64

(b)

NORMALIZED

Employee # + skills code	Skills description	Skills level
0021 + L1	Spanish	Fluent
0021 + M1	Cabinet maker	Expert

(c)

Figure 10-6 Putting a Table in First Normal Form.

repeating dependencies and to call it third normal form, forgetting we ever heard of fourth normal.

Assuming we have no repeating domains, as we first had in Figure 10-6, we are in first normal form. Figure 10-7(a) shows a table in first normal form. However, two problems with this table will tend to make the information in it unstable. One problem is that all the attribute data items in (a) are not fully dependent on the full primary key. It is not necessary to know item # to identify purchase order description or the vendor information. Purchase order # by itself does that. To put a table already in first normal into second normal form we must make all attributes *fully* dependent on the primary key, whether concatenated or not.

In Figure 10-7(b) we achieve second normal form by decomposing the one table into two tables. To know vendor information we only need purchase order #, as in table (b)(1). To know quantity cost information we need purchase order # + item #, as in table (b)(2).

PURCHASE-ORDER TABLE

Purchase-order # + item #	Quantity	Quoted-cost-per-unit	Quantity-price-at quoted-cost	Purchase order description	Vendor #	Vendor-name	Vendor-address

(a)

Purchase order #	Purchase-order description	Vendor #	Vendor name	Vendor address	(1)

Purchase-order # + item #	Quantity	Quoted-cost per-unit	Quantity-price at-quoted cost	(2)

(b)

Purchase-order #	Purchase-order description	Vendor #	(1)

Vendor #	Vendor-name	Vendor-address	(2)

Purchase-item + order #	Quantity	Quoted cost per unit	Quantity price at quoted cost	(3)

(c)

Figure 10-7 From First to Third Normal Form.

Now we are in second normal form but not yet in third normal form. To put a table already in second normal form into third normal form we must have the attribute items dependent on *only* the primary key of the table and *not* on another attribute field. Note that in Figure 10-7, table (b)(1), vendor name and vendor address are fully identified by the vendor # without need of the purchase order #. Therefore, in order to finally realize third normal form, we must decompose (b)(1) into two tables, which are shown as (c)(1) and (c)(2).

Now we have three tables in third normal form, as shown in Figure 10-7(c), rather than one table in first normal form as shown in (a), and we are ready to fold these tables into a conceptual schema with some assurance that the table data is stable and very flexible, since the tables are cohesive and easily joined for application use.

To sum up normalization, we require:

- flat tables (without repeating domains)
- attributes fully dependent on the whole key
- attributes dependent only on the whole key and not on each other

THE ELEMENTS OF USER VIEW BUBBLE CHARTING

User-view charts are a very satisfactory way to show and manipulate data and data relationships. Combined with either the neoclassical flowcharting method of Chapter 3 or the data flow diagram method of Chapter 4, by which process flow is arrived at, the physical system charting touched on briefly in Chapter 8, and one of the system design charting methodologies which we have not yet discussed, we have a complete charting methodology of considerable power of communication. Figure 10-8 shows the elements of the user-view chart. While there are only two forms in this method, the bubble and the arrow, the combinations of these two forms are somewhat involved. It takes a while to gain skill in drawing user views.

The Bubble

The bubble, as shown in the first row of Figure 10-8, contains either a primary key or an attribute. The primary key can be either a single data item or concatenated data items, such as ABLE + BAKER. In either case the primary key is underlined. CHARLIE is an attribute.

The Arrow

The arrow shows direction and relationship. Unidirectional arrows have an arrow-head at one end only, implying that the user view is not interested in the reverse direction. Bidirectional arrows imply that the user view, or combined user view, needs flow in both directions. An example is shown in Figure 10-8 in the relationship between ABLE, BAKER, and CHARLIE1 on the third row where the key ABLE points to BAKER, CHARLIE1 but CHARLIE also points to ABLE.

Besides direction, the other attribute of arrows pointing to bubbles is the *one*, *many* relationship. As row 2 in Figure 10-8 shows, we can have one-to-one, one-to-many and many-to-many relationships. These can occur in either or both directions. A many-to-many relationship must be resolved into bidirectional one-to-

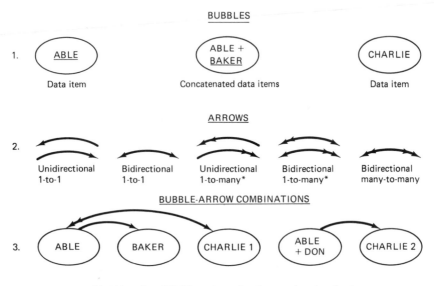

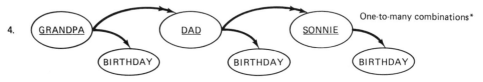

A bubble such as <u>ABLE</u> is a *primary key*, because there is at least
one single-headed arrow leaving it. The same is true for ABLE + DON.

A bubble such as <u>BAKER</u> is an *attribute* (of a primary key). It has
no single-headed arrows leaving it.

A bubble such as <u>CHARLIE</u> 1 is a *secondary key* because it has one
single-headed arrow entering it and one double-headed arrow leaving it.

*Many = 0, 1, many can also mean none or one only.

Figure 10-8   User-View Charting Elements.

many relationships in order to be implemented in a physical database, as shown
in Figure 10-9. When drawing these arrows connecting bubbles, we will consider
only directions and one/one, one/many relationships that actually pertain to the
particular user view.

Bubble-Arrow Combinations

We can use the bubble-arrow relationship to identify primary and secondary keys
and their attribute data items.

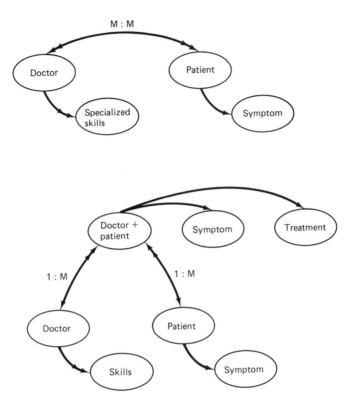

Figure 10-9   Resolving Many-to-Many Links.

In row 3 of Figure 10-8 the bubble <u>ABLE</u> is a primary key because there is at least one single-headed arrow leaving it. The same is true for <u>ABLE + DON</u>. The bubble BAKER in row 3 is an attribute, an owned data item of <u>ABLE</u> because it has no single-headed arrows leaving it. CHARLIE1 is a secondary key because it has one single-headed arrow entering it and one double-headed arrow leaving it. (Secondary keys are owned data items that always point to primary keys in a 1 : M relationship.)

When doing data modelling with user-view diagrams, where the end result is to be implemented on a tree-type database with predefined pointers, we will eventually try to work our user view around to the parent-child 1 : M segment view typical of this database physical structure. This is shown in row 4 of Figure 10-8, where we have three 1 : M "segments" from GRANDPA down three levels to SONNIE.

Note: When we speak of "*many*" we assume that at any specific point in time *many* could include 0, 1, or *many* occurrences but that *many* can and will occur.

EXAMPLES OF USER-VIEW DIAGRAMMING

Let us go back to Figures 10-6 and 10-7, where normalization was demonstrated in table format, and put those tables into user-view diagrams the way we would first draw them after analyzing the data in the view.

In the case of the repeating domains in Figure 10-6 this would be immediately apparent in the user view, which is shown on top in Figure 10-10. It would be simple to make two flat normalized tables from this user-view diagram. Of course, the analyst must be smart enough to recognize that the table has repeating domains and is not flat. Every subsequent step in data modelling is dependent on the elegance

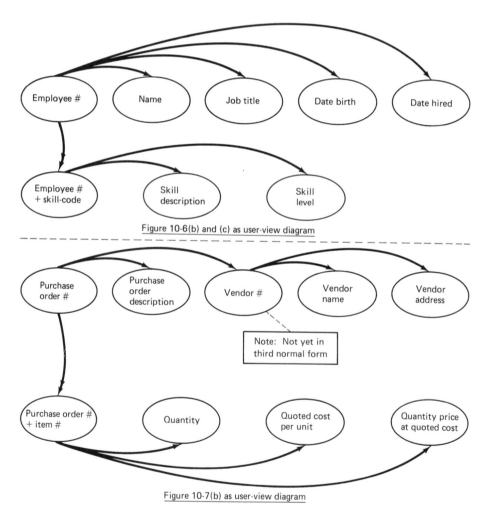

Figure 10-6(b) and (c) as user-view diagram

Note: Not yet in third normal form

Figure 10-7(b) as user-view diagram

Figure 10-10 User-View Diagrams of Table Views in Figures 10-6 and 10-7.

of the user view. This does not mean presynthesizing the user view into larger views; it means accurately catching the independent view.

The user-view diagram at the bottom of Figure 10-10 shows the way the problem from Figure 10-7 would very likely be drawn by the analyst. The problem concerning the dependency of the quantity and amount information on only part of the key would be resolved during the diagramming. The problem concerning the dependency of the vendor attributes on a non-key field (vendor #) would probably be left in the user view. The problem is, however, obvious from the diagram, since the bubble VENDOR # with a single arrow line out and a single arrow line in does not fit any of our descriptions just given for either an attribute field or a key field; therefore it is ambiguous and must be resolved. The resolution is the further decomposition of the user view in the data modelling process. Again, success in data modelling from this point on, whether done with or without support software, depends on the success of the analyst in capturing the true user view.

TAKING USER-VIEW DIAGRAMS FROM DOCUMENTS

Figure 10-11 shows a monthly bank statement regarding a checking account. Many of the user views made by the analyst will come from similar documents. Making these user views leads the analyst to begin looking at documents in a new way. The analyst quickly learns to decompose the document into flat tables and be sensitive to other problems of normalization. Looking at the Figure 10-11 bank statement, we can quickly see three levels of $1 : M$ relationships: customer tax Id. # (or social security #), account #, and deposit/withdrawal code. Assumed in the solution shown in Figure 10-12 is that while a customer could bank in more than one branch and have more than one type of account (checking, money market, IRA, etc.), for any given account number there would be only one branch and and one type of account.

Check # in Figure 10-12 is shown as $1 : M$ because deposits and withdrawals have been combined in this user view, and deposits have no check #. Also cash withdrawals from automatic teller machines would have no check #. Therefore we have a 0 or 1 situation, which we will show in this user view as $1 : M$, resolving it later on when we normalize. What is important for now is to capture the data relationships in the document for synthesizing with other user views of the data.

USER-VIEW SYNTHESIS DIAGRAMMING: THE DATA MODEL AS MINIMAL SUM OF COMBINED USER VIEWS

Our aim is to define the middle layer in our three-level database, the conceptual schema from which all applications are drawn and from which we will build our physical database system. This layer is independent of both other layers and, being so, gives us the benefits of our modern database design. Modern database design

<u>NORTH SHERMAN BANK AND TRUST CO.</u>

| Social Security No./Tax ID No. |
| 122-84-6666 |

William W. Westfield
RR 1, BOX 2620
No. Sherman, CT 06785

For any information call 203-394-0001

	Branch 22	Account-number 18489696	Starting date 12/22/84	Closing date 01/24/85
Deposit		Withdrawals	Date	Balance
		STARTING BALANCE	12/22	1,346.50
	11.21	CHECK NUMBER 466	12/23	1,335.29
	23.84	CHECK NUMBER 465	12/23	1,311.45
1,000.00		DEPOSIT	12/27	2,311.45
	110.12	CHECK NUMBER 471	12/28	2,201.33
500.00		DEPOSIT	12/29	2,701.33
	101.33	CHECK NUMBER 468	12/29	2,600.00
	200.00	CHECK NUMBER 467	12/30	2,400.00
		CLOSING BALANCE	12/30	2,400.00

Figure 10-11   Making User Views from Documents—Input.

is design for the best-case sharing of common data. If an organization's data has a high level of cohesiveness at the corporate level (e.g., sales, engineering, accounting, purchasing, warehousing, and manufacturing all using data items in common), and *given* that we want to come at the data from different 1 : *M* points of view, then resolved: we need a common database system, a common conceptual schema. The resolution of the conceptual schema into a relational, tree, or plex internal schema is a separate matter.

The idea of user views as the source of the conceptual schema can be reduced to water-cooler parlance by saying: "Why would we keep a data item on file if we couldn't get it into the machine and if we didn't need to pull it out of the machine for use?" Why, indeed? Nothing gets on the database system without first coming in as an add or update transaction. If we input and store a piece of data, it is for

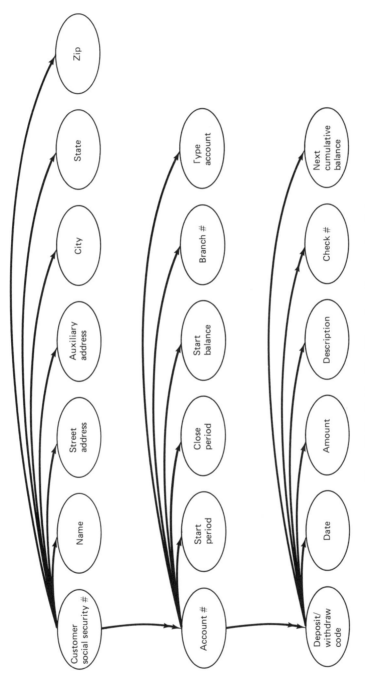

Figure 10-12 Making User Views from Documents—Output.

only one reason: because we intend to use it either now or sometime later in an output query, report, or feedback transaction.

Therefore, when we define user views as system inputs and outputs or some meaningful part thereof, we are saying that user views make the common data store—the conceptual schema.

Information specialists have been building data stores from inputs and outputs since before the internal-programmed computer was invented. In the beginning the user views were called card layouts, then transaction-record layouts and report-field descriptions. This was long before the terms "user" or "user view" were employed.

The advantage of user-view diagramming boils down to three things:

1. User-view diagrams may show errors of analysis.
2. User-view diagrams can be combined into one diagram.
3. The combined user-view diagram can be cleaned up, reordered, and minimalized.

The next group of figures illustrates this advantage of user-view diagrams. We will fold together three user views into one common user view and rearrange this common view in a way suitable for physical implementation on a tree-structure internal schema.

Figure 10-13 shows user view 1. One way to do user views is to ask a user question which the user view answers. A comprehensive output report could therefore contain a multitude of user views, an ad hoc query perhaps only one. We will define our user views by asking the question first: What is the number and color of all the phones in all the households in a town?

We will assume that phone number includes extension number. This user view 1 is shown in Figure 10-13. Given a town, there are many households, each of which could have zero, one, or many phones. Any phone will have one phone color. The modeller decides to draw a bidirectional arrow between household and town.

Figure 10-14 introduces a second user view, which responds to the question(s): What are the phone types and the street address for any person's household and what are the sex, occupation and age of that person?

User view 1 showed relationships between towns, households, and phones. User view 2 also deals with households and phones, adding new attributes in both cases, and introduces a new data group: persons. As we draw these user views, we show the degree of normalization by the way we connect the bubbles with the arrows, and this normalization process becomes part of the user-view drawing process.

Figure 10-15 shows the combining of user view 1 with user view 2, thus starting the conceptual schema. Now we see four keys, their attributes, and their relationships to each other: town, household, person, and phone. Town has no attributes.

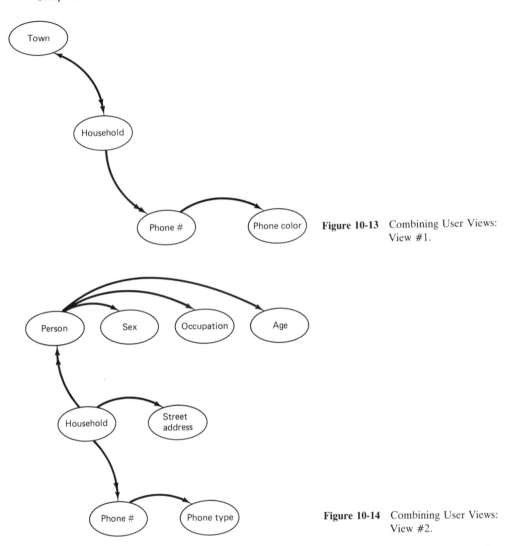

Figure 10-13 Combining User Views: View #1.

Figure 10-14 Combining User Views: View #2.

Figure 10-16 introduces user view 3. This view does not add any keys but does add attributes to the existing keys. Town still has no attributes. The question asked is: What are the town, household, residence type, street address, sex, occupation, and age of a person? This question implies bidirectional arrows. Given a person, what is their town? Given a person, what is their household? We already have 1 : M arrows going from town to household to person.

Figure 10-17 shows the result of combining user view 3 with the other two user views. In this little example we have said that the current known user views from which we wish to design our database system are the three views shown. This, then, is the data which we wish to bank. This will become our conceptual schema.

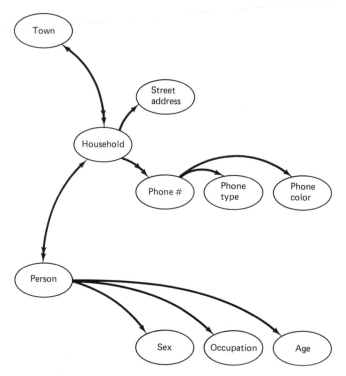

Figure 10-15 Combining User Views #1 and #2.

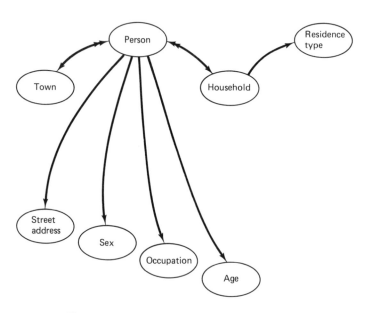

Figure 10-16 Combining User Views: View #3.

But first we need to clarify our combined user view, eliminate redundancies, and rearrange the bubbles for clarity of schema.

Note on Figure 10-17 that two redundancies have appeared. Town has two single arrows pointing to it. We can get to town from person via household and so do not need a bidirectional single arrow from person directly to town. Also we can get from person to street address via household and so do not need the single arrow pointing directly from person to street address. Eliminating these redundancies, we get the combined user view shown in Figure 10-18.

Examining this combined view, we start to ask some final questions: Town, which has a $1 : M$ relationship to household and person, is, nevertheless, starting to look less and less like a primary key. It has no owned attributes. It is an attribute of only one other primary key (household). Town is a good candidate for a secondary key, owned by household and pointing $1 : M$ to household and person. Otherwise, household, person, and phone # are shaping up as flat tables, where the attributes are owned by the keys and only the keys. Phone color, for instance, has been found by the analyst to be determined only by the phone # and not the phone # + the phone type or just the phone type. If either of these conditions were true, we would not be fully normalized to third normal form, and the combined user view would be incorrectly drawn. It is easy to see that the credibility of the conceptual schema we will arrive at is predicated upon the analyst's skill and on the users represented in defining the original user views. The process of synthesizing and normalizing user views can be automated, but if the original view is not the whole truth and nothing but the truth, the schema will be flawed.

Figure 10-19(a) shows the results of making town an attribute of household and a secondary index pointing to the primary keys household and person. Figure 10-19(b) shows the combined user view better arranged for implementation as a tree-type conceptual schema. We can see household as the root "segment" with phone # and person as children "segments." Figure 10-20 shows this implementation into a conceptual schema.

We could also have implemented this combined user view as a relational database, a single-path database, or even as a one-level redundant file. The important achievement is to get a true combined user view. Instead of doing user views, we could have done table layouts (as we always did in the past), but it is easier to do user views using normalization techniques. It may be hard to see in this simple little example, but given the ordinary and enormously complex combined user view for a common database system the advantages of this methodology become more obvious.

The question is, does the organization really need to arrive at a common primary database of a great order of complexity to get the information which must be obtained? Which is the greater mistake, putting up with the long-term creeping disaster of the application development point of view (which we have already detailed) or going to the elegant quantitative analysis of data modelling, based as it is on high levels of analytical skill? There are horror stories about using this approach: organizations who have found their development effort brought to a halt

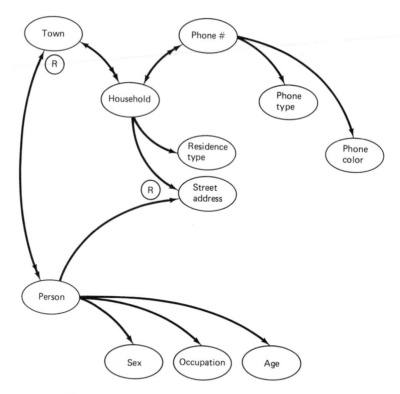

Figure 10-17 Combining User View #3 with #1 and #2.

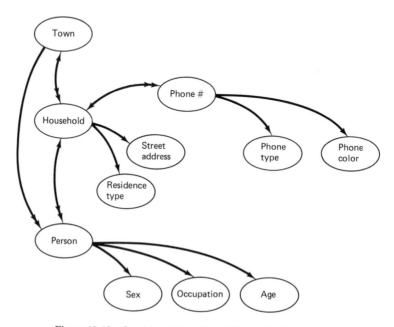

Figure 10-18 Combined User View: Eliminating Redundancy.

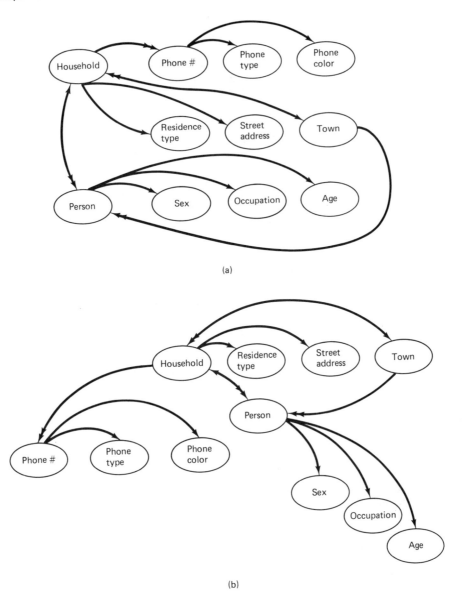

(a)

(b)

Figure 10-19   Combined User View: Creating Secondary Key and Rearranging.

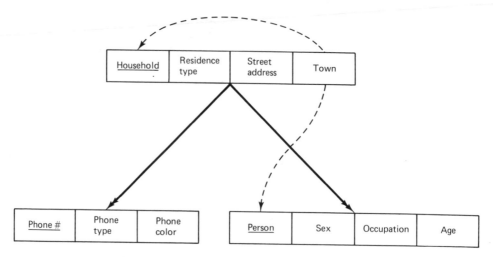

Figure 10-20   Conceptual Schema for Tree Database.

while major resources concentrated on the development of the efficient common database without achieving the expected results.

A SKEPTICAL VIEW OF THE COMMON, PRIMARY DATABASE SYSTEM

By a *primary* system we mean one in which we build and update the tables. We could retrieve information from a primary database or, if it is suitable, from an extracted (uploaded, downloaded, derived) database or database system.

Technical Problems with the Common, Primary Database

Doing high-volume queries and reports (decision support) from the common, primary database system is like drawing your drinking water just downstream from where you throw your slops and water your cows. It may be necessary but it sure isn't desirable.

The organization information experts and the users need to question whether it is necessary to know real time when an update takes place or whether the inquiry can use data aged a few hours. In cases like an airline or hotel reservation system the converse question must be asked. Can we afford to have the whole organization sending transactions into the reservation system database and slow down response time? The answers to these questions are usually: yes, we can use information two hours old for decision support in most cases, and no, when we do need real-time systems we want that application dedicated, not available across the organization.

Then there is the problem of updating a system of databases with predefined pointers linking many databases together in order to make a conceptual schema to support the external schemas of the applications. There are two ways to enter a transaction to update a table or segment in such a database. One way is in the *conversational* mode, the other, in the *pseudoconversational* mode. Another set of expressions describing the exact same update is *running long* and *running short*.

When we are running short (in the pseudoconversational mode), we select a record for update as we would do an inquiry, without tying up the record until the very last millisecond. This means that anyone else can retrieve the same record for either update (changing) or inquiry while we are still working on the record. We may select the record, have to go to the john and come back and complete the update in that very last millisecond. This could and does produce the possibility of breaching data security. Two person-terminal dialogues could update or make decisions using two different views of the same information. Sophisticated defenses against the problems of running short include comparing the original retrieval against the preupdate retrieval at the moment the postupdate retrieval is returned and nullifying the update if they do not compare.

The problems of running long (in the conversational mode) are what really concern the skeptics of the common primary database system. When we run long, we lock out from our internal schema any other user until we get the table, go to the john, and finish the update. This gives maximum data security. The problem is that the internal schema does not define one database affecting only one group of users who perhaps sit next to each other. No, the internal schema consists of ten or twenty or fifty physical databases *linked together* with predefined pointers. The conceptual schema supports many applications, and the path we lock up will likely weave its way through several physical databases.

The results are horrible. A purchasing agent or a sales product manager makes a change to a purchase order or a sales plan. These two databases have pointers to a location database where raw materials are stored. The product manager just wants to change the name of the copywriter, or the purchasing agent just wants to change a delivery date. As a result of these little changes' being made in a suburb of New York City, a nationwide network of locations is locked out, and somewhere outside of Cleveland a truck cannot be unloaded even though the materials on the truck are needed immediately on the shop floor at that location. Rule 1 in information system development: don't let your reach exceed your grasp of the consequences.

Politics Problems

Another serious consideration is the amount of organization politics that goes into the development of the common database. Many different user communities are involved. Within the data processing department itself the Database Administration, Data Administration, Systems Support, and Applications Development (not to mention all the other role-players such as the Applications Maintenance Groups) have to interact with each other in a much more collectivized manner than they used to. Sometimes born politicians rise to the top, and the hard-working professional who has no stomach for this level of wheeling and dealing becomes disgusted and longs for the old days when a smaller group worked to get systems on the air when promised.

Systems analysis, much to the discomfiture of the superior programmer who migrates into this work, must involve political skills. Politics is not used here in a pejorative sense. Politics is the art of arriving at consensus in a democratic environment—the smaller the group the better.

Regarding large-scale operations, no modern organization anywhere can survive by ramming decisions down the throats of powerful subsystem leaders such as vice-presidents or department chairpersons or cabinet members. However, the more one forces these power figures to share scarce resources such as good response time, the more likely it is that disorganization will result. Enter the political compromise. The systems analyst must know how to function in this political arena.

Other Considerations

Furthermore, we can see as we combine user views that if we don't really know our data and call the same thing by two different names or call two different things by the same name, combining many user views is going to result in one big mess.

It is usually possible for one application area to know its data. Information science up to now has been predicated upon that. But it is by no means clear that common data is clearly understood to the high degree of accuracy necessary to combine user views across functional or application areas.

In most organizations such a data item as PRODUCT-QUANTITY-NEEDED is not only not commonly used functionally but is not known clearly through *development time*. Sales may forecast a PRODUCT-QUANTITY-NEEDED two years before the warehouse is concerned. This quantity may be quite slushy indeed and go through many revisions, particularly at decision forecast time, when it is necessary to make a commitment regarding such items as purchasing raw materials. Even during production the PRODUCT-QUANTITY-NEEDED may be revised . . . and so on. Does manufacturing have the same perception of this data item as does sales?

All these different perceptions and interests have to be worked out much more seriously than is deemed necessary while doing fragmented applications development.

THE FAITH OF THE OPTIMIST

So then, caveat emptor, are we seriously proposing to build a common database system linking together (either when built or when collected for retrieval) perhaps fifty or a hundred subject databases through the development of user views? Do we have the hubris to suggest doing this across functions and time?

Yes. We do seriously propose to do this. It can be done with the help of the data dictionary and with a very strong data administration function—if the leaders of the organization are committed to it. But it is not going to be easy or quick to accomplish and it is going to require investment of money and other resources in research and development that will involve all the key people in the organization. It will also be necessary to hire and train the very best people. Many organizations that might need this type of data clarification and integration are not going to be able to make the necessary commitments. This explains why so few organizations have undertaken data modelling for integrated common data.

We will discuss in the next chapter organization strategic planning, and we will see that before the data modelling covered in this chapter can be commenced, the strategic plan must be developed. This monumental effort is the type of task which is undertaken in an organization only once in a generation. Should it be done? Yes—for the large organization it really must be done in order to make proper use of what is now becoming available in the way of information and consequent knowledge in what is being called the Information Society or the Information Age. Otherwise the very large organization may go the way of the dinosaur.

Meanwhile, for the smaller organization or the larger organization with other more compelling priorities, at least two alternative strategies have merit for the short-range and perhaps even the mid-range future.

ALTERNATIVE STRATEGIES FOR ARRIVING AT COMMON DATA STORES

1. Uploading from Application-Oriented Data Stores

First of all, no alternative is possible without some type of comprehensive review of organization data. This is best accomplished by organizationwide strategic planning, such as we will cover in Chapter 11, and through the use of a data dictionary, which we will cover in Chapter 12.

Figure 10-21 shows the strategy for an application derived common database. Insofar as is possible, each application database is kept separate from the other application data stores. Access to these smaller primary databases is available for the purposes of the application but is not encouraged across the organization, so that response time to the primary databases is not degraded. Problems within each application database remain circumscribed. If the database is changed, the application programs might need to be changed and recompiled. Redundancy of data, which is inevitable as each new application comes on the air, is encouraged.

Most importantly, the data dictionary mediates the data confusion that results, exposes the problems, and provides for a solution. Note in the example of Figure 10-21 that PRODUCT-QUANTITY-NEEDED from application-1 is found to be the same as PRODUCT-QUANTITY-NEEDED + ESTIMATED-DESTROY-

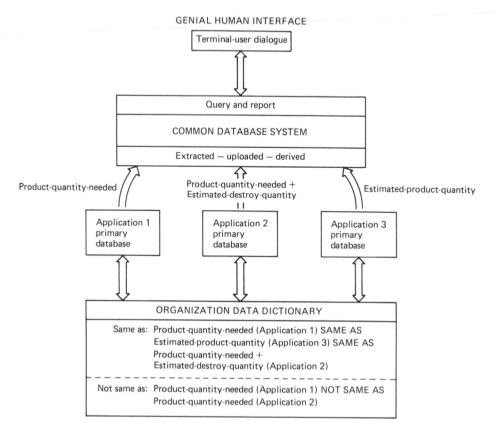

Figure 10-21   The Application-Derived Common Database.

QUANTITY from application-2, which is found to be the same as EST-PROD-UCT-QUANTITY from application-3. Only a well-maintained, seriously used data dictionary could mediate this confusion.

Data-dictionary mediation aside, the main solution is to make available a table from one application view for use with any other application view so that the table is updated only once by one application, though retrieved for all applications.

As far as decision support across the organization is concerned, the *upload* is used. Once or twice or more times a day a series of extractions to high-speed, read-access-only, shared databases are made to refresh these data stores. These databases are part of a state-of-the-art database management system which excels at providing decision support data store systems with all the retrieval tools. IBM offers DB2. The author is familiar with Mathematica's RAMIS II and Information Builder's FOCUS. Both are sophisticated fourth-generation systems which are programmer friendly and user congenial. Any user who can work a spreadsheet, word-processing software, or a file system on his or her PC can deal with this software, given adequate support. Meanwhile, the DBMS dialogue interfaces are expected to become much more congenial for all users. There are several other powerful, integrated DBMS packages not mentioned here which are surely also satisfactory.

Figure 10-21 shows the concept just described of separate application databases organized as a data administration function with the use of the data dictionary, uploaded for query and reporting and driven by a genial terminal-user dialogue.

Sooner or later, for some, this simple solution may be overwhelmed by the developing complexity (in addition to uploading there will be an increasing amount of transactions output from one database and input to another, for instance) or by less-than-first-class data administration support. In the meantime, many applications will have been quickly and simply developed in a decentralist atmosphere conducive to quick big hits for the good of the organization. After all, the name of the game for many excellent organizations is "get the project out the door—make it happen—we'll pick up the loose ends later."

2. Uploading from the $i - 1$ Functions

If we say that the whole system, the organization, is the ith level, then what is the $i - 1$ level? It depends on the organization. For a giant business corporation formed of ten or twelve divisions, each the size of many large organizations and run by an

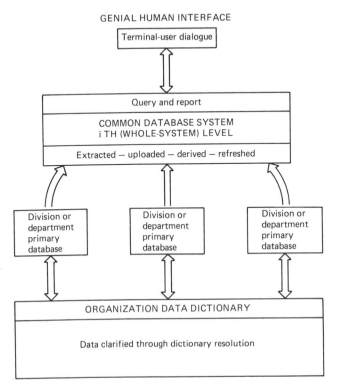

Figure 10-22 The $i-1$ Functional-Derived Common Database.

executive committee that essentially makes investment decisions and does personnel work on the fast-track managers level (deciding who gets moved up and who doesn't), the $i - 1$ level might be the division or the function within the division. For many other organizations it will be the large functions in the business environment, such as sales, manufacturing, accounting, legal, purchasing/warehousing.

In the $i - 1$ plan shown in Figure 10-22 data modelling through user views, normalization, and synthesizing to a conceptual schema are all done as depicted earlier in this chapter. The difference is that the primary database system is *not* common across the entire organization but only across the divisions or functions at the $i - 1$ level. Above that level the treatment is exactly the same as in the decentralized applications system shown in Figure 10-21. For decision support at the ith level a database is uploaded. All anomalies resulting from not giving the whole system one whole database are resolved—as already touched on—through the data administration function with the data dictionary. The very big difference between this $i - 1$ strategy and the application-derived database strategy is that modern techniques of data modelling are fully employed and the door is pulled open to the future without an attempt to bite off the whole job as one enormous indigestible chunk.

The reader is left to mull over these different strategies of database organization. Whatever is decided by an organization, it would be wise to develop a strategic plan and make use of the data dictionary and the concept of data administration. This is the subject of the next two chapters.

STRATEGIC PLANNING FOR INFORMATION SYSTEMS

This chapter on strategic planning which involves very high level (VHL) systems analysis, is placed at this point in the book for several reasons:

1. As shown in Figure 11-1, this subject is the natural companion piece to the application data modelling discussed in Chapter 10, where we developed a conceptual schema through user-view diagramming. We developed these user-view collections into tables and put these tables into the third normal form. Our database system was formed from these normalized tables. This was a bottom-up inductive method of arriving at an information base by putting together all the little pieces of data views to build a whole data view. These user views resulted from the process analysis undertaken at the application development level.

But how did a particular application come to be selected for detailed specification? What did we already know about the data and how the data is shared before we started the application? Surely the perspective at the single-application level is not broad enough to answer these questions. It is clearly necessary for the best development of information systems to support the goals of an organization to have a long-range plan of systems development that takes all contending needs for information into consideration. That is the top-down part of the job of systems analysis shown in Figure 11-1. Like all long-range planning, it allows for proper capacities management in addition to data sharing. Also, we shall see that certain

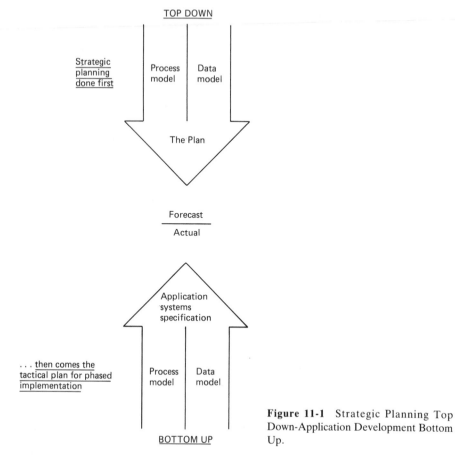

Figure 11-1   Strategic Planning Top Down-Application Development Bottom Up.

basic problems with systems that become apparent as systems mature can be observed and dealt with only through a methodology based on a view from the top.

2. Another reason for introducing this material now is the new perspective this chapter will give on the best use of the data dictionary, which is the subject of the next chapter. A better definition of the data dictionary, as we shall see, is: An Encyclopedia of Subjects and Their Relationships.

3. The other reason for introducing this material here is the encouraging fact that those readers who have struggled through the methodologies introduced in Chapters 1 through 10 are fully prepared to understand VHL systems analysis. It is just a matter of adjusting one's perspective to analyzing the whole system rather than some lower-level subsystem—and presenting specifications in terms meaningful to the senior managers of the organization. (Subsequent chapters on research and interviewing techniques are a requirement for doing VHL analysis but they do not introduce any new charting methods.)

STRATEGIC PLANNING AS AN ORGANIZATION FUNCTION

The planning and control process in an organization is inevitably divided into a *strategic development plan* and an *operating plan*. The development plan must precede the execution of the operating plan by a length of time determined by the necessities of capacities management. It could take a considerable length of time to finance a project to build a new manufacturing facility. We would have to acquire a plant site, get the necessary local government cooperation, and build, equip, and staff the plant. It is, of course, very important to be able to place costs and benefits on a long-range development project, since at any given time there may be several worthwhile development proposals in contention for financial and other resources.

Examples of Strategic Planning Processes

The development of policy, the gaining and disposing of resources, the evaluation of basic goals that control policy, the allocation of resources, the development of an outward image, investment strategy, adversary strategies, organization planning, and information resource planning are some of the major processes dealt with under the title of strategic planning.

Examples of the Operating Plan

Examples include product/service planning, financial planning, staff planning, facilities and materials planning, information systems planning, and measurement control plans—all guided by the objectives broadly set by the strategic planning process. This may also be called the *tactical plan* in support of the objectives of the strategic plan. Once the operations plan is established for a short- to mid-range period, the operations management is charged with the plan's successful execution and evaluation.

SPIRACIS

Profit-making or not, all organizations may and should have a concern for the quality of life in the local and world community and for the welfare and development of employees and their suppliers and contractors, but there is an overriding consideration of economy which helps to make planning quantifiable. That is SPIRACIS. SPIRACIS stands for *Strengthen Position, Increase Revenue, Avoid Costs,* and *Improve Service.** It is possible to put a benefit and ascribe a cost to most but

---

* SPIRACIS is the offspring of IRACIS, which is an acronym that has been around for a while. See Chris Gane and Trish Sarson, *Structured Systems Analysis: Tools and Techniques* (Englewood Cliffs, N.J.: Prentice-Hall, Inc., 1979), p. 160.

not all endeavors that can quantify SPIRACIS. Making the SPIRACIS business case for a project is part of both strategic and operations planning.

STRATEGIC PLANNING FOR INFORMATION SYSTEMS— THE BUSINESS CASE FOR DOING IT

Having put information systems development in its proper context regarding an organization's strategic planning (one of several concerns), let us say that useful information when needed is becoming of preeminent importance to the organization. Previous chapters of this book have established that information is an intrinsic quality of every process, including the process of life itself. Some observers have called what we are now living through the *information revolution*, equal in importance to the agricultural and industrial revolutions. This is, to many experts, the flowering of the age of information. If this be true, then the strategic planning for high-quality information systems is a task of the highest priority. We must build our future development of information systems upon a sound current information base.

Let us now look at three cases where information systems have had or will surely have a major impact on the health and welfare of the organization.

Case 1: Changes at the Beatrice Food Company

The facts in this case are taken from the *Wall Street Journal* of Tuesday, September 27, 1983.[1] Beatrice was, until recently, an unusually decentralized company marketing a wide range of products. The company grew by using its strong purchasing-power capability to buy companies in many different fields. Management was then often left in the hands of the entrepreneurs who were the former owners. Beatrice grew to be a sprawling empire of 430 different businesses, each performing the usual business functions such as purchasing, research and development, and marketing without consolidation of such functions. At the high point of the decentralized operation, for instance, Beatrice employed 140 different advertising agencies.

As time passed, many of the original entrepreneurs retired. Also Beatrice's smaller companies found themselves competing not only with each other but with such centralized operations as Procter and Gamble. It was a case of gunboats shooting it out with battleships. As a result of the difficulties imposed by this handicap, Beatrice, which has always been a well-managed company with quality product lines, decided not too long ago on a major change of strategy. Their sprawling, decentralized empire is being (or has now been) merged into six operating groups, which will then plan strategy closely with the headquarters operation (as of 9/83). "Beatrice wasn't capitalizing on its strengths as a big company," says a former Beatrice executive, according to the *Wall Street Journal*. Such functions

---

[1] Sue Shellenbarger, "Slimming Down," *Wall Street Journal*, 9/27/83, p. 1.

as advertising and purchasing are being consolidated as part of the reorganization. But nowhere in the lengthy article does the *Wall Street Journal* reporter mention anything about consolidating and sharing information and information systems. We can assume this must be taking place with the usual difficulty.

It is one thing to draw up and execute the reorganization plan for centralization. It is another, very complex, process to develop and carry out a new strategic plan for information systems, and without sharing common information it is very hard to carry out any reorganization.

The same problem with regard to information systems would result from the decentralization of functions in such a large organization. It may be even more difficult to share common information in that case.

Case 2: The Great Stock Market Information Fiasco[2]

Starting in the fifties and picking up speed in the sixties and seventies, the number of trading transactions handled by the New York brokerage houses increased in volume from below ten million to over two hundred million transactions on a busy day. This trend was clearly observable in the market, but these observations did not extend to changes in the so-called back office, where antiquated information facilities continued to try to keep up with the deluge of transactions. Insufficient attention was being paid to the need for strategic planning for the information systems. DP managers were probably tolerated as chronic complainers when they started predicting apocalypse now. Then one day in the mid-seventies "everything hit the fan."

Many of the brokerage houses simply couldn't process the transactions. Buy and sell orders were executed under changed price conditions. Large sums of money were lost in a vast float, and some of the large brokerage houses faced financial ruin. Only then did the whistles start to blow. Old systems analysts still talk in awe of the type of heroic measures it took to quickly specify and build the information systems necessary to correct this gross error in strategic planning.

Case 3: How New York City Almost Went into Bankruptcy Without Knowing Just How Bad the Problem Was[3]

During this same period that the brokerage houses had their day of reckoning the same thing happened to New York City. The information systems necessary to run an organization of the size and complexity of New York are impressive. For many years NYC had been slowly changing. Many poor, unskilled people came to the city because of its opportunities and compassionate support of the needy. At the

---

[2] Based on interview with Felix Rohaytn by David Susskind. David Susskind Television show, 3/4/84

[3] *Ibid.*

same time the middle-class city dwellers took advantage of inexpensive, available mortgage money to leave the city in a historic and major migration, settling in their own homes in the suburbs. At the same time, too, the small manufacturing operations for which NYC was so well known left for more favorable locations as New York began its metamorphosis from a manufacturing to an information-processing center.

Poor people with and without jobs created a need for many more municipal workers such as policemen, firemen, teachers, and sanitation and welfare workers. These workers were effectively organized into powerful unions which did what they were supposed to do: greatly increase the wages and fringe benefits of their membership. On top of all this, a recession and gas shortages "fueled" inflation.

As all these pressures mounted, the city leadership began to borrow money from the banks in ever-increasing amounts to meet current operating expenses. No one seemed to know how much the deficit amounted to. City officials stated it was just seasonal borrowing that would self-correct.

When it became obvious to the financial community that the situation was indeed more serious than expected, the Governor of the State of New York was called upon to save the day. He appointed a commission of financial experts to try to figure out what the deficit really was. With a great deal of difficulty (since valid information was not available) this group of wizards managed to piece together the facts, which horrified them. Even they did not expect the deficit to be over six billion dollars. The city officials denied that such a deficit was possible, but it became evident that it was true. New York City (and the banks and New York State and probably the whole nation) were saved by the last-minute creation of the Municipal Assistance Corporation.

The question is: Would the banks and the State of New York have let these events proceed so far down the road to disaster if reliable planning and information systems had been in place?

Hopefully, these three examples have made the point that strategic planning for information systems is a very worthwhile endeavor, no matter what the effort involved. The alternative to such planning is to pay off political debts or oil the hinge that squeaks the loudest and never build the type of database system that will allow common sharing of minimally redundant and program-independent data effectively related to resource and process management.

It is important to interject here that the reasons for conducting a strategic planning study are positive as well as negative. We will have more to say about the positive motivators later in this chapter.

UNDERSTANDING STRATEGIC PLANNING FOR INFORMATION SYSTEMS

Strategic planning for information systems (IS) is, first of all, done to determine the *application systems* that need to be developed for the whole organization and the *data entities* that are needed to support the information requirements for these application systems. These data entities, when grouped according to affinity with other entities, form subject databases that need development into functioning database systems. Therefore, the result of the strategic plan is a high-level information-system development plan, including what we called in the last chapter the conceptual schema or database organization.

Now that we have introduced the term *data entities*, we are obliged to define it. The meaning is somewhat different from that in Chapter 10 where in Figure 10-1 we equated *entity* to *data group*. Figure 11-2 introduces strategic planning terminology, showing the also-known-as (aka) synonyms much as in Figure 10-1.

In Figure 11-2 data entity is related to data group—a term with which we are familiar. In Chapter 10 we defined data group as one table row and in Figure 10-2 gave an example of it (08, Birdsong, M, Systems Analyst, 26). In the strategic plan study we will stick to this definition of a primary key with a set of attributes— with these addenda: the primary key becomes a meaningful class name, and the group does not have to be normalized. In the case of our example above, we are not interested in Id# (08) as much as we are in Customers (Birdsong). We may be interested in knowing Birdsong's hobbies, music and reading interests, and many other such attributes about Birdsong.

Column / Row	"Ann Arbor"	IBM'S BSP	This book up to now	Other definitions in this book
1	Strategic planning for information systems (IS)	Business systems planning (BSP)	Strategic planning for information systems (IS)	
2	Buisness function	Organization by leader	Jurisdictions, subsystems	
3	Process	Process group	Activity	i − 1 level
4	Activity	Process	Operations / Processes	i − 2 i − n levels
5	Information group	−	User view	Documents transactions
6	Data entity	Data class	Data group	Table row
7	Subject databases	Databases	Conceptual schema system	

Figure 11-2   Strategic Planning Terminology aka Table.

Other examples of data entities are:

- machines
- vendors
- employees
- workplace locations
- sales plans
- projects
- tax information
- vehicles
- sales schedule
- parts
- finished goods
- organization profile
- department profile
- employee benefit
- facilities plan
- class schedule
- class description
- students
- curriculum
- faculty

There are two other powerful reasons for conducting the strategic plan (the Plan) besides the development of information systems. For those looking for dramatic results from systems analysis the next few paragraphs may be the most important in this book.

The Plan involves building an organization model relating organization structures (functional areas such as accounting or sales department) to organization processes (provide employee benefits, pay accounts, make payroll, do taxes) to organization information groups (user views such as invoices and purchase orders) to organization data entities (such as vendor, parts, sales-plan).

When the final study report is presented, what usually shows up from the matrix tables that make up part of the report (if they are allowed to survive) is a significant and embarrassing redundancy of effort—a productivity drain significant in scope. This shows up in two areas.:

1. Most organizations that have been around for a while have significant duplication of processes related to functional departments. Sometimes this is deliberate, as in the case of competing sales operations within the same organization, but usually

this is a rationalization covering other motivations. Sometimes purchasing is done in several departments. Sometimes duplicate inventories are kept by several competing groups, entailing considerable redundancy of effort and resources. Very seldom in these situations is there any sharing of resources. What there is is duplication of resources. Financial information may be developed by the accounting department that is also developed by the market analysis group in the sales department. Often the information that should agree does not agree.

All this duplication of effort may have been all right when manual information systems were being used. New sales departments were set up for new products, and people had to be hired to deal with the workload even though the procedures of developing and executing sales plans used the same processes and information groups and data entities. With automated systems this formula is invalid. You don't hire new people if you add product lines that use the same information and generic processes. A computer-based system will handle ten times this kind of workload without any noticeable increase in computational load. After you get past the development costs, these systems handle information loads under an entirely different concept of productivity.

What does an automated information system care whether the product line is cars, dresses, ice cream, or airplane seats? It cares a little (perishable-inventory factor) but not as much as one would think using old-fashioned parameters. Certainly if the product line is cohesive (you would not want to build a new information system for each school or campus in a university system), it cares very little.

2. When we examine the relationships between functions, processes, information groups, and data entities, the numbers are amazing. Give or take thirty, there are only about fifty functions, fifty processes, and fifty data entities in a large organization.[4] (These numbers are mean averages and there may be some few exceptions, especially on the less-than side.)

The uninitiated, not familiar with results of strategic plan studies, are amazed at the small number of data entities. Given that each entity, group has five to twenty or so attributes fully identified by the primary key of the entity, that is not much data. So what is all this talk about converting over to the information society? There is not much data in that information!

The answer is that there are usually thousands of information groups (user views) thought to be necessary in the average large organization. After all, there are only 26 letters in the English alphabet, and look at the proliferation that has resulted from that small number of letters.

We expect to see a large number of activities one level down from the process level, maybe hundreds, and we expect to see information groups in the many hundreds, or thousands. But there are always more information groups than there should be. For every redundant process or activity there have to be many redundant information groups. For every nonredundant process in a mature organization there are many unnecessary information groups. It is very easy and rewarding for the

---

[4] Results of strategic planning studies participated in by author at Reader's Digest Association.

workplace to vulgarize information, to be uneconomical in the use of information. What is good for the department that has a life of its own apart from the system may be deadly to the system as a whole. This is the main problem with large, mature organizations. This is a problem not solvable by traditional applications systems analysis. It is solvable by VHL systems analysis, which is what we are about in this chapter.

It is not necessary to document here how organizations reward the mechanisms of their own downfall. Just ask yourself which is usually more profitable and enhancing—to manage large numbers of redundantly occupied information-handlers, or to lead a small team developing vital knowledge for the organization. Who uses more information groups?

The author is aware that there are other normative values besides productivity that drive societies. No one should want to return to the cruel Victorian age of exploitation in the name of productivity. Maybe the normative value of our world culture should be keeping every human happy at makework while the robots do the productive work. The issues raised here are explosive. Making a large organization employing 4000 information workers productive could damage the economy of an entire community. Information handling is that inefficient in some places. At least let us not delude ourselves about what is involved in doing a strategic plan study for information systems development.

Perhaps we should be motivated by three goals in doing the Plan: supporting opportunity and finding and curing pathology, to be sure—but also supporting community.

PLAN OVERVIEW—A GLIMPSE

The Plan is carried out as a total systems effort. The entire organization is involved. A study team is formed. The study team consists of information systems specialists and users, working together. The effort is sponsored by the organization's chief executive officer. All the key people in the organization, anyone who might make a contribution to the study plan, are interviewed. After the current system is analyzed and presented, organization executives are interviewed in depth to develop new requirements from their whole-system perspective. New requirements are further developed, perhaps resulting in new data requirements. A final report is made, outlining future development needs.

STRATEGIC PLANNING TERMINOLOGY AND RELATED MATTERS

Before proceeding to the details of conducting the Plan we should, as we did in Chapter 10, ground ourselves in an appropriate terminology. To do this we will turn again to Figure 11-2.

The author has labeled as the Ann Arbor Connection (for want of a better identifier) those firms located in Ann Arbor, Michigan,[5] and their doppelgängers in other places, who specialize in top-down strategic planning (our current subject) tied to bottom-up data modelling (the subject of the last chapter). These companies offer, in the author's opinion, one of the more comprehensive, well thought out methodologies now available in the marketplace in this subject area. We will make good use of the Ann Arbor terminology (which is or has become generic), as shown in the first column of Figure 11-2. Those readers who feel more comfortable with another terminology can select a synonym from another column on the same row.

The other major methodology available in the marketplace is IBM's Business Systems Planning (BSP).[6] This is an established, quality methodology used by many organizations, many of whom regard BSP as the method of choice, having produced effective results with it. The terminology for this methodology is shown in the second column of Figure 11-2.

In the third and fourth columns are the terms which have been used in this book up to this point—which, as we shall see, are quite valid, even though the author is deferring to another terminology.

It will be instructive for us to deal with Figure 11-2 row by row. Interesting differences in points of view will be revealed. Also, even though this is called an also-known-as table, there are row differences to be uncovered which will clarify further discussion.

Row 1: Strategic planning for information systems.   The use of the word *business* in the second column is too restrictive. Many organizations, such as government agencies and schools, cannot be called businesses.

Row 2: Business function.   It is very smart for Ann Arbor and BSP to make a distinction between jurisdiction and process. As we have already pointed out, this makes for some very interesting and productive analysis. BSP uses the word *organization* to mean the job title of the head of the business function—e.g., Vice-President of Sales or Vice-President of Manufacturing. This could get tricky. Why not just refer to the business function itself?

It is important for the reader to fully understand that a business function is a jurisdictional area with a geography and table of organization—not a process.

Row 3: Process.   A *process* is our largest category of what people do in the organization. Both Ann Arbor and BSP make a mistake in not putting this concept of process levels in an organization in the proper context. This goes to the heart of the whole subject of systems analysis. Ann Arbor defines activity within

---

[5] (a) Holland Systems Corporation, 3131 So. State Street, Ann Arbor, Michigan 48104; (b) Database Design Inc., A Doll-Martin company, 2395 Huron Parkway, Ann Arbor, Michigan 48104.

[6] IBM, Information Systems Planning Guide (Business Systems Planning), GE20-0527-3, 3d ed., July 1981.

process. BSP defines process within process group. There is not enough explanation that this is skimming the top two levels off a multileveled (recursive) process that goes down a variable number of levels per process until we reach the base level where it is pointless to decompose process any further.

Major sections of this book, particularly Chapters 3 and 4 which covered process charting methodologies, have been given over to stating the case for vertical decomposition (levels) along with horizontal decomposition at each level (partitioning). The fact is that if the whole system is the ith level (the whole process), then there are $i - n$ levels, and it matters not what you call these levels. Able-Baker will do just as well as Process-Activity or Activity-Operation-Process. The important thing to remember in this chapter is that both Ann Arbor and BSP have as an objective the blocking out, in a compressed period of time, of the major areas of function, process, and information and how they relate. This is a top-down skimming of just enough analysis to make strategic assessments and set a proper course.

Row 4: Activity.   We now come to rows which show the main difference between the Ann Arbor and the BSP approach to the Plan. When BSP describes Process Group-Process, it is dealing with a higher-level view of the organization processes than does Ann Arbor when it describes Process-Activity. BSP states that "most studies result in only 20–60 processes for the whole business"[7] and "there will normally be 4–12 process groups."[8] Ann Arbor might not agree with this scoping of the process.

Perhaps they would say that process group is not comparable to their process level and that BSP shows only one level: process. Their average activity level would probably show several hundred activities within processes for which BSP has no counterpart. This is the first instance that shows the study according to the Ann Arbor approach is scoped differently and more comprehensively than the BSP study—which is not to say that more is better.

Row 5: Information group.   You cannot arrive at data entities (or BSP's equivalent data classes) without first identifying and examining information groups. This is the most significant and time-consuming task in the study, since there can be thousands of information groups. BSP has no formal mention of information groups. It is stated that through document collection and the interviewing process data classes will be arrived at.[9]

Row 6: Data entity.   There is agreement here that data entities are modest

[7] IBM, Information Systems Planning Guide, p. 33.

[8] IBM, Information Systems Planning Guide, p. 36.

[9] IBM, Information Systems Planning Guide, p. 41.

in number. BSP states: "In most cases the study team will identify 30 to 60 data classes. . . ."[10] Ann Arbor might place that number higher but in the same ballpark.

Row 7: Subject data-bases.   The term *conceptual schema system* is preferred, but the author concedes that *subject databases* is more descriptive for many.

What can we make of the apparent difference in scope between BSP and the Ann Arbor method? This is a tricky question. BSP seems to commit less company resources and less elapsed time to accomplish the Plan. What if BSP is sufficient to solve the problem? The problem is to understand the whole system, spot problems, make new requirements, and do a high-level development architecture for new information systems. The benefits from the study are down the road. Applications ready to go on the air give direct benefits; long-range planning studies do not.

In many organizations BSP may solve the problem. These organizations have a good grasp of their operations and data already and need a combination review, company conference, and launch meeting to get the armies moving. Other organizations may not really have a grasp of their situation, may need deep quantitative analysis, and are not ready to launch any plan. For these organizations, unless experienced leadership is present to *make* BSP work, the study could and will degenerate into a meaningless dog and pony show—yet another useless study that senior executives have become cynical about.

The Ann Arbor method also has its pitfalls. It is sufficiently complex to beg for expert consultation. It is sufficiently comprehensive to require software support. It has the negative potential to drag on for a long period, using up the best systems brains in the organization, deferring important applications development. Yet if done with robustness and vigor and a sense of deadline it is, in the author's opinion, a good choice for many organizations.

And, of course, even as this is being written, both methods are being improved and expanded by both the vendors and the more sophisticated users.

Whatever choice is made by a particular organization to do the study, the methodology put forth in the rest of this chapter will be found complementary. BSP and Ann Arbor are both very good about telling what needs to be done and who should do it but are not so clear (in writing) about just how it gets done.

STRATEGIC PLANNING METHODOLOGY

The hard methodologies needed to do a strategic plan for information systems (the Plan) have already been shown in quite some detail in the first five chapters of this book and in Chapter 9. Those astute readers who have already observed this can pat themselves on the back. They have distinguished careers ahead of them as VHL systems analysts. A review of some of the illustrations in each chapter will

---

[10] *Ibid.*

show clearly that the tools are already learned. What is left after we review what we have already learned is to tailor the methodology to the scope of the Plan.

Note. We have not as yet discussed soft methodologies such as interviewing, document research techniques, documentation of findings, and relating to users, all of which are subjects still to be dealt with in upcoming chapters. We will deal very briefly with those subjects now, as it becomes necessary.

Glossary Note. A *hard* methodology, which in systems analysis is usually a charting technique, deals with modelling a process or a data collection in sufficient detail so that it can be replicated in some reductionist form sufficient to be useful. A *soft* methodology makes observations about processes and data and how we might learn more about the nature of the subject. Both are useful, but the history of baking tells us nothing about how to bake a chocolate cake with coconut icing.

The Plan is nothing more or less than a current system specification, followed by a requirements analysis, followed by a new system specification. We have done all that before in this book. We have already defined systems analysis as being just that, saying it could be done at many different levels. The Plan is a VHL flyover done with the active participation of the entire policymaking stratum of the user community. The user gets to play analyst for a time. The other differences between the Plan and the ordinary specification is that there is a concentration on relationship analysis, as we have already mentioned (such as processes to data entities), and, of course, a difference in scope. Matrix reports are used to show relationship. The whole organization gets together to take a hard look at the whole system, so that all the needed information systems can be identified and scheduled.

Figure 11-3 gives a simplified picture of the vertical process and data layering in a system and how the Plan skims the top two layers, along with the high-level view of data at those levels to do the study.

Let us now review what we have learned already about doing system specifications. We will do this by taking another look at illustrations in past chapters.

Figure 1-1.   This figure shows activities containing operations containing processes. We have now agreed to call these layers of a natural system such as a social organization *activities* within *processes* of a natural system. Who cares what we call these levels; the point is that the system model is best thought of as hierarchically leveled. We want only the top two levels to meet the objectives of the Plan.

Figure 1-2.   This figure shows the analyst creating the specification document in flow chart form and then reworking the charting to communicate the findings to the user groups. We will use matrix diagrams as well as other techniques to show Plan findings.

Figure 1-3.   Chapters 1 and 2 outlined the entire system development life cycle (as this figure shows). Everything we now intend to do in the Plan is covered

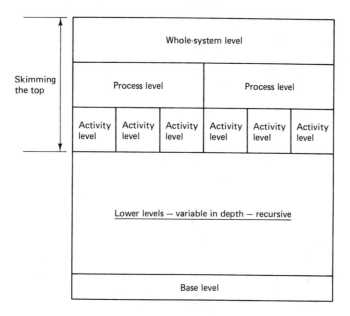

Figure 11-3   Skimming Vertical Levels for the Strategic Plan.

in the development life cycle. We won't dig any deeper than is necessary to arrive at a long-term development architecture. We will cover the following subjects, shown as boxes on Figure 1-3:

- get and publish a letter of authorization from the executive sponsor
- do overview current system study
- present study
- set inclusion-exclusion boundaries for further study
- do requirements analysis
- do new system specification alternatives
- define and present a conceptual new system showing automated information systems needed plus support data needed
- present recommendations and decide on development plan at the whole system level

Figure 2-1.   This figure shows an example of how requirements relate to the general and specific goals of the organization. The first objective of the current system study is to understand these goals.

Figure 2-2.   This illustration shows that we have not only covered this material before as part of the development life cycle overview but that we are clear as to what we expect from the strategic plan study. We want, as the figure shows, an identification of subject databases and a system development architecture, maintained on data dictionary software (the subject of the next chapter). From this architecture we intend to define a development schedule, including resource allocation by date, from which we will launch detailed analysis, design, and implementation of system applications in an orderly, phased plan. Once these applications go on the air, we will maintain them and develop them.

Figure 3-9.   Here we see the concept of leveling and partitioning. We will partition the whole system into a set of processes and down-level those processes into a set of partitioned activities. In this example the whole system, which is a process called Have Dinner Party, is down-leveled and partitioned into five processes: Invite Guests, Greet Guests, Serve Drinks and Munchies, Do Dinner, and Do After Dinner Entertaining. We will take these processes down one level to Make Dinner and Serve Dinner activities for the process Do Dinner. However, since this is a high-level effort, we will not go down one more level and decompose Do Dinner into Make Soup, et al. We will diligently resist this impulse.

In terms of information we will have an entertainment schedule, a guest list, and several different recipes—perhaps four or more for the entree alone. These are all different information groups (user views). However, the single word "recipe" describes the data entity sufficiently. Recipe stands for a list of ingredients and a processing description. What we want for the purposes of the study are the data entities Entertainment Schedule, Guest List, and Recipes, with a brief description of each. We resist the urge to delve deeper into interesting data, such as that unique recipe for chocolate mousse.

On the other hand, while resisting the urge to delve deeper into activity and entity, remember that we want it *all* at the specified level of detail: no missing entities, no missing activities.

Figure 3-11.   Parts (a), (b), and (c) of this figure show a more realistic example of leveling and partitioning a system and relating process to needed data. Here we are specifying the current system for a major league baseball team. We see that at the process level we have partitioned the flow into seven processes from Do Spring Training (1) to Do Intraseason Operations (7).

Associated with these seven processes are the data entities shown in the lower left-hand corner of part (a). There are many more information groups—scorecards, tickets, etc.—that have to be reviewed to arrive at these entities. We then make subject databases from these entities, the sum of which are the conceptual schemata which are translated by the DBA into the physical database system in support of the one common process.

Part (b) shows the activities level of the Do Spring Training process, which involves ten activities. Part (c) goes beyond the scope of our Plan. At this and lower levels the data entities get referred to over and over again for different information groups.

Figures 4-3, 4-4, and 4-5. These figures show the bakery model using the other process charting method presented in this book. This is more like the real world in terms of process complexity. The three figures are all quite relevant to our task of doing the Plan. Only two levels are shown—and not all of the second level. The data entities are shown on the lines between the bubbles on this data flow diagram. The moving data is shown as it is being transformed, which is more than we want for the Plan. What we want to record as entities here are the data stores in the open-ended rectangles.

What we observe in Figures 4-3 through 4-5 is an example of a specification for both the current and the new system, which is the basic documentation from which we will derive our Plan study. In between is the requirements analysis shown in Figure 5-1. For the purpose of the study we want a high level requirements analysis defined by the senior management of the organization. Still, the methodology is the same.

How, then, do we arrive at this model, at the depth shown in Figures 4-3 through 4-5, for every process in the organization and from which we will make our study? Where do we get the knowledge to do this?

RESEARCH TECHNIQUES IN SUPPORT OF THE STUDY

We will deal with this subject in a subsequent chapter, so at this point we will only touch on some highlights.

Who Does the Research?

It is not recommended here that users participate in the systems analysis needed to create the organization model at the research level. An experienced VHL-grade systems analyst is worth about as much (and gets paid about as much) as a corporation lawyer in a corporate legal department. We should strike a reasonable deal with users such as the organization's lawyers: we won't try to research the law or settle litigation out of court or draw up a contract and they won't try to build a model of a large organization with its automated information system support.

It is very popular now, as this book is being written, to believe that users can carry on their own work from their own backgrounds and still do systems analysis. It just isn't so. Putting skills aside, users are not always objective regarding problems.

What the user community can do is:

- welcome the researchers into their territories (functions)
- serve as a committee of oversight to see that their interests are being served
- become part of the study team after the current system research has been concluded

- participate actively in the group process of working up the new system requirements
- understand and participate in the development of the new plan for system development

Of course, if the organization doesn't have enough analysts to do the project, it will be necessary to hire consultants or train users as analysts competent in this one area. Our goal should be to spend no more than about two months doing the study.

It should be obvious from the list just given that we anticipate an increasingly active role for the user as the study plan develops beyond the research stage.

How Do We Conduct the Research?

We interview in each of the organization functions. (Remember that there are functions; i.e., departments and functions do processes, and processes can be partitioned into activities.) We interview, if necessary, at three levels and in two functional areas for each function.

The three levels are manager, supervisor, and mission worker. The mission worker may be a Nobel-prize-winning chemist or a drill-press operator. This is the person who actually does the work. They all have different stories to tell, which may not hang together. For a high-level study most of the interviewing is with the manager, unless this person turns out to be unsatisfactory.

The two functional areas are the function itself and the set of functions that provide input and receive output from the subject function.

If we know in advance that the manager of the function under research either is going to be unfriendly or is non compos mentis, we interview in the other functional areas first and then bottom-up in the function after a first obligatory interview with the manager. By the time we get to the unfriendly manager, if we keep our mouths shut, this individual will tell all about where the gold is buried plus much more. The function manager has no chance whatever of holding back information or distorting information when dealing with an experienced systems analyst backed by the organizational leader.

If this approach to interviewing sounds cynical, please be assured that its purpose is usually to check honest confusion. Operations management does not often live in the same world of precision in which the systems analyst who started as a programmer has been raised. These managers are often primarily concerned with resource management, particularly human resource management, and are not always sharp on the process logic in their function. However, let us not turn away from the fact that many managers have ambivalent feelings about change, not always for less than good reasons.

Before going beyond the first interview with the manager, we confront the second major area in which we do research. We request from the manager *all* the documentation available in the department. This documentation is of two kinds:

- user views
- explanatory material

First we attempt to learn as much as we can of the system from the explanatory documentation. Then we turn to the user views, which are the action documents that drive the process, such as invoices, orders, schedules, workload reports, purchase orders, and sales plans. These user views are our information groups and are a vital element in the study. Unlike the full-fledged system specification, we only glance at these information groups to identify data entities.

A final word about interviewing. There are all kinds of interview techniques to be learned. For now let us just advise to keep the interview informal and warm and mutually participatory. Most managers want to improve their operations (when assured their own self-interests will be served) and have lots of good ideas about how to do it. Do not use more than one interviewer or use a notetaker or even carry a tape recorder. Don't do anything to stifle free and easy expression. Learn to write without always looking at the notebook. If the reporters for the *New York Times* and the *Wall Street Journal*, two sources for future history, can do this, so can we. Just like these reporters' stories, our notebooks are important source documents which are permanently saved.

How the Research Process Actually Goes

"*What do you do in this department?*" We will get back a combination of processes, activities, and information groups. Separate these by category.

"*What documents do you use?*" We will get a partial list of information groups and some more processes.

Since we have already reviewed all the documentation, we ask: "*What about document A? Why haven't you mentioned document A?*" We will get back information about this document, other documents, and more information about processes and activities within processes.

"*How can you do process X without knowing something about . . . ?*" At this point we mention a data entity. We get back more information about information groups at the least. We keep this up until we are satisfied that we have all the facts. We do the multiple interviewing already covered as it becomes necessary to do so. We also record ideas and gripes for the time when we have completed the current system study and are looking into needed changes.

This all must be done very quickly for this high-level study. We do not get in too deep unless the research data looks bogus.

What to Do after the Research Is Concluded in a Function

Chart it as in Chapters 3 or 4. Check charts with the sources. Get them approved.

Then extract from the charts the necessary relational lists: Function DOES Process CONSISTING-OF Activities WHICH-USES Information Groups MADE-

UP-OF Data Entities. Enter these relational lists onto the data dictionary or other documentation medium and move on to the next function. Figure 11-4 illustrates what we have discussed in this paragraph.

Sounds easy? It is not easy. The more thoroughly the analyst knows the organization and the more experience the analyst brings to the task, the closer the model will be to the truth. It is particularly hard to arrive at the few data entities that make up the information system. Remember, we are working under deadlines, since to be useful this project has to be short-lived. From an impressive stack of information groups we have to cull out data entities.

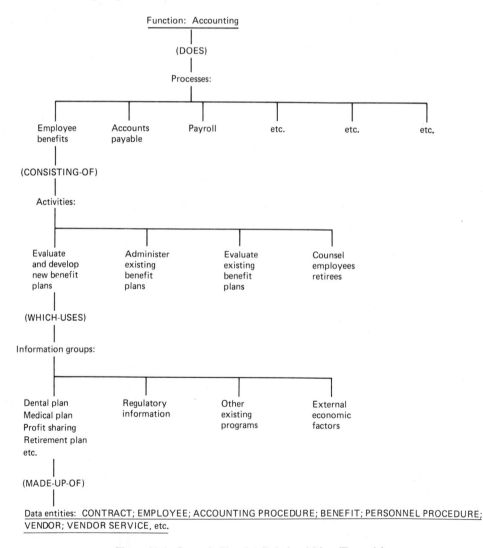

Figure 11-4 Strategic Planning Relational Lists (Example).

Figure 11-5 shows an example of a document that represents an information group and the data entities that must be quickly identified. Do not think, because there are several data entities on this one group, that we will end up with more entities than information groups. The opposite is true. There are many information groups that use these entities. MATERIAL, for example, will appear on sales plans, engineering specifications, warehouse inventory reporting, purchase order changes, receiving slips, invoices, and so forth.

Sometimes it is good to work as a tandem team on this. One person drives through the documents, banging out the data entities and ignoring redundant documents. The follow-up person verifies the selection that was made. We are looking for that new data entity. We pick up a purchase order and see VENDOR, LOCATIONS, REQUISITIONER, PARTS, and SCHEDULES as the data entities. We pick up a purchase-order change and learn nothing new.

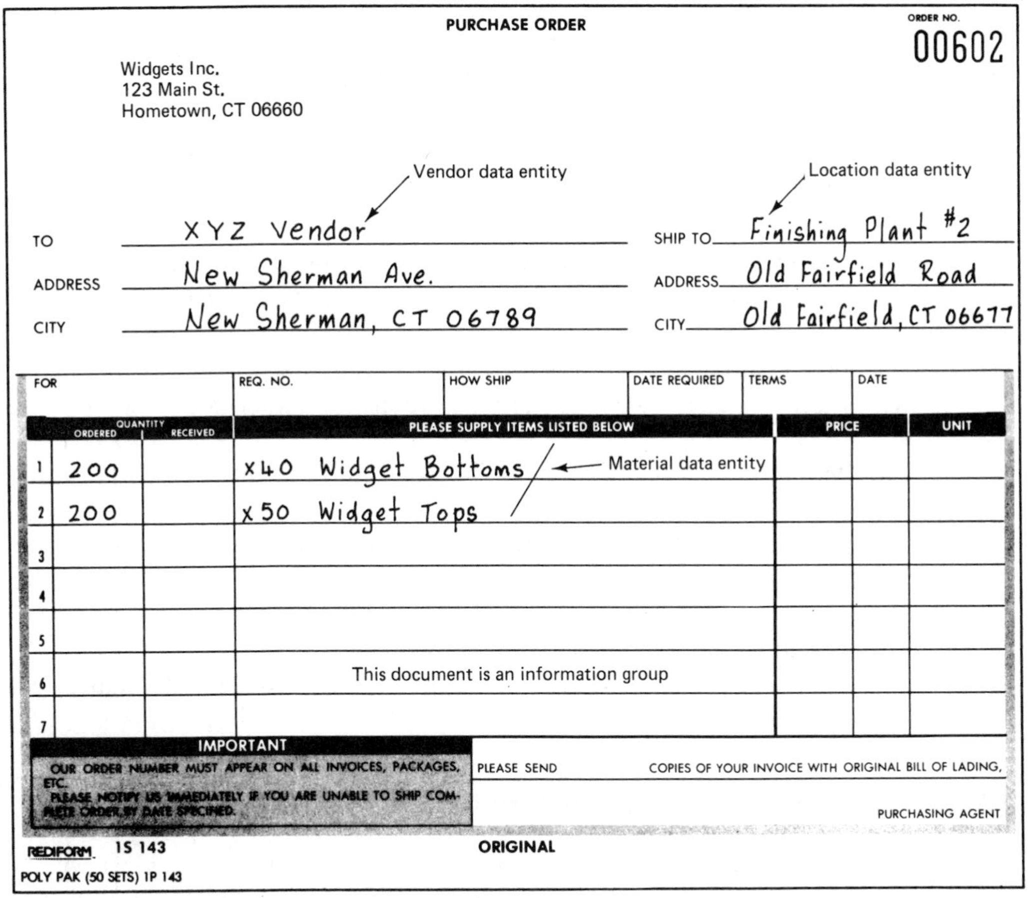

PURCHASE ORDER

ORDER NO. 00602

Widgets Inc.
123 Main St.
Hometown, CT 06660

Vendor data entity

Location data entity

TO XYZ Vendor

ADDRESS New Sherman Ave.

CITY New Sherman, CT 06789

SHIP TO Finishing Plant #2

ADDRESS Old Fairfield Road

CITY Old Fairfield, CT 06677

FOR	REQ. NO.	HOW SHIP	DATE REQUIRED	TERMS	DATE

	QUANTITY ORDERED	QUANTITY RECEIVED	PLEASE SUPPLY ITEMS LISTED BELOW	PRICE		UNIT
1	200		x40 Widget Bottoms ← Material data entity			
2	200		x50 Widget Tops			
3						
4						
5						
6			This document is an information group			
7						

IMPORTANT
OUR ORDER NUMBER MUST APPEAR ON ALL INVOICES, PACKAGES, ETC.
PLEASE NOTIFY US IMMEDIATELY IF YOU ARE UNABLE TO SHIP COMPLETE ORDER BY DATE SPECIFIED.

PLEASE SEND COPIES OF YOUR INVOICE WITH ORIGINAL BILL OF LADING.

PURCHASING AGENT

REDIFORM 1S 143 ORIGINAL

POLY PAK (50 SETS) 1P 143

Figure 11-5 Data Entities in Information Groups.

STEPS IN THE STUDY

Here, in the form of a check-off list, are the necessary steps to be undertaken in the study:

1. Decide on the scope of the study.
2. Make a proposal.
3. Get top executive commitment to back the study.
4. Top executive assigns executive review committee.
5. Decide whether to include users on initial research projects.
6. Sign on study team.
7. Sign on user oversight group (designate users to be phased onto study team after current system study).
8. Make sure software support is ready. (This is data dictionary support, discussion of which is deferred to Chapter 12.)
9. Do current system study charting (high-level).
10. Prepare conclusions—especially matrix reporting of relationships.
11. Present conclusions to executive committee and user oversight group.
12. Work with both groups to arrive at new goals, new requirements.
13. Work up new requirements into a requirements analysis.
14. Present requirements analysis to both groups and revise to gain consensus.
15. Revise current system charts to become new system specification done at a high level.
16. Identify affinity groups of data entities.
17. Select candidate data-entity affinity groups for automation as subject databases.
18. Decompose information system development tasks into phased application projects, setting scheduling priorities.
19. Write up and otherwise prepare presentation material concerning the study proposal. Include cost-benefit.
20. Gain consensus with the user oversight group.
21. Present the final study to the executive committee.
22. Gain support for the study if possible. Be sure to present alternatives concerning development strategies.

PRESENTATION TECHNIQUES: MATRIX ANALYSIS

When we present our study conclusions (for the current system study) we will find that the matrix presentation of relationships will shed the most light (as well as the most heat) on the subjects of the study. We have already discussed this in general and should now look into these matrix paired relationships in some detail. We have, as a minimum, the following sets of paired relationships to deal with:

- FUNCTION/PROCESS
- PROCESS/DATA ENTITY
- PROCESS/PROBLEM

Beyond this minimum we will want to see *all* the relationships between all the categories of subjects: function, process, activity, information group, and data entity. Also we want to relate our findings regarding problems to these categories. This is why we need the data dictionary software.

When we present our final conclusions regarding the new system specification, we will want to see as a minimum the following sets of paired relationships analyzed:

- SUBJECT DATABASES/FUNCTIONS
- SUBJECT DATABASES/PROCESSES
- SUBJECT DATABASES/DATA ENTITIES
- INFORMATION SYSTEMS/SUBJECT DATABASES
- INFORMATION SYSTEMS/PROBLEM SOLUTIONS

Beyond this minimum we will want all the relationships between all the categories of subjects we were interested in during the current system study plus the new subject categories, such as subject database and information system. We will want to see this information in several different ways besides pair matrix reports. Hooked to this data store of subjects and relationships we will need a database such as is provided with some data dictionaries plus a good access language for queries and reports.

Matrix analysis reports are used to show how the row subject and the column subject affect each other. If we are comparing data to process, we are interested in what data is needed to support what process and what process is responsible for creating the data. If we are relating problems to processes, we are interested in which processes have such problems or present opportunities. If we are pairing functions and processes, we are interested in observing duplication of process in different functions.

This sounds as if much of what we are interested in is finding pathology in the organization. This should not be true, although it is one of our considerations. More important is the support this analysis will give to the long-range organization planning to strengthen position and to increase revenues, avoid costs, and improve service. One important subject that should come up at the beginning of the study and also during the requirements analysis is what plans the executives have on their drawing boards.

EXAMPLES OF MATRIX REPORTS

Figure 11-6 shows the PROCESS/DATA-ENTITY matrix report for a large university. It is at a high level, showing 31 process rows as they relate to 21 data-entity columns. At this level of penetration into the organization the data entities are really subject databases. STUDENTS or CLASS SCHEDULES are legitimate data entities but certainly seem reasonable as databases also.

A university system is an unusually complex organization, and it would certainly be legitimate to derive twice as many data entities and processes. Also the subprocess or activities level would divide into hundreds of subjects. The level dealt with in Figure 11-6 is appropriate for showing clearly what analysis can be derived from this important matrix output of the study.

We have pulled our processes from the flow diagrams (such as the data flow diagrams shown in Chapter 4. The data entities were the data stores on those same charts. Now we arrange this data into the matrix. There are various ways of using a matrix of this kind. The author has decided to mark with an M (for Make) the processes that make the data and are therefore responsible for it. Of course many other processes retrieve the data, but retrieval is a very diffused situation and doesn't lead anywhere in helping us develop the information systems and associated databases. This development is the goal of this particular analysis.

It seems appropriate to use this chart for another purpose—to record opportunities and/or problems. We will see, as this analysis develops, how this charting is used. We will circle these high-opportunity intersections.

In this matrix we are assuming the following two opportunities:

1. The university wishes to make computer terminals, including personal computers, on-line to a common database and available to a large segment of the university population. This is not now possible without some development work, which will involve upgrades of both the information system and the physical system that undergirds the information, such as the local and remote area networks that will be involved in this multicampus operation. To support this opportunity a group of data entities associated with resource management have been circled on the matrix.

2. Another group of M's have been circled, all related to one process. The university has begun to image an ambitious plan for regional development of high-technology industry in the geographical area surrounding the main campus. This would be somewhat similar to the development of Silicon Valley with the close support of Stanford University. The university has developed the necessary knowledge and skills base to support such an important goal. The data needed is associated with the process COMMUNITY RELATIONS. Such entities of data as PROJECTS, CLIENTS, FACULTY, and ALUMNI are involved.

The university has decided also to try to make some order out of its many aging application systems, each with owned files. The motivating impulse to conduct the study was the intuition that the time had come to integrate to a common, shared, and private system of databases and programs.

Figure 11-7 shows the same information as Figure 11-6 plus something else. Boundary shapes have been drawn around groups of data that form natural information systems. This is really quite easy to do, if the processes and the data are initially arranged to separate the obvious clusters. With a little experience (not much) at matrix manipulation, one can do this intuitively in a short time. After putting the M's in these elliptical shapes, numbering and labeling them, we arrive at the following list of information systems as shown in Figure 11-7:

1. RESOURCES
2. STUDENTS
3. CURRICULUM
4. PROJECTS
5. STAFF
6. COMMUNITY
7. ALUMNI
8. FINANCIAL PLANNING
9. ACCOUNTING
10. SCHEDULES

It is now possible to relate our information systems (which have good relatedness to process) to the data entities. We see this accomplished in Figure 11-8. Since our data entities are high-level enough to be suitable for subject databases, we now have a set of subject databases related to major information systems, and we have taken a very long step toward the goal of developing an architecture regarding strategic planning for information systems. It is very normal to have duplication of databases across an integrated information system. The system that gets done first develops new databases, which are then used by information systems developed later in the phased development plan. The last information systems to be developed find most of the database development already done.

In our university strategic plan we can be sure that the COMMUNITY and the RESOURCES information system along with the necessary databases as shown in Figure 11-8 will be developed first. The STAFF (faculty, professional nonfaculty, support, etc.) and PROJECTS information systems will find much of their data on-line, and development will then proceed much more quickly than was possible for the first systems done.

These are the kinds of results one should expect from such a study. The reader should be creative about designing matrix reports and not just copy routine examples. But, however inventively this is done, the relationship between process and data entity and the relationship between organizational function and process are really obligatory.

Process \ Data entity	Facilities	Housing	Vendors	Materials	Development	Student prospects	Students	Curriculum	Courses
Do accounting									
Do financial planning									
Develop resources	Ⓜ	M	Ⓜ	Ⓜ	Ⓜ				
Provide materials				M					
Plan facility	Ⓜ	M			Ⓜ			M	
Maintain facility	M	M	M	M	M				
Maintain resources	M	M	M	M	M				
Provide housing	M	M	M	M	M		M		
Provide food	M		M	M					
Provide other	M		M	M					
Materials management	M		M	M					
Purchasing	M	M	M	M					
Recruit students						M	M	M	
Admit students		M				M	M		
Register students							M	M	M
Collect student fees							M		
Measure student performance							M		M
Keep student records		M					M		
Assign students to classes							M	M	M
Make programs							M	M	M
Make curriculums							M	M	M
Develop projects									
Do projects									
Maintain alumni contact									
Maintain endowment									
Community relations									
Schedule classes									
Schedule events	M		M	M					
Plan facility and staff								M	M
Recruit faculty and staff									
Support faculty-staff needs	M	M							

Legend: M = Make, Ⓜ = Opportunity

Figure 11-6 Process/Data Entity Matrix.

Class schedules	Projects	Clients	Library system	Publications	Community relations	Faculty	Staff	Academic planning	Financial planning	Alumni	Accounting systems
											M
								M	M		
								M	(M)	M	
						M				M	
						M	M				
							M				
							M				
							M				
								M		M	
					M			M	M		
M			M								
											M
						M					
M											
	M	M	M	M						M	
	M	M	M	M							M
										M	
									M	M	
	(M)	(M)			(M)	(M)	(M)		(M)	(M)	
M											
					M				M		
						M	M	M	M		
						M	M				
						M	M				

Figure 11-6 *(continued)*

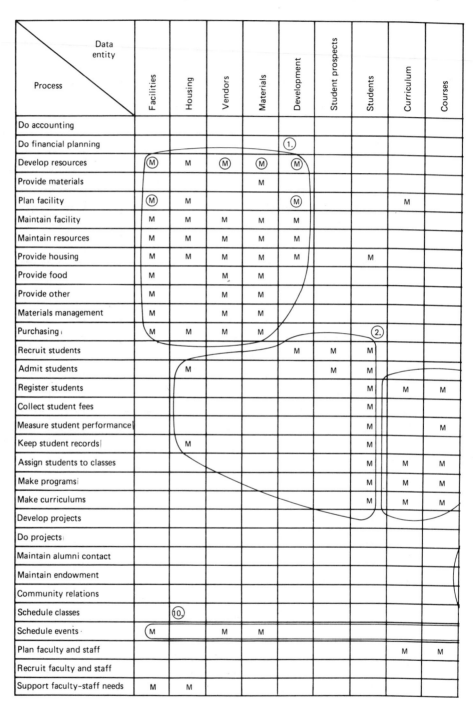

Figure 11-7 Process/Data Entity Matrix—Developing Information Systems.

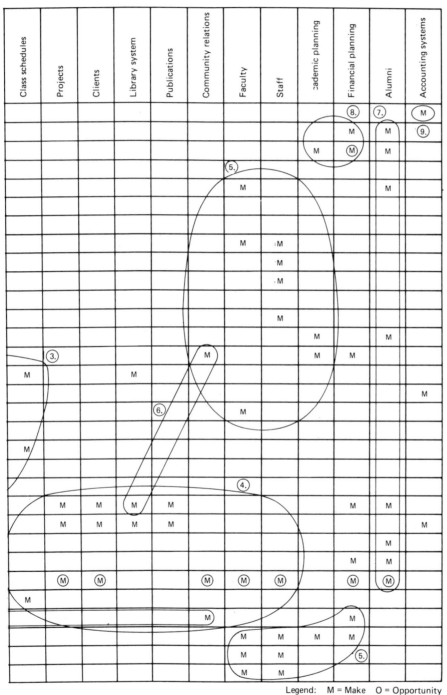

Legend: M = Make   O = Opportunity

Figure 11-7   (*continued*)

Information systems

1. Resources
2. Students
3. Curriculum
4. Projects
5. Staff
6. Community
7. Alumni
8. Financial planning
9. Accounting
10. Schedule events

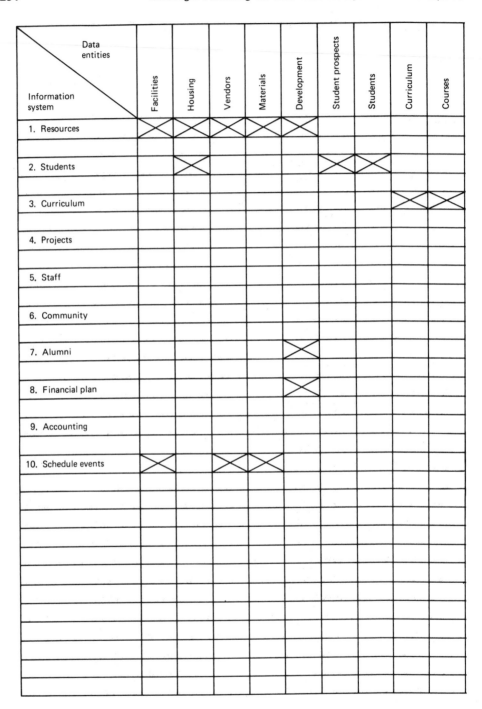

Figure 11-8   Relating Information Systems to Data.

Class schedules	Projects	Clients	Library	Publications	Community relations	Faculty	Staff	Academic planning	Financial plan	Alumni	Accounting systems
X											
	X	X			X	X	X		X	X	
			X	X		X	X	X	X		
		X	X	X	X	X	X				
									X	X	
									X		
											X

Figure 11-8   (*continued*)

STRENGTHENING POSITION—THE SP OF SPIRACIS

We cannot leave a chapter on strategic planning for information systems without a discussion of adversative strategies for using information systems to advance an organization's competitive goals. The competitions of the future, whether peaceful or violent in intent, will involve the strategic and tactical use of information. We will close the chapter on this interesting subject, giving two examples of adversative strategies with which all of us are familiar. But first we should look again at the university example and the university's goal to be the knowledge hub of a regional high technology industry. This is a case of adversative strategy if ever there was one. Regions of the United States are competing fiercely for a finite number of high-technology enterprises, not only with other U.S. regions but in the international arena. What would the economy of the New England region be without the great universities in and around Boston?

Our first example of a successful adversative strategy through exploitation of an information system is the case of a bank that is the first in its community to offer on-site tellerless banking to business corporations and their employees. The bank provides a local-area network that handles automatic payroll transfer to employee accounts and most other routine banking services, which can be done without customers' leaving the shop or office. No more two-hour lunches on payday!— which includes standing in long lines at the bank. This is a saving to employers and a service to the employees. What the bank gets in return is most of the banking business in the community—to the point where the other banks that couldn't compete are dealt a serious financial blow.

Our second example is the computerized checkout available at some supermarket chains but not others. The computerized checkout does optical scanning of the groceries, resulting in much faster checkout lines, an itemized listing of purchases, accurate billings, and real-time inventory management (important in the supermarket business). It really is hard for the supermarket without this information system to compete with such an operation.

This is just to scratch the surface on the area of adversative strategic planning. This approach to information systems is becoming more important and puts a new light on the importance of the professional developers of such systems and their worth to the organization. The information system is becoming a major profit center. Good strategic planning is the necessary base from which we can develop the information systems which will improve the organization's competitive performance.

<table>
<tr>
<td>

12

</td>
<td>

DATA ADMINISTRATION, THE DATA DICTIONARY, AND DOCUMENTATION

</td>
</tr>
</table>

The data dictionary used to be thought of as the database about what is in the common database—or, put succinctly, *metadata*. Metadata is data used to describe data. The data administrator was the keeper of the metadata, the arbiter concerning data standards, and the user's friend in the inner sanctums of information-system database technology.

This metadata was, of course, important documentation concerning some of the organization's machined data stores, to be used along with all the other documentation needed to manage the organization.

The data dictionary told what moveable quanta of data were in the database (e.g., segments) and what data elements made up each moveable quantum. The attributes of the data elements were given—such attributes as type and size of element, name in COBOL or BAL (basic assembler language), synonyms, and a full language description of the element.

This data dictionary could be manually maintained, although it soon became clear that it was complicated enough to be maintained on a computer system.

The modern data dictionary is this and much more, having become the best-integrated, most coherent documentation concerning the organization's operations, particularly concerning the information resource. Correspondingly, data administration has become, in some organizations, one of the key roles in the management of the enterprise.

In order to understand this new, more expansive view of the data dictionary, we will have to overcome some semantic difficulties concerning the differences between *data*, *information*, and *knowledge*. These differences were covered in Chapter 6. In subsequent chapters the author has had to obliterate the clean lines between these subjects in order to relate the ideas in the book to the current naming conventions in the "industry." For this chapter we will need the naming precision of Chapter 6.

DATA, INFORMATION, AND KNOWLEDGE— A REVIEW OF CHAPTER 6

In Chapter 6 we spoke of a datum, or a piece of data, as a fact with a recognizable code system. If someone hands us a piece of paper with the number 182 on it, we recognize this as some kind of a data fact. Since there is no sign, we assume it to be a positive number, no doubt in the decimal system—some number larger than 181 and smaller than 183. It is not really information. We are as uninformed as we were before we received this data fact on the piece of paper. We have apprehended nothing and are surprised by nothing.

But when John gets on the scale in the morning because he has been having trouble getting into his clothes and the SCALE REGISTERS 182 POUNDS, a real piece of information has been apprehended with some surprise, because a relationship has been established. To put this in data-dictionary terms:

- category–PEOPLE REGISTERS-WEIGHT-ON SCALE
  - subject–JOHN REGISTERS-WEIGHT-ON FLOOR-READOUT-SCALE
    (registers-weight-on is the forward relationship)
- category–SCALE SHOWS-WEIGHT-OF PEOPLE
  - subject–FLOOR-READOUT-SCALE SHOWS-WEIGHT-OF JOHN
    (shows-weight-of is the reverse relationship)

In this instance John registers a weight of 182 pounds on the scale.

Therefore, in order for data to be considered information it must have:

- forward and reverse relationship to other data
- been apprehended by an interested party (if the cat steps on the scale without interested people home to observe, there is no information)
- surprise (if the scale is broken and always registers 182 no matter who gets on, there is no information

So-called data dictionaries do not deal with data about data; they deal with information about data. Network communication systems deal with just data. Data dictionaries are interested in category relationships, subject relationships, attributes of subjects (John is 5′1″ tall and 13 years old and lives in Sweden), and text description about subjects. The text is all about relationships but expressed in a less precise and formal system.

Relationship can be either explicit or implicit. In the 182-pound example REGISTERS-WEIGHT-ON is an explicit relationship. In information systems, particularly advanced information systems, it is as important to be able to manipulate relationships as it is to manipulate subjects. We may want to know all the relationships that deal with Swedish overweight children and all the subjects these relationships point to.

Implicit relationships are not as easy to manipulate by relationship-list processing but are very efficient for most other purposes. Implicit relationships in information systems are parent-child and key-attribute relationships. Vendor # has an implicit relationship with Vendor Name and Vendor Address. The reason we want to put database tables in the third normal form is to make these implicit relationships unambiguous. It is not necessary for many purposes to state these relationships explicitly. In the database we do not usually store IDENTIFIES or IS-IDENTIFIED-BY between the key Vendor # and the attribute Vendor-Name. Neither do we store IS-THE-CHILD-OF between two segments in a tree database. In a full-service data dictionary we do store relationships between categories and subjects as explicit.

We can make all sorts of inferences, once we have information. Inferences are formed by dealing with multiple relationships, relating the information at hand with the knowledge base of information, as we have defined information. A number of inferences could be made from the information that John weighs 182 pounds. That much weight is a physical and emotional health problem. John's diet is at fault. He eats too much sugar. Maybe those long Swedish winter nights are causing John to be sedentary. Does he gain more in winter than in summer?

Knowledge is information that is:

- understood
- retained
- connected
- evaluated
- formed into rules
- made retrievable as solutions

For systems analysts (and data administrators) not involved with artificial intelligence systems the data dictionary is an entry point into this type of information processing. The full-scope extended data dictionary is the first step towards an intelligent knowledge base that is concerned with every aspect of the organization. We discussed the structure of this knowledge base in Chapter 11 when we dealt with strategic planning for information systems and we related organization functions (departments) to processes to *information* groups to *data* entities. That is one reason why the chapter on strategic planning was placed between the chapter on modern database design and the chapter on data dictionaries. This chapter will build on the previous chapter.

DIFFERENT ROLE PLAYERS IN THE MANAGEMENT OF INFORMATION RELATED TO THE DATA ADMINISTRATOR

Figure 12-1 shows the relationship between information-resource objectives and the information-system managers or stewards. This figure points up the amount of responsibility vested in the data administration group. Data administration is systems analysis in so far as the information resource is being specified. The custodial role of the data administrator regarding the information resource and data-dictionary maintenance is not particularly systems analysis. Whether an organization

Responsibility / Objective	DA	DBA	Communications Physical-system support	Application development	System maintenance	Very high level analysis	Users
Strategic planning for information	R					R	R
Tactical planning for information	C	C	C			R	C
Information standards	R	C	C	C	C	C	C
Development or selection of information systems	C	C	C	R		R	C
Maintenance of information systems	C	R	C	C	R	C	C
Database design	R	R		C		R	
Physical system support	C	C	R	C		C	C
Database resource management	R	R		C	C	C	
Organization information documentation	R	C		R	R	R	C
Information methodology selection and training	R	R	R	R	R	R	

R = Responsible, C = Consults

Figure 12-1 Information Resource Management.

formally designates a data administration group or not, the function of data administration is a vital system process and somebody is doing data administration.

The data administrator shares responsibility with the systems analysts (shown in Figure 12-1 as very high level analysts as opposed to applications level analysts) and the users in the development of the strategic plan. Before the strategic plan can be performed, the data administrator should work out the documentation support system, including software, for the strategic plan information storage and reporting. It is recommended that the data dictionary be used as the documentation-software support for the Plan. The data administrator, as the custodian of the dictionary and the expert in documentation software and information analysis, works alongside the VHL analysts to architect the long-range strategic and tactical plans which will determine information-system policy for years to come. It is important that these two groups—the systems analysts and the data administrators—learn to cooperate comfortably with each other. They should both report to the same management.

In addition to responsibilities in support of strategic planning the data administrator (DA) has the following important responsibilities, as shown in Figure 12-1:

1. The DA has primary responsibility for information standards, such as naming conventions. Beyond this, data administration should actually do the naming of data items that are going onto all organization information systems. This means language-naming in COBOL, BAL, PL1, FORTRAN, Pascal, BASIC, LISP, PROLOG, C, et al. It also includes names on fourth-generation data dictionaries and common data names which are input to software packages. It is important that this be centrally accomplished because of the redundancy of data names concerning the same data items that can occur as applications are being developed. Ideally, when an application needs a data group it should be delivered electronically as an export from the common data dictionary database to the recipient system in a format acceptable by the recipient.

2. The DA has shared responsibility for database design. The database administrator (DBA) has primary responsibility for the construction of the internal schema or physical database. (See Figure 10-4 for a review of the three-level database schema.) Otherwise the data administrator (DA) has shared responsibility with the analysts and/or the DBA for the selection of database software and hardware, for data modelling as described in Chapter 10, for the development of the conceptual schema, for support of applications development needs for data structures exported from the conceptual schema to the recipient programs, and for relating the reality of the developing database system to the real needs of the user community. This last responsibility is the quintessential DA concern, to which all other concerns must be subordinate. This is what it is all about. In the matter of relating the database information to the users' needs the DA has primary responsibility. As a result, in a highly computerized organization, there is rarely a major change in

operations which can be contemplated without having the DA present to participate in the planning.

3. The DA shares responsibility with the DBA for database resource management insofar as the physical database system relates to the users' need for storage space and good response time. The DBA is responsible for the physical tuning of the system to reach these goals and for making recommendations for adjustment of hardware capacities. Otherwise, the DA has primary responsibility for the performance of such database functions as data security, auditing, recovery planning, data independence from applications, redundancy problems, database transportability, conflict resolution (how come *our* response time is worse than theirs?), clarity of documentation, training in information use, and data-information quality.

In addition, the DA is responsible for reviewing these factors as they concern all data stores, both manual and automated. Quite a large order of tasks! Many of these tasks used to be under the hegemony of the systems analysts or the generalist managers. It is still unclear how these task divisions are going to finally end up in terms of information-system department organization. In the view of the author, data analysis and system analysis cannot be divided in terms of process, only in terms of organization. It is, after all, the processes which use data stores as a vital resource and transform moving data.

4. The DA is responsible for the data dictionary. Insofar as the data dictionary is a documentation tool, the DA has primary responsibility for documentation in the organization. (We shall see that the dictionary often has other functions besides documentation.)

We will be presenting the case in this chapter for the use of the data dictionary as the primary documentation methodology in the organization. The dictionary either stores the information directly or acts as a directory, pointing to the information repository. The rationale for this point of view is the need to automate the specifications for the entire organization, including the information systems. Once automated, this information can be maintained in one place, retrieved in many permutations and combinations, attached to inference-handling programs, and otherwise manipulated.

The dream of the systems analysts née data analysts is to thus specify the whole system and to avoid information pollution and confusion in the process. We can then more clearly address a major task of our age: how to take our greatly expanded ability to store and retrieve information and convert it into knowledge. A lot of information does not even a little knowledge make ipso facto. Some very smart people have made a good case for the proposition that, right now, knowledge has become inversely proportional to information volume. (Dealing with this is the subject for another book.)

5. The DA has a shared responsibility for the selection of information systems methodologies and training. Everyone gets into this act. The DA is the expert on

data modelling, database logical functionality (What does the user expect from the database? Does database X meet this expectation?), and the data dictionary. In these areas the DA has primary responsibility.

6. The DA serves in an important consultative role in numerous other areas, many of which are shown in Figure 12-1. Tactical planning, whereby the general conclusions of the strategic plan are further examined, staffed, and scheduled, requires the participation of the DA. Maintenance of the whole information system concerns the DA, particularly in the area of conformance to standards. The development of information systems scheduled by the tactical plan very much involves the DA, as the database system needs constant rethinking and tuning in the light of real implementation. And, finally even the physical-system support requires DA participation. Such matters as distributed processing versus centralized processing greatly affect the organization approach to data and data storage, transmission, and presentation to the users. The DA is interested in the physical-system support for relevant information as contrasted to the physical-system support team, who are essentially interested in the data problems unconnected to information relevance.

Figure 12-2 shows a simplified organization chart for a data processing organizational subsystem, indicating where the various role players we have been discussing, especially the data administrator, may be hiding out in terms of function. It is interesting to note how little the usual organization reflects what players are actually involved in the process. The DA, for instance, shows up redundantly in planning, documentation support, development (design and implementation), and maintenance. This type of distribution of responsibility, as we have already touched on, has the potential for causing trouble.

The worst thing that can happen to a group researching and developing information systems is the formation of little hegemonies or principalities that are competitive. It may help a sales department to have competition between product sales managers. It may help an organization to have the big spenders in competition with the tightwads (in the little town in rural Connecticut where the author lives, this would be the school board versus the finance committee). It is of no help at all to have the various functional groups in data processing, whom we have been discussing, in competition—friendly or otherwise. Add to this the potential for destructive behavior that results when these analysts interface with contending users, and we have a prescription for failure. Autocratic rule would be preferable to this type of first-century Roman republicanism where everyone becomes a conniver. Don't let it happen! The way to develop good information systems is to cooperate and leave egos outside the meeting rooms.

Where internecine striving for dominance has trashed the information-development process, so that nothing gets done right, the users eventually abandon the whole-system approach to common information development and revert to

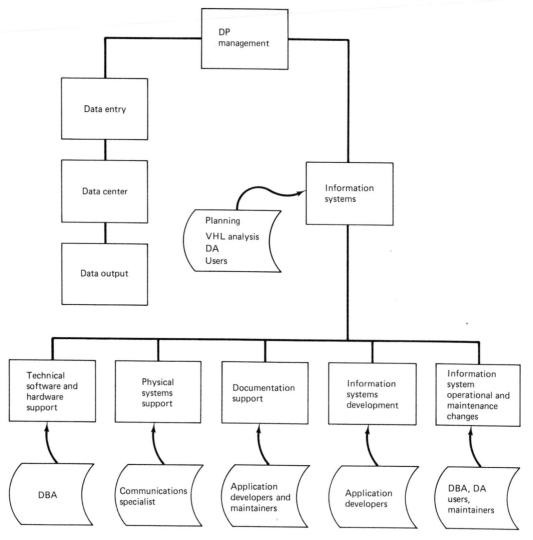

Figure 12-2   Information Management Participants in the Data Processing Organization.

either the old disassociated applications-development approach or to going it alone on the personal computer. Either way they soon hit a hard wall. The applications approach, as we discussed in Chapter 10, demands a very sophisticated documentation and upload support system, and the personal computer has only limited function in an organization without being able to download common information from a common database. The use of the personal computer is tied to the common-database approach. The PC is fine at combining the common data with private data and further manipulating the resultant data to produce private information.

DATA-DICTIONARY TYPES AND FUNCTIONS

Having placed the custodian of the data dictionary, the DA, in proper perspective in the organization, let us discuss the data dictionary itself.

Figure 12-3 illustrates the different types of data dictionaries and the functions each addresses.

There are two kinds of data dictionaries: integrated and stand-alone. The *integrated* dictionary is related to one database management system. To the extent the organization data is under this DBMS it is global or organizationwide. However, very few enterprises have all their data eggs in one basket, so the dictionary

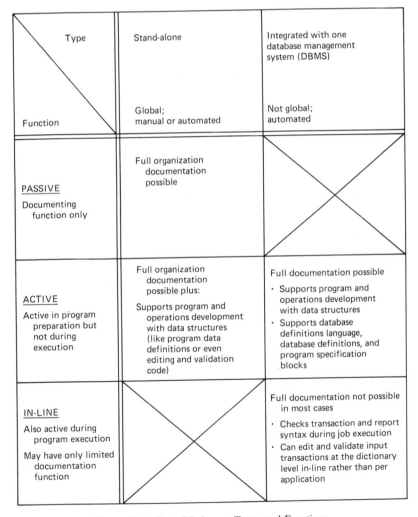

Figure 12-3 Data-Dictionary Types and Functions.

documentation (metadata) can be considered as local and fragmented. The *stand-alone* dictionary is not tied to any one DBMS, although it may have special advantages for one DBMS, such as the IBM DB-DC Data Dictionary,[1] which has special features related to the IBM IMS DBMS but is still a stand-alone variety of dictionary.

Both these types of dictionaries can be identified by function as either passive, active, or in-line. Viewed either way, by type or function, the differences are striking. Passive, active, and in-line dictionaries differ functionally as follows:

Passive Data Dictionaries

The functionally passive dictionary performs documentation only. This variety of dictionary could be maintained as a manual rather than an automated database. For more than limited documentation use, the automated passive dictionary has clear advantages. From the organizational view the documentation function is the most important dictionary service with the most potential benefits, so the passive dictionary should not be thought of negatively. It has more limited functionality but may perform its critical function of global documentation best of all. (See Appendix A for an example.)

Active Data Dictionaries

Besides supporting documentation to one degree or another, the active data dictionary supports program and operations development by exporting database definitions and program data storage definitions for languages such as COBOL and Job Control Language (JCL) for execution-time performance. The IBM DB/DC Data Dictionary already mentioned is such a stand-alone, active data dictionary. A dictionary such as this is not an in-line data dictionary as delivered, which is not to say that it could not be put in-line by a determined effort of major proportions. The author is not aware of anyone running an active dictionary in-line.

In-line Data Dictionaries

An in-line data dictionary is active during program execution, performing such feats as transaction validation and editing. Such a dictionary would always have some documentation value, but documentation across the organization about the organization functions and activities and all the organization information data stores is not likely. In-line dictionaries are associated with DBMS products such as Cullinet Software Corporation's IDMS-R or Cincom System's TOTAL, to name just two. The author would not be surprised to see in-line dictionaries or information di-

---

[1] (a) OS/VS DB/DC Data Dictionary—General Information Manual, Release #5, GH20-9104-5, IBM Corporation; (b) OS/VS DB/DC Data Dictionary Terminal User's Guide and Command Reference, Release #4, SH20-9189-0, IBM Corporation.

rectories expanded to full documentation systems but is not aware that such is yet the case. It is dangerous to describe product specifications, since in our business products are constantly being enhanced. Suffice it to say that the reader who is thinking of purchasing any software should consult such periodically updated information sources as DataPro[2] or Data Sources[3] or the IBM product catalog before contacting vendors. We will discuss getting software in an upcoming chapter.

To review Figure 12-3, the stand-alone dictionary is either passive or active and is used to provide documentation of varying scope. The stand-alone active dictionary also delivers preexecution-time development data structures.

The integrated active dictionary can do both of these functions and also support the operation of a specific database at preexecution time by delivering DBMS data structures. The IBM data dictionary falls into this category and can export data structures such as program specification blocks and database descriptions in support of the IMS DL/1 language.

The in-line data dictionary or data directory is always integrated with a DBMS. At present, full documentation is not possible (as we have defined "full"). However, organizations that decide to go into partnership with a specific fourth-generation DBMS with an integrated dictionary and do all their information systems with this vendor's product do get a considerable amount of built-in documentation for their money. An enterprise rapidly growing from a modest organization into a large one and in serious need of information-system support without years of delay should (and perhaps must) consider this route.

The integrated in-line dictionary can edit and validate input transactions at the dictionary level rather that at the application level, which is a palpable advantage in the situation where more than one application program creates or changes any one data group (this condition is generally considered poor practice).

However, different products offer different degrees of sophistication in transaction refinement, validation, and editing. We saw in Chapter 7 just how complicated transaction refinement can get between presentation to the information system as a raw transaction and presentation to the database for update.

THE MAKEUP OF DATA DICTIONARIES: DATA-DICTIONARY INTERNALS

The minimum data dictionary is shown in Figure 12-4. We have a database system consisting of databases or files. These files consist of data groups or segments or records. These data groups consist of data items or fields. There is an implicit relationship here, which needs no additional comment. A certain amount of at-

[2] *Datapro—The EDP buyers Guide.* Vol. 3: *Software.* Datapro Research Corporation, Delron, N.J. 08705.

[3] *Data Sources—Software Volume.* Ziff-Davis Publishing Co., One Park Avenue, New York, N.Y. 10016. Published quarterly.

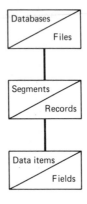

Figure 12-4 Minimum Data Dictionary.

tribute information is always present. In the case of data items we need to know if it is a primary or secondary key or an attribute field, if it has aliases, what are the field type and field size, what is the name in various languages, and what is the user description of the item. We need to know whether the data item or data group is in test, system test, or production status. We need to know the number of occurrences of this data item on the dictionary.

Addressing these last points, a data item (for instance) on the IBM data dictionary may look strange to the uninitiated.[4] It will look like this:

T,C,BALANCE-ON-HAND,0

We recognize balance-on-hand as an inventory quantity. The T is the status code, which we will say is T because the data-item balance-on-hand is on the test-data database. The C is the subject code, which in this case is the primary programming language: COBOL. The 0 is the occurrence number where duplication exists in the common information system. So, in terms of this dictionary, the full description of the data-item consists of the four elements mentioned above. This convention holds for all subjects defined on the IBM data dictionary.

Note. We have referred, and will refer again, to the IBM data dictionary in the discussion in this chapter. The reasons are simple. The author is familiar with this product and wants to discuss from experience rather than research alone. The assumption is that all full-service dictionaries are similar in function. The other dictionary with which the author is familiar is a DBMS integrated dictionary, which is not broad enough in scope to satisfy the discussion of documentation we are about to embark on. Please accept the author's apologies for not using as an example the reader's favorite dictionary product.

Before discussing the functions of the full-service extended data dictionary we need to review data-dictionary elements. Figure 12-5 shows these elements.

---

[4] OS/VS DB/DC Data Dictionary—General Information Manual, p. 15.

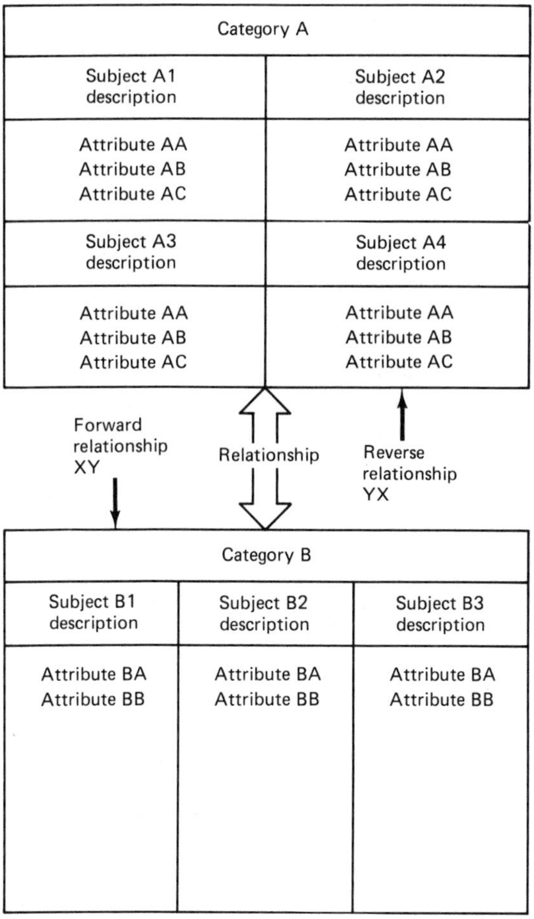

Figure 12-5 Data-Dictionary Elements.

We have already discussed categories, subjects, relationships, attributes, and descriptions on other occasions and now will do so again in the most appropriate possible context. These are the elements that make up the data dictionary. In Figure 12-5 Category A has a forward and reverse relationship to Category B. We have two-way relationships simply because we may want to examine these relationships in both directions. Data-Items is an example of a category. In a full-service dictionary some categories are predefined regarding attributes and relationships, but the dictionary has the capacity to handle user-defined categories. This, in IBM parlance,[5] is an extended use of the dictionary. This "extensibility" feature is the heart of the full-service dictionary, allowing documentation of the whole organization and allowing us to use the dictionary as the software support for strategic and tactical planning.

[5] OS/VS DB/DC Data Dictionary—General Information Manual, p. 10.

Category A in Figure 12-5 has four subjects. Each subject has the same attribute set as the others (attributes AA, AB, AC). For instance the category may be Projects. The four subjects are four different projects, described by name and description as unique. Perhaps the attributes are Project Leader, Project Due Date, and Percent Accomplished. All four projects would have identical attribute names.

Perhaps Category B is Information-Systems, with subjects and attributes defined in a similar fashion. The forward relationship might be Projects ACCOMPLISH Information-Systems. Reverse might be ACCOMPLISHED-BY.

The description associated with each subject can be as detailed as is desired.

Other minor elements are available on the IBM data dictionary. These are attributes for relationships, the same as for subjects, and user instructions regarding use of each subject screen in the dictionary. In the interest of grasping this concept of an information database about organization information the reader can ignore these details.

Figure 12-6 shows another example of the elements that make up a data-dictionary database. In this case we have the category Business Function (or department) related to the Processes of the organization such as Provide-Materials. Remember that the subject name looks like this: P,,Provide-Materials,0. Remember that the P is the status code, which in this case stands for Production. The two adjacent commas means the subject code is not used for this kind of category, and the zero is the occurrence (only this occurrence exists).

The IBM data dictionary, which is actually six linked databases, each with many segments, consists of standard categories and the infrastructure needed to "customize" installation categories. The standard categories have the attributes prebuilt and ready for the user to fill in. These standard categories are:

- DATA-BASE
- SEGMENT
- ELEMENT
- PROGRAM COMMUNICATION BLOCK
- IMS SYSTEM DEFINITION
- APPLICATION SYSTEM
- JOB
- PROGRAM
- MODULE
- TRANSACTION
- PSB

These categories are all related to servicing the data processing function and are not sufficiently broad in scope to support a dictionary for the entire organization. The strategic plan cannot be documented with just these categories. It is the ability

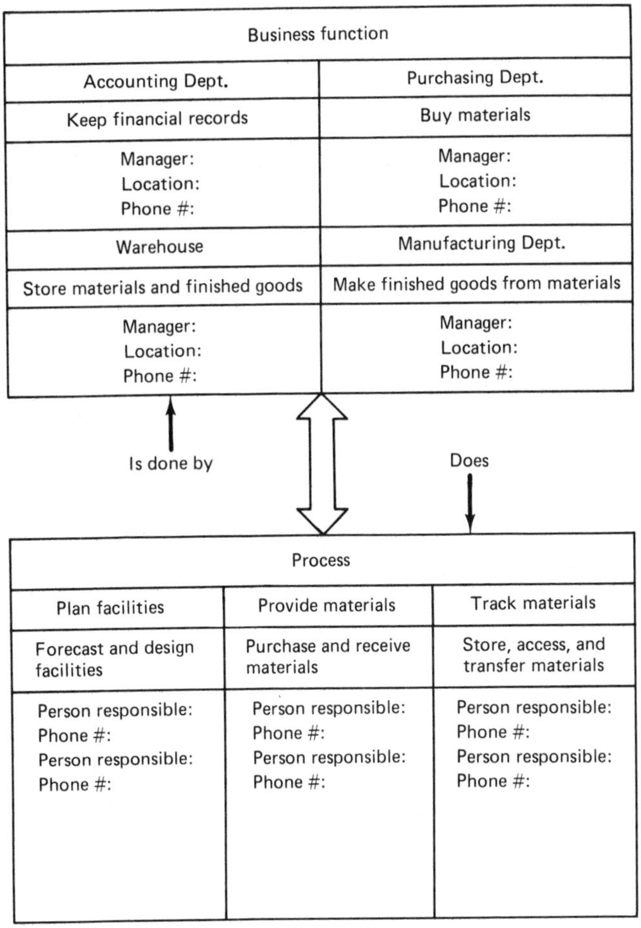

Figure 12-6 Example of Data-Dictionary Elements.

of this data dictionary to allow the creation of other user-defined categories that allows us to consider the dictionary a serious tool for systems analysis and documentation in support of the current and new system specifications.

Figure 12-7 is an example of the dictionary used as the central directory of organization documentation. The basic dictionary functions in support of data processing, such as exporting data structures for use in programs, are still present. In addition, the dictionary either is the repository for organization function, process, and information definitions or points to further documentation which is not now on the dictionary. The trend, which is still in an early stage even in organizations with an understanding of the power of the dictionary, is to actually keep the full documentation on the dictionary, including the charting.

We have defined the central documentation which describes the current and planned organization to be a charting method. For the dictionary to really be effective as more than a pointer to other documentation, we would need to au-

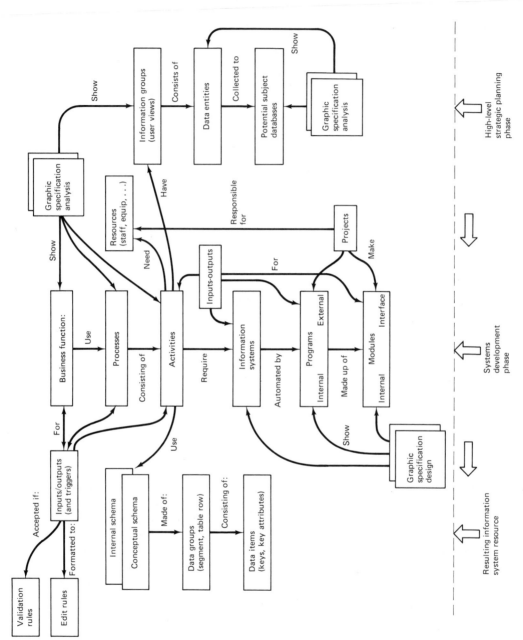

Figure 12-7 What Is Kept on the Extended Data Dictionary.

tomate the charting function so that the elements that make up the charts (such as the data flow diagrams discussed in Chapter 4) are kept in digital form on a database. This database would then be part of the dictionary family of databases.

We are now getting software products on the market which do just this. By the time this book is published, this software support should be in commonplace use in many organizations. This software will have word processing associated with it. The full-service dictionary of the future will be need to have not only digitized charting but also a word-processing ability which is not present in current dictionaries. Then we will have a full documentation tool automated for flexible retrieval.

USING THE EXTENDED DATA DICTIONARY TO DOCUMENT THE ORGANIZATION

In the extended data dictionary we set up our own categories while still retaining the categories provided with the dictionary. These categories are all related to each other, so that the subjects within each category can be related to subjects in other categories.

In Figure 12-7 we can see how this extended dictionary can support the effort that begins with the high-level strategic plan for information systems development covered in the last chapter, proceeds through the systems development phases, and documents the accomplished integrated and fully defined information systems resource. This is shown in Figure 12-7 going from right to left. Let us go through the details, keeping in mind that our purpose is to document the whole organization and the information systems that support it, or as a minimum to point as a directory to the actual documentation source, much as a library reference system gives information on where to find the source documentation and gives an abstract of what will be found.

We are familiar with the categories covered by the strategic plan, and we see these categories in Figure 12-7 as the boxes covering Business Function, Processes, Activities, Information Groups (such as documents), Data Entities, Potential Databases, and Information Systems. (The reader who is not familiar with these categories should review Chapter 11.) This documentation created in the strategic plan should not end there, but should be enriched with detail and updated with changes. The same can be said for every category on the dictionary. This is the basic information on the dictionary and is of interest to all who labor in the organization. Under the aegis of the data administrator the information is kept precise and is the product of consensus.

The users should not only learn how to extract category documentation but also become comfortable with using the category and subject relationship information, which in Figure 12-7, to simplify the illustration, is shown as the forward relationship only. The fact that Business Functions Use Processes or that Activities Require Information Systems can be as important to some users as the documentation within a category.

Another example of the use of relationships (in this case in support of the strategic plan) is given here by two illustrations reprinted with the kind permission of the *IBM Systems Journal.*[6] Figure 12-8 is a business systems planning (BSP) data model similar to the model shown in Figure 12-7. Nine categories are shown with their forward and reverse relationships. Figure 12-9 shows a summary report for the category Process, giving all the relationships this category has with the other categories in the model. These illustrations are used to indicate how the IBM data dictionary can support the business systems plan. We are suggesting here that this dictionary documentation can be used as the basis for the entire organization documentation methodology.

Returning to Figure 12-7, we see that graphic specifications are linked to all documentation topics. We want to establish a documentation methodology that uses graphics as the basic specification. These graphics must be integrated with clear text material. In some cases the graphics must be convertible to text for those many users who prefer a more traditional reporting method. The reason the systems analysts prefer graphics is that it is a rigorous notational form which best expresses the real life of an organization, which is nonlinear. Many processes go on at the same time in the real world, and this is hard to model in serial text. Also a chunk of processes, captured in a flow diagram, can be grasped better as a whole chunk, which is different than the serial sum of its parts.

The following are the graphics, or charting methodologies, which we wish either to keep on the data dictionary or to relate to the dictionary as a directory:

- process flow charts
- data flow diagrams
- decision tables and charts
- physical-systems charts, including data transmission systems
- transaction layout tables
- user-view diagrams
- database data-group tables
- strategic-plan architecture
- matrix charts
- design structure charts
- design flow charts
- user-terminal screen flow diagrams

Many of these charting techniques have applicability beyond the development of information systems and are useful across the entire organization.

[6] J. G. Sakamoto and F. W. Ball, "Supporting Business Systems Planning Studies with the DB/DC Data Dictionary," *IBM Systems Journal*, Vol. 21, No. 1 (1982), pp. 61 and 72.

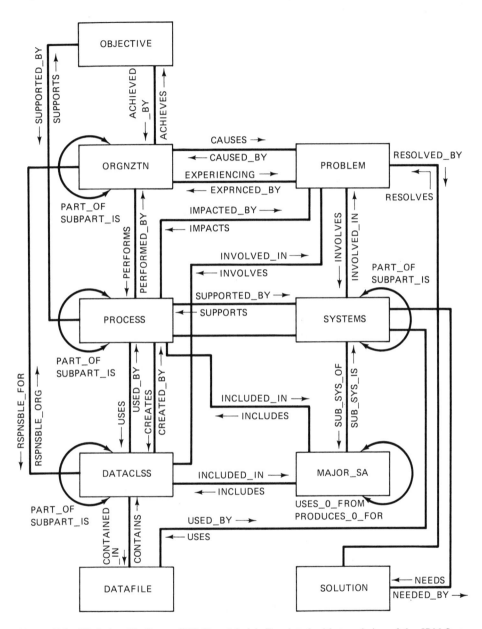

Figure 12-8   "Relationships" on a BSP Data Model. Reprinted with permission of the *IBM Systems Journal*, IBM Corporation, Armonk, New York.

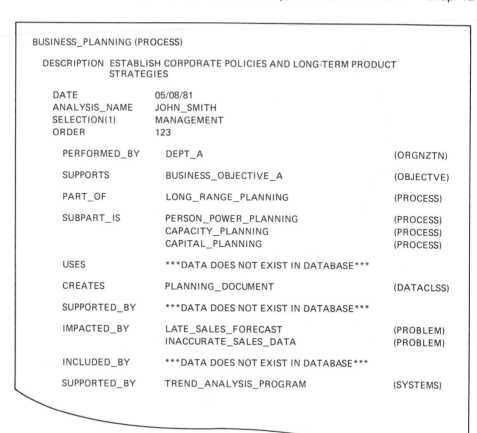

Figure 12-9   Reporting Relationships. Reprinted with permission of the *IBM Systems Journal*, IBM Corporation, Armonk, New York.

APPROPRIATE DOCUMENTATION FOR INFORMATION SYSTEMS

No book on systems analysis can be complete without establishing an approach concerning how to document the planning, development, and maintenance of information systems. The problem of documenting the information system has never been resolved satisfactorily and is the critical risk path in automation.

The type of documentation that suffices for most engineers who are developing devices will not do for documenting a complex human-based social organization. In terms of general systems theory, devices are small-number systems and natural systems are medium number systems (disorganized "heaps" are large-number systems). The type of documentation used to describe a device, such as the building plans and contracts for a house, will not describe how a university is developed and maintained.

It is, perhaps, just about impossible to fully document a complex natural system. It may even be vulgar to do this, since in the process we could reduce the natural system with all its surprises to a useless device. However, we must *appropriately* document our model of the natural system in order to make planned changes to it and in order to insure the survival of the system when it breaks down. The prehistoric tribe that keeps an oral record of where the back-up waterhole is located may find the information unreliable. This doesn't matter until the day the primary waterhole dries up.

Appropriate documentation means just enough documentation and no more. Nothing is as polluting and discouraging as too much documentation. Too much documentation cannot be kept accurate as the system changes. It is hard enough to persuade the enterprise managers to pay the costs of any documentation. It is hard enough to get the systems analysts and programmers to faithfully maintain documentation. We have to make this task as easy and simple as possible and keep the operating costs as low as possible. This is another reason for investing in automated software support for documentation, particularly the maintenance of charting.

System Maintenance and Use Documentation

What, then, is appropriate documentation? We will now show the documentation subjects required at each stage in the system development life cycle, starting with maintenance and use and working our way backward to the original systems analysis. Figure 12-10 shows what must be developed—or at least set up for use—by the previous analysis, design, and implementation effort to support the maintenance and use of the information system.

1. Program source code.   In some computer installations the source code is the only surviving documentation. This is an improvement over "the good old days" when sometimes all that was available was the object code (fortunately not in hexadecimally coded binary). Today, with the advent of on-line systems that cause severe disturbances and even danger when down, and with the constant accumulation of the program base, no one can afford the luxury of trying to bring up a crashed system using just a source listing and a dump. This can work only if the person who wrote the system is there to find the inevitable bugs that develop. However, the source code is still the primary document.

2. Run book type documentation.   This includes the program input (transaction) layouts, the output (report) layouts and specifications, and the I/O triggers that cause things to happen. Also included are

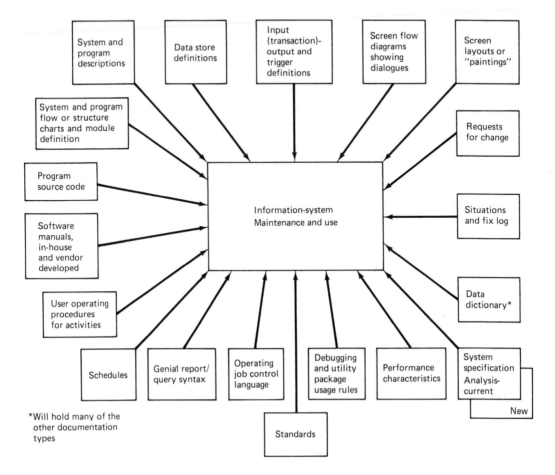

Figure 12-10   Documentation Needed for Information-System Maintenance and Use.

- requests for changes
- situations and problems and resolutions of same
- change and fix log
- data store definition (the external schema view seen by the application)
- maintenance and change schedule
- program and system flow charts or structure charts and module definitions
- system and program descriptions
- screen flow diagrams and paintings (what each screen looks like to the user)
- performance characteristics (compute limited? I/O limited?)

3. Other support documentation.   These documents include software package manuals and in-house summaries of such manuals, the data dictionary and documentation on how to use the data dictionary, and such other software support items as debugging packages and utilities packages.

4. The new system specification.   This big picture is necessary on occasion—such as when doing a major operational change under the general heading of maintenance.

Note the inclusion of software package documentation in this list and in Figure 12-10.

Again, we must stress that most systems in the future will incorporate application packages and the customization of such a package in the information-system development. (Not since the earliest days of automated information systems has any organization done sorting and physical access of data with in-house routines without having a very special reason for doing so.) A major documentation problem concerns getting a quick grasp of how to use and fix an application software package.

5. User documentation.   This consists of the very important user operations manual. Users need to have concise information concerning how to enter transactions on-line, how to prepare transactions in the batch mode, how to audit the system for accuracy, and how to do queries on-line. The user needs the query syntax (to be used with data definitions from the data dictionary) to allow retrievals to take place under user initiative.

System Implementation Documentation

This is the documentation the implementation team must receive from the analyst and designer; it is shown in Figure 12-11.

Like the maintenance team, the implementation team needs the screen flow diagrams and paintings, the system and program descriptions, the system and program flow charts or structure charts, a schedule, the data dictionary, debugging and utility packages, program I/O, software package specifications, system specifications, data store definitions, and debugging and utility documentation. Instead of performance characteristics we have performance expectations. Instead of a change and fix log we have a design change log. The situation and problems log is associated with the design change log.

Systems Design Documentation

This consists of the documentation received from the analysis phase plus the system maintenance documentation (we are now more often then not developing a new system from an existing computer solution), the data dictionary, and the software package specifications. This documentation is shown in Figure 12-12.

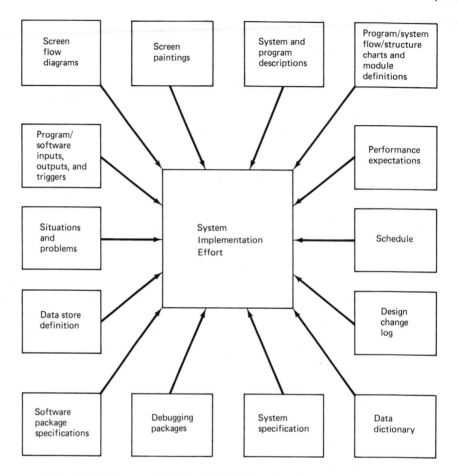

Figure 12-11   Documentation Needed for Information-System Implementation.

Figure 12-12 shows that the system designer needs the following documentation from the systems analyst:

- system specification process charts and text
- performance expectations
- data store definition
- data-in-motion definition (transaction handling, query and report specification, triggers)
- process and information requirements analysis
- financial estimates (cost-benefit analysis)
- a conceptual view of how the new system will work

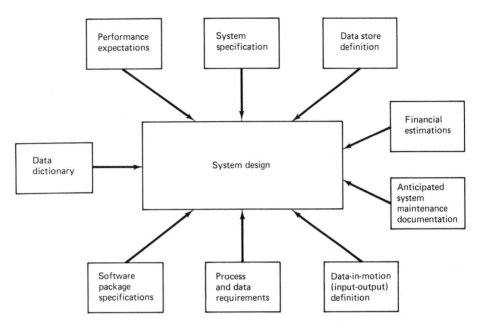

Figure 12-12   Documentation Needed for Design.

All this documentation just mentioned comes under the general heading of The New System Specification.

Systems Analysis Documentation

Figure 12-13 shows the documentation for systems analysis:

- organization statement of objectives and goals (and analysts' notes on actual objectives and goals, if different)
- the strategic plan for information-system development
- the tactical plan for information-system scheduling (and capacity planning)
- system resources and resource costs (available and needed)
- expected benefits
- forecasted development costs
- function, process, document, and data relationships in the organization
- data dictionary (particularly as extended to define the organization)
- software package specifications (what is available in the marketplace and what is available in-house)
- application-area documents; examples and notes
- interview notes

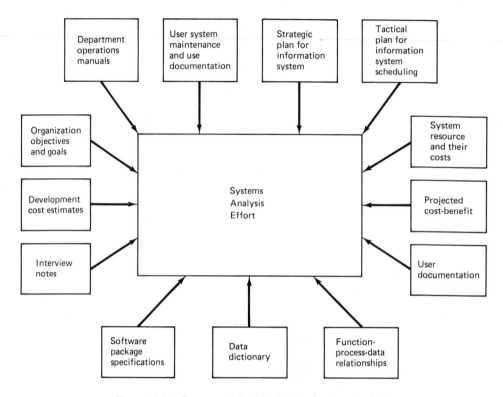

Figure 12-13 Documentation Needed for Systems Analysis.

Several of these items, such as document review, estimating, and interviewing, are the subject of the next chapter, which is about research techniques.

Before leaving the present topic we must come to some sense of what is appropriate in terms of the distribution of documentation. We have shown here that the type of documentation needed by all those involved in the development and usage effort concerning the organization information system is an order of magnitude more complex than what is usually involved in engineering, building, and using a device. In fact many devices will be involved in the building of the information system. The documentation can really pile up. With that in mind, two points must be made:

1. Not all people need to see all the documentation. It is not fair to burden the users with the details of program development or to introduce users to design discussions that are still tentative. Users should not be burdened with trying to keep up with revision after revision of paper documentation that takes hours just to file. The same holds true with other concerned parties in the development process, including information-system management.

2. All the documentation should be kept—and kept in one place, if technically

possible. This is one reason why the data dictionary is becoming so important in our thinking about how to do this whole business of information-system development. Keeping our documentation in one place means keeping it on a database, such as the dictionary, where it can be manipulated and where it can be viewed by all concerned people who have access to a terminal. Pending the time when charting documents and documents that need word processing can be incorporated into the data dictionary, we must be able to use the dictionary, or some less-than-satisfactory substitute, as a directory, which integrates all this great amount of necessary documentation about the information system—and about the whole organization.

13
RESEARCH TECHNIQUES

This is a chapter about the raw material of systems specifying—about listening, entering into dialogues, reading, and inventing for the purpose of doing the current and new system study. The main purpose is to make sure that the system accomplishes its objectives in the best way possible.

This book up to now has concerned methodologies for organizing and manipulating observations once made. We will now attempt to deal with the methods found to be most useful for the capture and integration of the raw observations.

We will be dealing with invention in a separate chapter. For now we will say that one of the two major purposes of research is to allow the mysterious process of invention to take place. The other purpose is to record observations in an orderly way so as to be retrievable by all those concerned.

Often we seem to stumble into invention. The information needed for the inventive process is meaningless unless we are prepared to apprehend it and convert it into knowledge. This process of changing information into knowledge is much more likely to happen if we have some idea of what we are looking for—if indeed we know we are looking for something. Research is looking for something about which we have some advance ideas.

It is generally accepted as unarguable fact in our "business" that some programmers are many times as productive as an average programmer. Ratios of 20 or 25 to 1 have been suggested. Barry W. Boehm's well-known COCOMO model, dealt with in his book *Software Engineering Economics*[1] (wherein COCOMO stands for COnstructive COst MOdel), plugs in personnel/team capability as by far the most significant factor in predicting system development effort—twice as important as any other factor.

The same importance can be placed on the relative effect on project performance of the quality of the systems analyst. The excellent analyst is worth many average analysts. We would further add that the systems analysis is the most important stage in the development life cycle. If the analysis is second-rate or worse, the best design, implementation, and operations teams in the world will not make a quality product. In fact, development failure is a possibility. *Inventive research* is the critical factor. Almost anyone can be taught the methodologies covered in this book. It will take a talented individual to use these methodologies to the best advantage. Researching is where this does or does not happen.

TWO KINDS OF RESEARCH

One kind of research is done in advance of assignment to a specific project. This includes special education, reading the technical magazines in the field, reading the books that might relate to future work assignments, attending conferences and user groups, and working on projects ongoing in professional societies such as the Association For Computing Machinery (ACM).

The other kind of research is the system studies which have been the entire subject of this book. We have discussed in detail the two types of project studies: (1) the strategic plan for developing information systems, and (2) the system specification in support of application development. Both are conducted with the same techniques. The strategic plan analyzes the top levels of the system. The full system specification analyzes the system from top to bottom, including every process which is transformative. Variations on listening, talking, reading, writing, and charting make up the techniques involved. Doing these seemingly simple generic functions involves some "tricks" which can be taught.

WHAT IS TO BE RESEARCHED?

Figure 13-1 shows the generic systems model. A discussion of this model will allow us to better understand how we will bound our research and what form it will take.

First, however, we must once again resolve the problem of potentially confusing terminology. An earlier discussion presented the system terminology used in Figure 13-1, which the author feels is best suited to a discussion of systems at any level of complexity. Chapter 11 introduced another set of terms, covering

---

[1] Barry W. Boehm, *Software Engineering Economics*, (Englewood Cliffs, N.J.: Prentice-Hall, Inc., 1981), p. 642 (illustration).

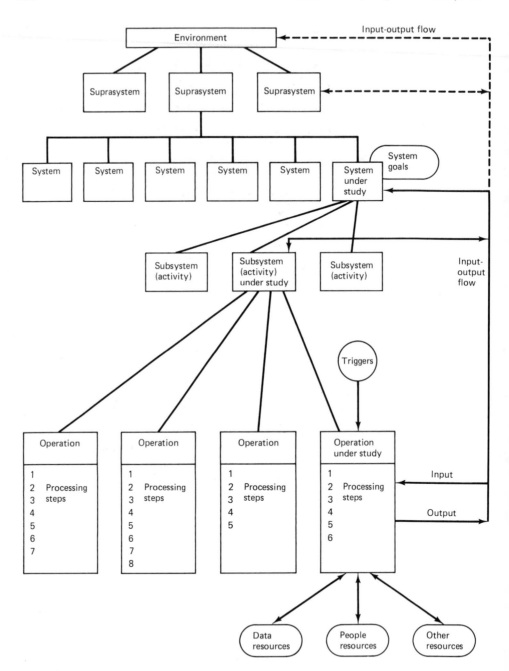

Figure 13-1 Generic Systems Model. From *Computerworld*, copyright 1982 by C.W. Communications, Inc., Framingham, MA 01701.

roughly the same systems model, so that we could introduce to the reader the work done in the field by such methods as IBM's Business Systems Planning. Figure 13-2 gives another terminology equation table. We now return to describing a system as existing in an environment (which is, at the least, a geosphere, a biosphere, and a social milieu). This environment (unless our study system is the universe) always has suprasystems in parent-child relationship to our system. If our area of concern is an oil company, there is always the energy industry and the international system of energy producers and consumers. If our area of study— our system—is the sales department of a corporation, then the suprasystem is the whole corporation. In this game of bounding a social organization as a system we are allowed to pick our own level of what is suprasystem, system, and subsystem.

To further review generic system terminology, we have below the top or subsystem level of activity the levels of recursive operations and processes. This is to say that any subsystem can have a variable number of descending levels, each of which we will call an operation within an operation. Eventually we will reach the base level, which is the most detailed operations level, below which it would be unnecessary to decompose the system. At this level, the base-level operation, we deal in serial processing steps: Mary opens the envelopes, John takes out and distributes the contents to like bins, Joe counts and puts rubber bands and a header card about each bin batch, ... Even at this level we have recursive processing. While John takes out the contents of each envelope, he may discover money or a check enclosed and have a special series of processing steps to perform at an even lower or recursive level.

It is this system which the researcher must fully understand upon the completion of the research task.

Chapters 1–4 and Chapter 13 terminology	Chapter 11 strategic planning terminology
Environment	
Suprasystem	
System	
Subsystem (activity)	Function
Operation*	Process (activity†)
Process*	

*Recursive (operations within an operation or processes within a process can occur).

†Activities occur recursively within a process.

Figure 13-2 Generic System Terminology Table.

STEPS TO BE COVERED WHILE RESEARCHING

Researching in this context will consist of interviewing and document analysis. We are interested in tracking the various inputs as they are triggered to become outputs through process transforms. We will be dealing with processes where information is the entire input-output flow. The material flow, if there is one, is simply a physical carrier for the information (such as a banking system involving check handling) or a product I/O flow system which is under information control (such as a bakery with flour and recipes among the vital inputs). In human organizations all systems have some vital information-control component.

Assuming we already have management support for the project, have identified goals and objectives, and have bounded the system, the following are steps to be covered during research:

- know all outputs
- know all data store resources
- know all other resources (people, equipment, etc.)
- know all triggers
- know all activities (subsystems)
- know all operations and their recursive relationships
- know all process details within operations and their recursive relationships
- know the balance between the need for internal equilibrium and the need for throughput efficiency, which is the basic and inherent systems dilemma
- know the role players' responsibilities and problems
- know the problems and opportunities present or absent in the system
- know how the subsystems fit together to create a whole system different from the sum of the parts
- know how to be prepared to deal with the sudden insights about the system which will come to the analyst who avoids premature closure—information not available through quantitative analysis or performance of uninspired methodology
- know how to arrive at a working hypothesis regarding the system specification and system problems and opportunities
- know how to test and reformulate the working hypothesis
- know how to record and present the study

INTERVIEWING TECHNIQUES

Interviewing involves listening, dialogues, group discussion, and getting feedback to our monologues. Our goal in all of this is not only to record the system accurately but also to recognize opportunity for invention. We may be given an accounting in advance of potential opportunities for improvement, or we may have to discover them during the research. The types of interviews we will discuss are shown in Figure 13-3.

There are three stages to interviewing:

- gathering
- understanding
- confirming

Gathering

The first piece of business conducted at this stage is to request and receive *all* the documentation available in the study area. We will cover this in detail in the section of the chapter on document analysis. No second interview should occur until this documentation is received and digested. The importance of this act of absorbing the available documentation before continuing the interview process should not be underestimated. Sometimes, if the documentation is good (and documentation is getting better and better), the interviews that follow can be confirming interviews and the process shortened.

During the gathering stage, they talk—we listen. During the understanding stage we all interact. During the confirmation stage, we present—they listen. All three stages require skills that are extraordinary. We must listen with an expectant

Emphasis on . . . / Interview Type	Listening	Dialogues	Group discussion	Monologues
Opening interview	X			
Interview with hostile user		X	X	
Concealment		X	X	
Listening to user	X		X	
Conflict with user		X	X	
Collaborative		X	X	
Three-level I/O	X	X	X	
Confirmation				X

Figure 13-3   Interview Type Matrix.

openness and quietness of mind that impels the interviewee to give forth information as if filling a vacuum. Most people being interviewed in this type of two-person situation love to explain what they know and are amazingly forthcoming unless they feel threatened or have something to conceal.

It is far better that the routine interview for gathering information be between two people: one analyst and one user. Do not tape-record the interview. Use only the ubiquitous yellow pad or notebook. Do not use a notetaker. Do not use a second interviewer. Conduct the interview on the user's turf, with the user behind his or her own desk, buying coffee for the analyst. The very worst type of information-gathering interview is a group of users and analysts. The ambience we wish to create is destroyed in this setting. Also we wish to interview users separately to pick up differences in how they interpret their system. These differences are little gold mines of opportunity.

The next-to-worst situation, which is recommended by an important source which we will tactfully not reference here, is the interview conducted by two analysts and one or two notetakers, all focused on one user. To make the situation even more threatening, this interview is conducted on the analysts' territory. This is the type of "interview" conducted by the New York City or the Los Angeles police department to extract a confession from a mass murderer. One interviewer is brutish and the other interviewer is a nice guy. The nice guy protects the felon from the brute and is usually the one to get the confession. In very few organizations will this achieve any results other than the dismissal of the analysts without severance pay. What we want in the usual gathering interview is to establish or maintain a friendly relationship between two people to enable maximum communication.

What we want communicated is clear. For the hundredth (and not the last) time, we want to know the operations in the subsystem that convert inputs to outputs using triggers, using resources. We want particularly to know the information and people resources, and we want to know the serial processes that make up each operation. As we gather this very specific information, we want to get a candid accounting of problems involved and we want to record the user's wish list of desired enhancements. Above all, we want to keep our periscope constantly scanning to pick up problems not mentioned and opportunities not dreamed of by the user.

It is not arrogant for the experienced analyst to expect to find problems and opportunities that the local system managers are unable to find in their own organizations. The analyst goes from place to place like a whale, eventually seeing the amazing similarity in all systems of social organization, and having the gift of finding past solutions applicable to the current system after extrapolations are made. The user manager is like the oak tree rooted for a season on his or her turf, doing the same operations day after day, constantly putting out fires, mostly dealing with personalities, and almost always developing blind spots regarding subtle changes or subtle opportunities.

What should keep the analyst who has the task of developing information systems humble is how often the best information system is less important than other functions with which the user manager is concerned. Such human-oriented concerns as team functioning, individual personality problems, and general morale and such other concerns as materials quality and availability very often are far more important to the manager than good, timely information. The best-written restaurant menu is meaningless (and is in fact a travesty) if the meat is tough, the potatoes cold, and the salad limp.

So the gathering phase is not just gathering information as a camera gathers information; rather, it involves interpretation and a plan to follow. The basic process, though, is listening to the user. What information is needed? Are they getting it? And, by the way, are they making the best use of their other, noninformation resources? Do they need training as well as better information? That is definitely also the type of question a systems analyst is entitled to ask, and must ask. What good is the automated inventory system if the warehouse workers bypass or otherwise misuse the system when moving materials around?

Understanding

After the gathering interviews (and the documentation gathering and analysis), the systems analyst retreats to a quiet place and struggles with all the information that has been acquired. A first attempt to chart the operations-process flow might be undertaken at this time if the analyst feels sure of a reasonable grasp of the system model. The data is modelled from the documents (and, if new data, from the interview notes). Either formal data modelling from user-view diagrams is undertaken or intuitive modelling is conducted. Both methods result in data definitions for all operations, showing transactions and data stores. These data definitions are in the form of data tables (layouts).

At the same time, the analyst develops a tentative working hypothesis or set of conclusions regarding the functioning of the system and how it should be changed. These are written down as a scenario for systems development. Several versions of this scenario are tried, running from a minimum-cost, minimum-benefit, quicker "hamburger" solution to a maximum-cost, maximum-benefit, slower-to-develop, prime-ribs-of-beef solution (or, if the reader is nutrition conscious, from brown rice to fresh fish).

These are rough notes in a very liquid state, ready for changing, not a hardened position to be defended without good cause. These rough notes are presented to the analyst's own colleagues and management through walkthrough sessions and further revised.

At this point the analyst is ready for interview stage two: the Platonic dialogue with the user. This dialogue should be conducted, if possible, with just the primary user with whom the analyst has been working. This is usually the activity manager. After working out the understanding with this user, the analyst can extend the interview process to the group scene. This should not take place until the analyst and the primary user contact have worked out their differences and are united in one concept of the model.

The Platonic dialogue involves a dialectical confrontation and should be carried out only in an atmosphere of mutual trust and respect. The user should be convinced that the analyst's efforts are beneficial to his or her concerns. The analyst must be convinced the user has no hidden agenda that forecloses on honest debate.

Dialectics involves the triangle shown in Figure 13-4. The analyst advances a *thesis* concerning what the system is like and why and how it should be changed. The user offers the *antithesis*, which shows where the analyst has missed the boat on understanding the system and what the changes should really be. An adversarial but friendly confrontation ensues as the parties go to the mat on each point of difference. Somewhere in this process of give and take an entirely third, different model starts forming, usually in a few brief moments of qualitative change, and the *synthesis* rapidly emerges. This syntheses is far more impressive than the net change wherein wrong facts are dropped. In addition, new ideas have been developed based on the melding of the skills of the whale and the skills of the oak.

The novice reader, such as the student, should have faith that this process really does take place. It is not a romantic notion. It is the product of skilled, inventive interviewing in the proper atmosphere. There is no further fixed methodology that can be used here to enable this process. What is needed are pleasant people who have done their homework, are well grounded in the details, and exercise good communication skills (probably learned by being brought up in the right family environment). The systems analysis will be second-rate if this type of interview is not done, no matter how good the methodology. What we can do here is at least introduce the reader to a possibility.

The one-on-one dialectic friendly confrontation of two points of view is, of course, not the only possible interview situation at this stage in the research. It is just the most certain to produce quality results. It is certainly possible for a small work team of from three to five individuals to produce some very good research.

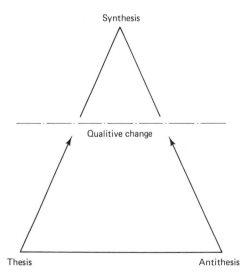

Figure 13-4 Dialectics in Interviewing for Understanding.

However, it is not easy. The more participants, the tougher it is. Once the inventive ideas are on the table, the small affinity group works at its best, since more than two minds are at work improving what has already been invented. Even so, the user-analyst duo, with the user reacting, is in the long haul the best method if the analyst is truly creative.

At this point in the understanding phase of interviewing, the model arrived at is no longer the product of the talent of the systems analyst but is now the result of the collaborative effort of the analyst and the user. This is usually a rewarding experience for the user, who has learned just as much from the dialogue-type interview as has the analyst. The expected result of the collaboration is that a united team has been formed.

At this point the user-analyst team will usually need to expand the interview process to include group discussion with other concerned parties. The analyst should try to make sure that this expansion of the dialogue does not destroy the gains that have been made. In order to insure this, the analyst needs to know something of group dynamics.

Group-Meeting Dynamics

The issue as to whether the project will be advanced or will deteriorate is often decided by who sits where at the conference table. At every group meeting held since King Arthur's round table there have been certain meeting role types that occupy certain seating positions. The position where the user-analyst collaborative team sits at the group meeting will probably determine the outcome if a conflict develops. (It seems lately that a conflict always develops among bright, high-achieving, ambitious meeting sitters.)

Figure 13-5 shows meeting table dynamics. There is always a leader at a meeting. Where the leader sits determines where the other role players sit. Ninety degrees in both directions from where the leader sits are the hot seats where the meeting antagonist and protagonist usually choose to sit. This is shown in Figure 13-5. Opposite the leader is a very strategic position for active participation for a conciliatory but active role. From this position a knowledgeable role player can evenhandedly mediate differences between the adversaries.

A more passive but very strategic position is to the right or left of the leader, as close to the leader as possible. Any attack on this player is construed as a challenge to the leader. The leader thinks of the person on the right or left as an adviser. In this position the analyst is diffident and low-key.

The strategy is to preserve the team invention from mindless ravages of a powerful and critical group player who could not possibly be as close to the problem as the team but is determined to play the antagonist's role. This is not an unusual situation!

If the user member of the team is capable of carrying the ball and has the respect of the players, the analyst should be sure to be positioned beside the leader and talk as little and as softly as possible. If a shoot-out is inevitable, the analyst

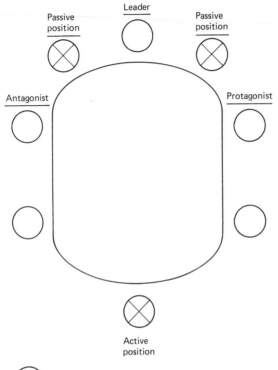

Figure 13-5 Sitting Strategies at Group Meetings.

should sit opposite the leader. If the user suddenly develops flop sweat, the analyst may have to assume the protagonist role, but this should be avoided except in unusual circumstances.

If the group approves of the research results, perhaps with minor modifications, the understanding phase has been accomplished and the analyst may be considered lucky. If, as often happens, the ideas are either emasculated or sent back to the drawing board, the analyst must be philosophical and by all means not become bitter. In a large organization the journey through committee acceptance is a rocky road. It goes with the territory. The people who succeed and rise to the top in large organizations are usually the "closers" who have somehow developed the ability to get someone else's ideas accepted. These are valuable people. They can sell refrigerators to Eskimos.

Confirming

At this stage the analyst develops the current or new system specification into polished documentation. If the entire process is accomplished, there will be current system charts and text—the requirements document showing process and data requirements and which requirements were YES (to be done) or NO (don't do). There will be new system specifications in charts and text showing information, system process changes and new data store resources to be built. There will be an estimate analysis, which will require charting of the proposed physical system.

This large amount of documentation must be boiled down to a presentation which will give the brown rice–fresh fish alternatives. Being in possession of clearly described alternatives expressed in scenic terms (what it will be like) and accompanied by cost and benefit estimates, it is very satisfying to nontechnical, non-operations management, responsible for investments, to make the circumscribed decisions now available to them. They pick blue or grey and get to participate in the decision process without going through the weeks or months of struggle to arrive at these choices. Their skill is that they are always dealing with important choices and must pick and choose from contending investments what is best for the organization.

In this environment, which is sometimes the organization directors' board room, the systems analyst should be businesslike and should radiate a quiet confidence. This is not the place to express doubts about the project, even though it is always perfectly reasonable to have doubts. If the analyst has substantive doubts about whether the project is a good investment, the analyst shouldn't be advocating the project and striving for confirmation.

Validity Checking of Interview Information

The analyst is fortunate in having two kinds of checks on the validity of acquired information. Given enough time and the energy to conduct thorough interviewing, there is only one reason why all the tangible facts about a system cannot be captured accurately. That one reason is that nobody knows what is going on. The two sure checks are:

- three-level interviewing
- I/O interviewing

Three-level-interviewing involves interviews conducted at the manager level, the supervisor level, and the mission worker level, all covering the same material. The mission worker is the person who actually does the work of the process. Do all three types involved in the process have the same story to tell? Do their stories seriously diverge? Can the differences in stories be accounted for by different perspectives? When the primary interviewee is the manager and the supervisor and mission workers tell a different story, it is necessary to return to the manager and ask why the difference. Sometimes there are breakdowns in communications (or nonexistent standards) in a department, and the systems analysis underway has the beneficial effect of pointing to easily corrected problems.

I/O interviewing means interviewing in those areas that send the inputs to the area under analysis and receive the outputs from the analysis area. Figure 13-6 shows what is involved.

The important first step in understanding how an operation works is to know what inputs go in and out of the process. The function of every process is to transform the inputs to the outputs. By interviewing in process area A in Figure 13-6 we learn about data X. Doing the same in process area B we learn about data Y. Process area D receives the only output from process area C. By understanding data X, Y, and Z we understand what the process function is in area C. When interviewing the manager of process area C we had better be learning about the transformation of X, Y, to Z.

The Negative Interview Situation

Almost all interviews are pleasant experiences conducted between people with a common goal—to improve organization functioning and increase productivity, mostly through automation. Some jobs may be eliminated, but other jobs will be created which are usually more interesting. Very few people, and certainly no managers, are going to have their jobs threatened by this in-house interviewing. Even so, the analyst will always come across individuals who are unsatisfactory to interview. Some of the more common reasons for this are

- the interviewee does feel threatened and is hostile
- the interviewee has something to conceal
- the interviewee has no confidence in the system development process owing to poor experiences in the past
- the executive sponsor of the systems analysis effort is too low in the command chain and has no authority over the activity area under scrutiny
- a hostile situation exists higher up in the organization between the data processing management, and this particular user management and the interviewee has not been encouraged to be forthcoming

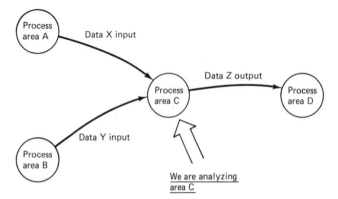

Figure 13-6 I/O Interviewing.

At the root of all these problems is usually the basic dilemma of all systems of human social organization. Question: Is the subsystem in existence to process inputs or to maintain a dynamic homeostasis or equilibrium? The answer is that a duality always exists and that both are true. Given no controls and a fixed input load, a subsystem will always keep growing and supervisors will be promoted to managers and then group managers, still doing the same work. The number of mission workers and supervisors will correspondingly increase, giving the manager some justification for this upward movement in career path. Managers often are not rewarded for the accomplishment of objectives but instead are rewarded for managing what is often unnecessary complexity. This can become—in some sick organizations—one gigantic back-scratching operation.

The subsystem that the organization leadership does not control and encourage to achieve not size but objectives will always develop a pathological whole-system identity. This identity will eventually develop a different set of goals from the real whole system. This may be related to why cancers develop in biologic systems.

What can the analyst do to survive in a hostile interviewing environment, whatever the reason? Plenty. In a duel between a reluctant, stone-walling operations manager and an experienced systems analyst over whether the analyst gets the proper information, all the odds are in favor of the analyst.

First, the analyst can learn nearly all that needs to be known by the published documentation.

Second, the analyst is dealing with a subsystem that is not that much unlike other subsystems the analyst has encountered in a long and checkered career. There are really not all that many ways of using information to transform inputs into outputs.

Third, the analyst can practice three-way interviewing, starting at the mission worker level and working up (after the initial hostile interview).

Fourth, the analyst can practice I/O interviewing, interview the sending departments and the receiving departments that surround the department in question, and nail down what goes in and what goes out.

Finally, the analyst can usually find a manager who had prior responsibility for the offending department and interview this individual.

Armed with the results of all this research, the analyst returns for a second interview with the unfriendly party. Most of what needs to be known, is already known but there are usually missing elements. Getting this information is the reason for returning at all.

The hostile interviewee is already aware of what has been going on and is not in a position to give deliberately misleading information. No matter what has been going on, *that* is taboo. It soon becomes apparent to this individual that the only recourse is total cooperation. As a general rule, hostile interview subjects, when painted into a corner as we have just described, make the best interview subjects. They always end up telling all about where the gold is buried—if the analyst is both patient enough and nonabrasive.

The analyst should never complain about hostile interview subjects. It is hard to prove and very poor form. Besides, as we have just said, they make the very best of subjects when handled correctly.

Having discussed at such length these tactics concerning the unfriendly interview, we must emphasize that this experience is not at all the rule in most organizations. If it *is* the rule, the organization is in serious trouble and will probably fail. If an analyst cannot get information, information probably is not flowing up to the senior management, the whole-system management. Unless changes are made from the top down, it would probably be better for the analyst to find somewhere else to practice.

DOCUMENT RESEARCH TECHNIQUES

One purpose of the initial interview is to establish the analyst's requirement for all documentation in the research area. The research area may be an organization subsystem or activity that is responsible for a group of operations, or it may be one department jurisdiction at the operations level responsible for a series of processes. Whatever the bounded area, several kinds of documents associated with the area must be put to the most exhaustive analysis if the research area is to be understood. This is solitary work. The analyst is supposed to be some kind of dual personality who can live well in the world of complex communication (as in the case of the types of interviewing we have just finished discussing) and also retreat into the world of the scholar. No expert in Sanskrit ever had to delve more deeply into the dense and obscure than the analyst who strives to be excellent.

It is essential that the analyst be clear about what kinds of documents are available for the study. Most documents will be found in the study area, but some will exist in such other source areas as the accounting department or the data processing systems and programming area, where the data dictionary, analysis and design specifications, run books, and program listings are kept. The more complete the retrieval of documents, the more successful will be the study.

Types of Documents Available from Other Sources

The accounting department usually has a wealth of information about operating departments in an organization—for instance, cost-accounting information about a manufacturing department. The accounting system usually carries information about labor costs, including fringe costs, overtime, and premium time. Also available from this source are the machine costs, material costs, and fixed costs of facilities. The organization needs this information to determine the cost of manufacture, which is one of the operating costs that offset the gross sales revenue to arrive at real profit in a profit-making organization. The analyst should review such general organization material as this before conducting an interview in the research area.

In the example just given, another valuable piece of information can be picked up in the sales department. In many businesses the inside manufacturing operation has its parallel in outside competing vendors who do the same manufacturing job. Often the sales manager may push for a vendor selection based on cost of manufacture. It is instructive for the analyst to know just how competitive the inside operation is and what is the management plan in this area. Needless to say, most of this information is highly confidential, and the analyst must have a reputation for being close-mouthed. This is the type of information that is not passed on by the working analyst, even to the systems department management.

In this case the question should occur to the analyst: What is it that will make the in-house manufacturing operation more attractive to the sales manager who lives by the profit-and-loss statement and still be acceptable to the people in the manufacturing operation? This goes into the analyst's ubiquitous notebook.

The main source of documented research information outside the research area itself has become the data processing department, where, as we have mentioned, the systems analysis and design specifications, run books, and programs are kept. We are heading in the direction where this area is the main source for current system information. In very few instances in large United States organizations has an activity that cries out for automation not already been to some extent automated. The question is: how deeply do we have to go into the automated information system to understand those aspects of the operations which are not available in the documents in the research area?

Now we can see one of the more important reasons for good information-system documentation in general and good system and design charting specifications in particular. Going into the program code is like falling into a deep hole with no observable bottom. The program code will sometimes be in several different languages, perhaps a mix of second, third, and fourth generation. Part of the information system will be software packages with mysterious, proprietary internals.

What we must have, in order to do a systems study in what remains of the twentieth century and into the twenty-first century, is the logical specification and design charting and either a flow chart or a language description of the serial process modules.

We are fast coming to a situation where no new organization can be created to enter into a complex information-intensive industry.[2] *The start-up information system is too staggering to contemplate. Only the existing organizations can start up new enterprises by replicating existing information systems.* If this is a true statement, then what happens to the organization that loses its information system due to poor documentation so that change becomes impossible? Answer—this organization loses its membership in the existing club. It is in this light that we deal with problems of system documentation and document research.

We can see then that the program listings will be referred to only to settle

---

[2] Gerald Cohen as told to Paul Gillen, "The Fourth Generation Moves into the Mainstream," *Computerworld*, 4/2/84, p. 17.

fine points that come up during the later stages of the interview process. The same goes for the run books kept in the computer center. What we want to do is place the computer operation within the context of the entire subsystem operation. When we discuss research-area documents, we will see that the inputs and the outputs of the computer system will be identified. To see what we have to learn about the computerized internal information system, please turn to Figure 6-1, where the overall or generic computer solution is illustrated. Look also at Figure 7-4, which shows internal input transaction processing.

Figure 6-1 shows us that we must learn what happens to an input transaction created in the research area before it is acceptable to the database updating system. Often the transaction becomes many transactions, as shown in Figure 7-4. We must understand this internal transform processing. Much of it has an efficiency orientation, in that the final refined transaction set is more what the research area understands as logical than the entering transaction, which is minimalized for cost and handling reasons and picks up vital information internally.

The other vital internal process to be understood, as shown in Figure 6-1, is the transform by which the set of all input transactions becomes the changing database system which is at the heart of the operation—and the process by which the output reporting is triggered out of the database system to become the documents we will see in the research area, thus completing the picture.

Document Analysis in the Research Area

Having gotten what we could from sources outside the activity itself, we now proceed to analyze the documents in the activity area. During the first interview, as we have already pointed out, the manager of the activity is asked to provide all the documents which explain or are used by the activity. It is important to emphasize that this document search must be exhaustive. Nothing, no matter how seemingly inconsequential, must be left out. The analyst must have the prerogative of determining what is consequential.

Documents will be of three types. One type explains how things work in this activity area and what standards are observed. A second type is the user view of the data store or data-base. The third type is the data-in-motion, the transactions that pass through the processes being transformed and the data retrieval in the form of queries and reports that provide the information support for the activity.

Regarding data stores it is important that the user produce a copy of even the most modest data store. This might be operating rules taped to the wall near a piece of machinery or a list of telephone numbers of vendors kept in the manager's front desk drawer. Any *record* qualifies as a potentially interesting data store.

In the case of the department transactions and output reporting, the documents must not be blank forms but instead must be filled in—in other words, actual used documents, as many as are needed for each document type to show all the variations of document data use.

What the analyst does with this complete set of activity documents is as follows:

1. Documents about the activity area and the standards in practice, along with the interview notes, form the basis for the process flow diagrams which are the system specification.

2. Documents describing input processing are included with the process flow diagrams just mentioned, as explained in Chapters 3 and 4. For instance, transactions are written on the arrow lines between transform bubbles in the data flow diagram method described in Chapter 4.

3. A great deal of time is spent analyzing output reporting documents. In Chapter 10 we described how user-view diagrams are constructed by formal data modelling using information on the user views. Report documents are the most important user views. These user-view diagrams were synthesized into a conceptual database schema, thus making the reporting documents the input to the process that develops the database. Using this technique, there is software available, as described in Chapter 10 to arrive at the conceptual schema from the user-view diagrams.

This software also produces database information as tables or relations in the third normal form. It is possible to produce these tables without software support and intuitively put tables together into predefined database structure (such as the tree or the plex) if that is desired. We actually were doing this during our discussion of strategic planning studies in Chapter 11. In that chapter we gathered all our documents and scanned each one for data entities. We did not specify all the fields in the entity—just the entity name. The difference with the full study is that, as we go through each document, we specify all the fields in the entity—the key and the attributes.

So, to sum up, we do document analysis either through formal or informal data modelling to build our database scheme, and we also analyze documents to build our system specification charts.

Previous chapters have described how these charting methodologies work. Now we are discussing the sources that are input to the methodologies: interviews and documents.

Two helpful hints may be offered to the systems analyst working with documents in the research area:

1. Put the report documents in order—the most important on top and the least important on the bottom. Most work areas have key documents that give all or most of the useful user views and data entities. By the time the analyst gets halfway down the stack, the job is just about done. At this time make a judgment whether the next document contains information already captured or that will never be automated. Not every obscure user view belongs in the data or process model.

2. Be on the alert during the interviewing to references to invisible documents.

When an invisible document is discovered, the analyst must write a document for it. An example is a query that is made in a fourth-generation reporting language for a screen display but never saved as a document or as a program procedure. Another example is the verbal agreement wherein a periodic call is made by another manager to ask about department backlog or accomplished quantity or whatever. The information is given regularly over the telephone or by the water cooler.

USER RELATIONS

At present the relationship between the information-systems people and the users is at a low point in many organizations. There are some economic reasons for this. Information-systems management is generally not paid as much as management in other vital organization areas, such as the finance, sales, and legal departments in the private corporation. Perhaps this is because management of the information system is perceived to be not directly related to profit-center responsibility but rather a support service such as house and grounds maintenance. This is an injustice that should be addressed by the senior management. Information systems has become itself a vital profit center and much more involved in the operation of other profit centers, such as sales, than anyone would have dreamed possible 25 years ago.

On the other hand, certain unsavory practices on the part of information systems departments have had a serious negative effect on organization viability. The three- and four-year application backlogs (and applications that never even get submitted) are not caused only by a great increase in quantity or complexity of new applications. This is a common misunderstanding, although it is true that there has been an increase in application complexity, mainly because of the development of on-line systems to replace batch systems. If the traditional in-house systems development route using procedural programming languages is followed, on-line systems are many times as expensive as batch systems.

As of the mid-1980s the following may be considered unsavory practices.

- designing systems that must be operationally maintained by the information systems department (example: the table-driven system where the users do not control the tables)
- using labor-intensive, obsolete programming languages under the guise of operating efficiency (example: building user-terminal dialogues using COBOL instead of non-procedural application development software[3])
- designing systems around obsolete, skilled-labor-intensive databases that require constant tuning and debugging by experts

[3] For instance, Oxford Corporation's UFO or IBM's DMS—both on-line application development tools using a nonprocedural language to which procedural languages can be attached as needed.

- producing great volumes of esoteric and redundant documentation, arcane to the user and always inaccurate, at significant labor costs and meanwhile failing to produce usable user operations manuals

- designing systems that must be implemented by the in-house programming staff when an off-the-shelf software package would have accomplished the job

- misusing the common-database concept to eliminate the ability to use shared local database systems and private systems

- designing real-time systems when on-line systems that are not always real-time would suffice

- designing on-line systems when batch systems would suffice

- estimating costs of development at many times what it could be done for

- building a battleship when what is needed is a sloop

- not doing evolutionary development starting with a prototype

- creating a situation where the user cannot become to some extent independent of the computer-systems expert when it comes to problem solving, dialoguing directly with the information base without the expert as necessary scribe

This last point is the most important of all. The invention of the printing press allowed the development of a literate community (with independent access to the Bible). A profoundly important current task for the information expert is to allow the ordinary user access to the computerized knowledge base. This will result not in a loss of jobs for systems analysts but in a great increase in the need for the enabling expert. Meanwhile, systems analysis—which is perhaps the second oldest profession—will never become obsolete. As long as human social organization becomes progressively more complex (which is the nature of all such systems) there will be an increasing need for systems doctors.

Figure 13-7 shows the ideal relationship between the analyst and the user in the modern world of computer-literacy expansion. Notice in this "user-analyst problem-solving machine" how the analyst's role devolves as the user becomes more adept. What is shown here is the development of the user's ability to develop reports, queries, and local applications either in the mainframe time- and data-sharing mode or in the personal-computer mode, where the user may or may not be on-line to the mainframe.

What we are *not* suggesting is that the user get involved in such matters as the type of research needed to develop the conceptual database schema for common data. We pointed this out in Chapter 11 when we discussed strategic planning and advised keeping the user out of the preliminary research. The user still needs to be given a bounded framework in which to work and should not, for instance, be changing the structure of large common and shared databases.

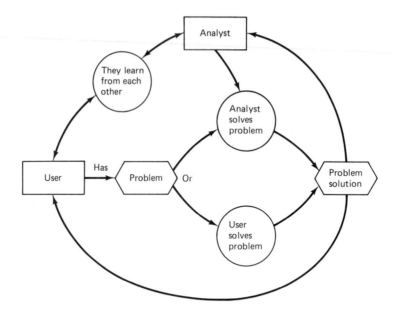

Figure 13-7   User-Analyst Problem-Solving Machine.

Professional Ethics in User Relations

No matter how clever the systems analyst, or how experienced, it will serve for naught if the analyst does not maintain a high level of professional ethics.

Most of these ethical standards are plain common sense, but it pays to discuss them because of their vital importance to the success of the project. On these ethical standards rests the analyst's reputation. Once damaged, a good reputation cannot be reinstated without the analyst's leaving the organization. Even then the grapevine may follow.

The analyst must, through good character, gain respect, be considered trustworthy, be capable of being a bridge over troubled waters, and be thought of as one who keeps confidences. In order to achieve this type of reputation with users the analyst must:

- let the user take the credit for good invention
- always tell the truth no matter how painful
- know how to keep secrets no matter how juicy
- never make a promise that is not kept—or, if not kept then explained
- treat everyone courteously whether humble or powerful
- never take sides between contending user groups—just give the facts and help the cooperative process not the conflict process

- avoid negative emotions on the job—such as hostility, envy and irritability—instead play Rugby or beat the dog
- don't talk too much or write too much (unless paid by the word)
- keep costs to the minimum needed to meet the requirements

The next chapter, which is about getting and using software, will give the user some reasonable alternatives in systems development to help users reduce costs and help themselves solve problems.

14 GETTING AND USING SOFTWARE

We will limit ourselves to a discussion of those types of software directly related to applications development. This means we will not be discussing operating-system and physical-transmission-system software. Neither shall we discuss (at length) utilities such as debugging and security software or sorts, as important as they are to implementing application systems. The reason is not that the systems analyst does not need to know about these types of software, but that such software is of less than primary importance to the practice of systems analysis.

One reason we discuss software at all in this section on analysis (after all, software could be considered part of design and implementation) is that the analyst, in most organizations, must know all the potential solutions to problems, even while defining the current and new logical system. Analysts defer solutions. Analysts cannot be oblivious to solutions. Any definition of systems analysis that does not include being able to conceptualize and cost-estimate a solution to a logical specification very early on defines systems analysis as a clerical job of little importance. We don't need such a limiting definition, because much more is expected of us.

Besides, as a practical matter, in many organizations the systems analyst becomes the systems designer becomes the implementation project leader. Not such a bad idea, really. A lot of knowledge falls between the cracks when the analyst and designer drop a project after their isolated task is accomplished.

Typically the analyst starts speculating on physical system and software solutions very soon after beginning the job of logical specification. The important point is deferring *closure*. The smart analyst never discusses an early solution or lets solution speculations affect the immediate job at hand. Solutions at this stage often come and go quickly. But eventually solutions come that don't go.

To put it succinctly, the more the analyst knows about how to solve a problem, the better the analyst will be at specifying the new system alternatives that are technically or financially feasible. The experience of the last few years has shown clearly that in-house development of computer solutions writing COBOL, BAL, or Ada programs is not always the best solution for many organizations (especially those organizations not involved in building software for device support—such as weapons systems). What is going to become the solution of choice, whenever possible, is off-the-shelf, easy-to-customize application software, hooked onto the framework of the modern database management system. We will devote this chapter to describing the use and acquisition of such software. For many of us this will be the main path for the systems development of the future.

We will be discussing the following two areas of getting and using software:

1. Getting and using the database management system (DBMS).
2. Getting and using already developed applications packages

We will have to work backward here and discuss using before getting. If we understand how to use, we can easily learn how to get. We can proceed from the assumption that the logical new system specification shows a need for an accessible data store, and a need for software resources that may involve generic applications such as word processing, workload and capacity management, purchasing and warehousing, statistical analysis, project management, shop floor management, and the various accounting functions such as general ledger, payroll, accounts payable, and many other applications packages that have been or will be developed.

Our first question is: how do we best use the DBMS to satisfy our need for information storage and management? We cannot discuss the problem of application software until we work out a resource management system which will deal with all our software as one system. The modern DBMS is the hook on which we can hang our software hat.

BASIC DBMS

Figure 14-1 shows the basic DBMS in all its bare bones. Database administration (DBA) nurses the database through a maintenance syntax and uses this language to construct and change the internal, conceptual, and external schemas as covered in Chapter 10. There are utility programs that can be called upon by the DBA to monitor performance, security, do back-ups, and so forth. The basic DBMS also

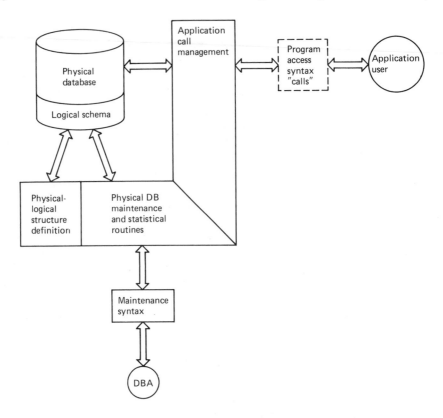

Figure 14-1 Basic DBMS.

allows "calls" to the database through use of its language. These calls are available in well-known languages such as COBOL, BAL, PL-1, and perhaps FORTRAN. A calling convention acceptable to the DBMS is converted into database access commands which retrieve, update, or add data groups (segments) that are the accessible quanta of the database. These operations can be in groups of data groups, such as children of a root-level data group, and can be qualified (retrieve IF age less than 30). On the basis of the flexibility of this call mechanism we can construct one kind of complex modern database system: the stand-alone basic DBMS to which all the necessary add-ons, each a stand-alone product, can be appended. We are going to be comparing this type of DBMS to the integrated, multiservice, single-vendor DBMS.

EXTENDED DBMS

It is now possible to take a basic DBMS such as IBM's IMS and surround it with modern application productivity software aids to achieve what is regarded as the essential DBMS extended package. Figure 14-2 shows most of the components that go to make up this "roll-your-own" system.

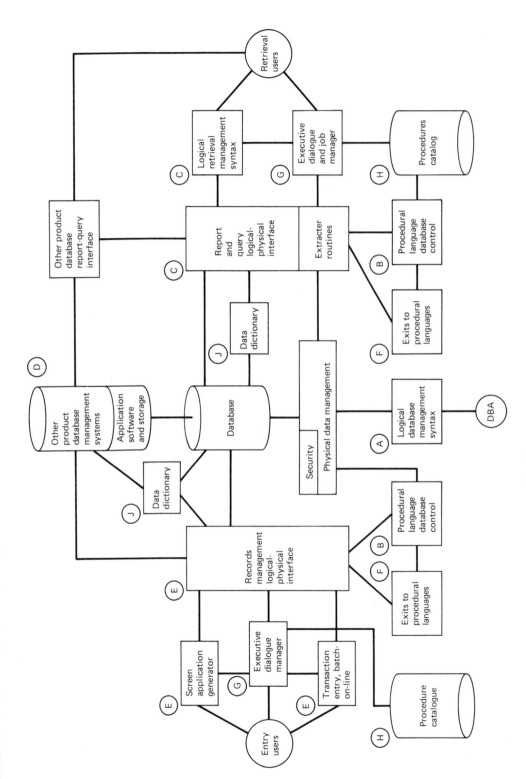

Figure 14-2 Extended DBMS.

In Figure 14-2, component A is the basic DBMS control syntax used by the DBA still intact. The DBA still loads the information for the three-level schema (see Figure 10-4). This problem exists for all types of databases, whether predefined hierarchical structures such as trees, or relational structures defined during execution, or a combination of both.

Component B is the access to the database through the basic call method, such as a call from a COBOL program, which is translated by the basic DBMS into physical reads and writes of the moveable quanta.

And now we proceed to the extended DBMS, which will allow us to contain all strategic plans for systems development as discussed in Chapters 10 and 11. Without such a system our strategies fail.

Report and Query

The first add-on to the basic DBMS is usually the report/query interface and the report/query syntax. This is shown in Figure 14-2 as component C. Sometimes it consists of two packages, a report generator and a genial query language. More recently a single package has been used for both functions.[1] In Chapter 7 we established the case that it is impossible to really separate reports from queries.

The addition of the report/query software brought a great improvement in productivity, since, as most people with any limited association with computing know by now, this report/query language is written in a reasonably easy syntax that is nonprocedural. It describes the report and the data in a few lines of "code" that perhaps replace several hundred lines of COBOL or BAL and have been written successfully by thousands of non-DP users. The argument that even this genial nonprocedural code is too complicated for the impatient user has been largely undermined by the alacrity with which many users learned spreadsheet and word-processing packages (in the office and at home). This PC software is every bit as difficult as the better of the fourth-generation, nonprocedural report/query languages.

The report/query add-on also brings us face-to-face with our first difficult purchase decision. DMS or DBMS? The DMS (data management system) is the software package such as report/query without a database or database management included.[2] It is a true add-on to the already existing DBMS, with no duplication of function. The DBMS report/query system is part of an integrated system, containing its own separate database as well as many other useful bells and whistles (as we shall see). Of course, the DBMS report/query can be purchased standalone, but this is seldom done (for reasons we shall explain as we go through Figure 14-2). The price difference between the low-end and the high-end report/query purchase can range from $10,000 to $300,000 and up.

---

[1] For instance: the RAMIS II report/query language (Mathematica Products Group, Princeton, N.J.) or the Focus report/query language (Information Builders Inc., New York, N.Y.).

[2] Examples of Data Management Systems (DMS) are: (a) Eastrieve (Pansophic Systems Inc.); (b) Dylakor 280 (Dylakor Software Systems Inc.); (c) Mark IV-Answer/DB (Informatics Inc.)

Other DBMS Satellites for Extracted Data Stores, Application Data Stores, and Software

This brings us to the components on Figure 14-2 marked as D. One feature that is essential in a build-your-own DBMS is a satellite DBMS containing a less complex, lighter-duty database for extractor processing of downloads from the heavy-duty primary database and also for small application processing. Typically this DBMS handles extractions for read-only purposes and for local and private data manipulations. The primary organization information is updated onto the major DBMS and then downloaded.

In this category of DBMS extensions (components D) we will also place our ready-made application software packages, which usually come with a proprietary database of their own. These are also satellites to the primary database. We will discuss this area in more detail later in this chapter. Along with the satellite database we have the storage of the satellite programs, program procedures, and directories or data dictionaries.

Now we can see why, when we make our first extended purchase of the report/query language, we have to think ahead to the other necessary extensions. Maybe buying the full-service, satellite DBMS is more practical in the long run than buying the less expensive stand-alone report/query. This is sometimes a hard decision to make.

Records Management

The next function that usually comes to mind when developing an extended DBMS is records management. By *records management* we mean the transactions that add, delete, and change logical records in the database. Component E in Figure 14-2 shows these functions. This is always the critical path in systems development, particularly in the case of on-line systems. It is in the area of records management that development costs can blow up to many times the original estimate. On this path, whether batch or on-line, transactions are validated and developed, applied to the database under conditions that can often adversely affect response time, and during application to the database go through the most complex system code and generate many outputs as a byproduct of records management. Report/query or downloading is routine compared to what records management can involve in a complex, operational (as opposed to decision support) type of system.

It is in the area of records management support that the wise selection of software is most critical. There is seldom a really bad report/query selection decision. Some packages are better than others, but they all work more or less. In the area of records management the demands on the software are much more stringent, and the inexperienced individual or group selecting this type of software can make truly harmful decisions. Let us make note that records management software includes on-line menu-driven application generators, user-terminal interactive dialogues with transaction generation, and batch-processing transaction generation.

Procedural Language Support for DBMS

In the case of a DBMS that has associated with it a nonprocedural language, component F in Figure 14-2 shows the required capability to descend to lower-level languages during either report/query, records management, or database update operations. There are input processing routines and complex output reports that are best programmed in a procedural language such as COBOL.

The Executive Job/Dialogue Manager

This is perhaps the most interesting and least-known package component making up the extended DBMS. It is shown as component G in Figure 14-2, and, as we see, it is the interface between the user and the rest of the extended DBMS. Here is a list of the functions performed by the executive:

1. Serves as a DBMS-oriented job control language (JCL), so that multiple procedures can be run as strings of subjobs or programs. Allows decisions to be made by the operator during the running of the strings, wherein different procedures can be selected based on the results of execution results to the moment. This can be done in a batch or on-line environment.

2. Allows variable parameters ("Parms") to be entered at execution time so that a procedure can be written in a generalized fashion and then particularized in real time by the Parm. This can be as simple as entering a time of day and as complex as making a major in-line modification to a skeleton procedure. This function is excellent for building dummy query programs, which are then particularized by the user's entering familiar parameters. This bypasses the whole issue of user-friendly query languages and is a nice midpoint between building hundreds of predefined user screens (many of which will never get much use) and leaving the user to the mercies of not-so-friendly query languages. The unfriendliest query language of all is that natural language in which the user just "converses" with the DBMS and never ends up getting the information needed (it always works great for everyone else but this hapless user).

3. Allows the user full access to all the data structures and procedures in the DBMS to which the user has authorized access, so that the user can browse what is in the various directories on-line to the executive before making a task performance decision.

4. Allows the DBMS and all the components of the DBMS to be manipulated by a built-in procedural language so that the DBMS can be programmed to execute the most complex tasks, making in-line decisions as mentioned in point 1.

5. Allows the inclusion of artificial-intelligence type procedures into the DBMS

so that procedures on the DBMS can be updated during a session to broaden the information base involved. These AI programs must be executed under the aegis of the executive dialogue manager. Now we can really make the natural-language query very friendly indeed!

6. Allows the construction of interactive real-time dialogues of any kind whatever, within the capabilities of the procedures and data stores on line within the DBMS, between the system and the user. The executive manages these dialogues.

Not a bad little list of functions. Unfortunately, to the author's knowledge, such a package is available only as part of an integrated DBMS. The main disadvantage to making an extended DBMS from a free selection of components available in the marketplace is that not all DBMS packages have such a software component as the executive just described here. Those DBMS packages that do have such a component tend to use it to build many of the utilities sold with the package.

Procedures Editor/Library

The procedures or programs shown as component H in Figure 14-2 should be able to be maintained within the DBMS by way of a procedures edit and a procedures library. With many integrated DBMS products it is easy to catalog, store, and retrieve a procedure. It is often necessary to edit a procedure, both during development and during operation. Procedures are often changed even after they are debugged. In either case the editor/librarian is an important part of the DBMS product.

Other Utilities and Functions

Every DBMS must have a security package, which includes logging on and off the system and access passwords to procedures, databases, files, records within files, and fields within records. Security must exist at all these levels. In each case the authorized use should be levelled to include either full access, no delete, update only, and read only. To protect data in transmission there should be data encryption and decryption capabilities (data scrambling and unscrambling when in transmission) and terminal clearance.

Another important function of the extended DBMS is the ability to create an extracted new database file exactly as we would create a report. This file or database is then held either for the length of the session or held permanently. This is a must function with many uses, such as those mentioned in the discussion of other product databases—namely, extracting derived, mostly read-only files for uploads and downloads.

The Data Dictionary/Directory

Finally—last but definitely not least—is the DBMS dictionary or directory, shown as component J in Figure 14-2. Chapter 12 was devoted entirely to a discussion of the extended data dictionary. The dictionary we will discuss here is a subset of the extended dictionary concerned only with DBMS necessities. This is the directory of databases, tables (files), data groups (records), and data items (fields) on the database system. (Review Figure 10-1 if unsure of these name meanings.) This is also the directory of catalogued procedures or programs provided with the package and developed by the users. The roll-your-own DBMS will require a separate, stand-alone dictionary. The integrated DBMS is essentially integrated *around* the internal dictionary plus the executive.

This, then is the extended DBMS based on a primary database holding common organization data and tied to local DBMS satellites and local application software and data storage. The questions that we should keep in mind as we go over this material are: Is it possible for the organization to function competitively without such an extended, linked, software-supported DBMS? Is buying the DBMS in parts from many vendors better or worse than tying in with one vendor who has developed an integrated, extended DBMS with all the capabilities just covered? We will now look at the integrated DBMS alternative.

THE INTEGRATED DATABASE MANAGEMENT SYSTEM

Figure 14-3 shows the single-vendor integrated full-service DBMS. The illustration can be seen to have a more logical and stable appearance than the Rube Goldberg contraption shown in Figure 14-2. (Rube Goldberg was a cartoonist who specialized in drawing ingenious but funny contraptions.)

We must somewhat qualify the idea of the integrated DBMS as the product of one vendor. Some integrated DBMS packages have become popular enough to be supported by the products of other vendors, particularly with regard to utilities and even report/query interfaces. It is also good to keep in mind that most very large organizations that purchase the integrated DBMS have, up to the present, used this type of software as a satellite system to a nonintegrated information system that includes the extended DBMS as already described plus any number of other systems, including application-oriented batch-tape systems and free-standing smaller on-line systems based on mini- and microcomputer hardware. Most of these organizations have not got anything like the type of information-system integration we are discussing in either of these two examples.

It is not necessary to discuss again the generic DBMS components that make up the system. They remain the same in Figure 14-3 as in 14-2. The difference is the consistency and lack of interface confusion in the integrated DBMS. Please remember forever that it is in the system interfaces that most of the system problems occur. This is true in software as it is in hardware. It is not the integrated circuit

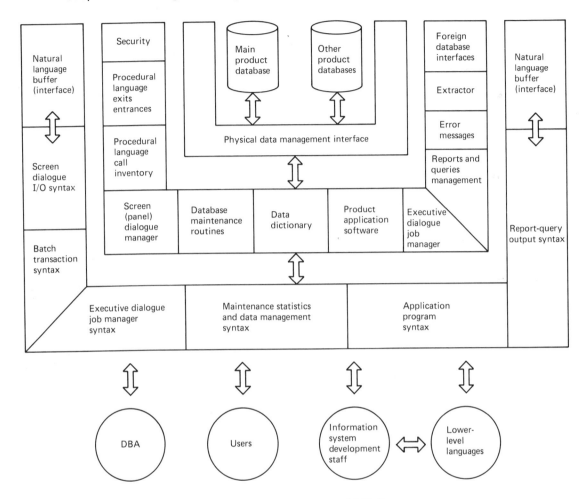

Figure 14-3 The Integrated DBMS.

that provides the main hardware development trouble; it is the connections between integrated circuits, both physical and logical. The same applies to software. Connecting all the components and maintaining zero-defect performance regarding logical and physical interfaces is the big software challenge.

How appealing it is, then, to have a reputable vendor provide a DBMS which combines in one package all the necessary system resources properly interfaced for efficient systems development. When the same vendor also offers applications packages written in the same language as the rest of the DBMS procedures, then we are really well served. If, on top of this, we find other vendors developing application packages compatible with the protocols of the integrated DBMS and easy to customize in the DBMS, then we can really knock off those huge application backlogs and proceed into the twenty-first century with a strengthened position.

Mainframe and PC hardware, other terminals, and data transmission systems all organized in support of the integrated, application-supported DBMS! Systems development revolutionized! Instead of developing disintegrated application-based systems in lower-level, obsolete languages, on each and every organization site, we get and use an integrated DBMS. We get and use application packages. When we have to program either to customize application packages or to develop highly personalized operational systems—which are often proprietary and provide a competitive edge—then we program in the DBMS fourth generation procedural or nonprocedural language and exit to second- or third-generation procedural languages only when compelled by process complexity to do so.

It has become clear that this is the way that system development must be done in the future. The development of thousands of application programs for the personal computer in just a few short years has shown what can be done. Some of these programs are very complex already and challenge existing software on the mainframe in terms of usability. The mainframe–PC connection is certainly a part of the integrated DBMS solution we are discussing.

Figure 14-4 gives a conceptual illustration of the integration of application software and data stores to the DBMS. It is important not only to integrate the application software but also to integrate the application data stores. When we acquire application software from the DBMS vendor, we get this data link as a given. The application internal data is also on the DBMS directory and available for all DBMS functions. Unfortunately, the same is not true for the independent-vendor data. Since it is unthinkable that the DBMS vendor could develop all the applications needed by the community of users, we have a serious standards problem, which tends to make application software packages less available to the DBMS as a whole.

Hence the first fly in the ointment. Let us review some of the problems involved when we tie the future of an organization to a one-vendor integrated

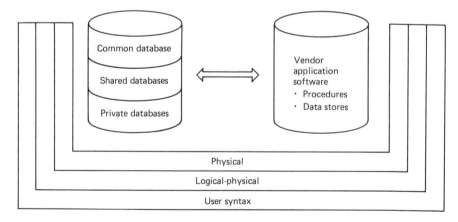

Figure 14-4   Integrating Application Software with the DBMS.

DBMS with perhaps some of the applications such as accounting or materials management being also a product of the same vendor.

1. It may be difficult to interface applications software from other vendors to the integrated DBMS.
2. The primary vendor may prove to be unreliable over time. This unreliability comes in several forms:

   - The DBMS vendor may not have the resources to keep up with the fast, changing business of developing information systems.
   - The inventive spark that made the vendor a front runner when the contracts were signed may die out, thus depriving the organization of a once-healthy competitive position. Remember, signing on with a DBMS vendor is not something we can back out of easily—not only are current systems maintenance and future systems development involved, but the documentation of the system and the training of the staff are tied up with the product.
   - The DBMS vendor will always be at the mercy of the mainframe vendor who controls the operating system—any time IBM wants to pull the plug, no one is "plug compatible" anymore
   - The archrival of the organization may decide to buy the DBMS vendor.
   - If some other vendor comes up with a great leap forward in DBMS component software, such as a really good applications generator for on-line systems, it will probably have to be passed by—thus weakening the position of the organization.
   - Requests for change to the integrated DBMS will go on the agenda of the users group and it may be years before they are acted upon—and then they may be rejected.
   - The vendor's prices will start to move up and up if the vendor is successful—there will likely be fewer possibilities to shop for good value.
   - The vendor may go out of business.
   - The vendor may become disappointing in terms of education and consulting support, even as the price increases.
   - The integrated DBMS is often a resource hog, usually running in the interpretive mode, unlike compiled COBOL and BAL type programs. Interpretive means that the source syntax is translated in-line into substituted blocks of machine code; among other unpleasant results, this may deteriorate response time.
   - The integrated DBMS may not have as good situation recovery performance as the stand-alone industrial chugger of a DBMS such as IBM's IMS—there is an increased risk of data loss.
   - Debugging facilities may be primitive compared to those which the programmer working in command-level COBOL is accustomed to. Good debugging

aids may not be available. Nonprocedural languages can be hell to debug, once past a certain level of complexity.

And the list could go on. It may be frustrating to the reader who wants a simple shopping list or scorecard for doing systems development to read through this book. There has not been a chapter so far where one solution or method can be stated as the best and only answer to the problem. For every list of yeas in information systems there is an equally impressive list of nays. All the author can say is, this is precisely the way it is in the real world of developing information systems. The important thing is to avoid making any really serious mistakes and to remember that nothing works as well as the vendor says it will (to which the vendors will reply that no user is as capable as the user claims to be). This is why the process of getting the software is so important. For any given software product this is where the worst mistake is made: inappropriate acquisition.

Figure 14-5 is a final review of the comparison between the mix-and-match

Criterion \ DBMS type	Extended DBMS	Integrated DBMS
Complexity	Higher	Lower
Achievability	High	Very high
Reliability	Lower	Higher
Development efficiency	Lower	Higher
For small system	Worst	Best
For large system	Best	Not best
Need for expert support	High	Low
Language sophistication	Lower	Higher
Cost	Higher	Lower
For decision support system	Not best	Best
For operational system	Best	Not best . . . yet

Figure 14-5 Comparison Between Extended and Integrated DBMS Development Criteria.

extended DBMS and the integrated DBMS. The following criteria are covered in this illustration for the two types:

1. *User complexity*. The extended DBMS is more complex.
2. *Achievability*. There is a high rate of achievability for the extended DBMS if (and only if) the proper experts are on board and don't jump ship—also a high rate of achievability for the integrated DBMS but not for every organization. In general terms, the larger the organization, the less wise it is to put all eggs in the one basket—the integrated DBMS is better used as a satellite manager and database.
3. *Reliability*. To the extent that the vendor is reliable and committed to research and development, the integrated DBMS is more reliable.
4. *Development efficiency*. The integrated DBMS wins hands down if the application is appropriately selected. As time goes on, more applications become possible on the integrated DBMS.
5. *Small system*. Extended is worst, integrated is best.
6. *Large system*. The extended could be best, but costs are high. The integrated DBMS could be worst when used as the primary DBMS, but the situation is brightening.
7. *Expert support needed*. Extended: absolutely. Integrated: desirable, but not a requirement if good vendor support is available.
8. *Language level*. Extended: It is easier to fall into trap of using lower-level languages at greatly increased costs. Integrated: It is easier to stay with a higher-level language when developing applications, thus providing cost avoidances which are often remarkable.
9. *Costs*. The extended DBMS is much more expensive over time. The integrated DBMS is much more expensive up front—but the cost savings begin to appear as soon as development work begins.
10. *Decision support systems*. The extended DBMS is a poor choice for this capture-and-retrieve type system. The integrated DBMS is an excellent choice for the decision support system (DSS).
11. *Operational system*. The extended DBMS is still the best choice. Use of the integrated DBMS in this environment is very tricky.

LOGICAL INTEGRATION OF APPLICATIONS SOFTWARE AND DATA STORES

By this heading we mean how the application programs and owned databases integrate with the systems analyst's logical flow diagrams. Figure 14-6 shows a logical specification depicting a high-level view of an organization. We see here in simple form the normal cycle of a product-oriented enterprise. Sales plans and schedules product development and manufacture. Materials management purchases and stores parts and finished goods. Manufacturing makes the products according to the sales plan and the engineering specifications. Customer service distributes the finished goods as ordered and handles customer-relations problems that are forthcoming. Sales and accounting reconcile the sales-plan forecast to the actual shipment and revenue. This is the flow of the enterprise shown in the flow diagrams in Figure 14-6.

In this case the organization has chosen to include on the common database the sales, customer service, and accounting information. It has also chosen to purchase application software for the materials management activities, which include material requirements planning, inventory receiving and tracking, and shop manufacture control. These application packages come with their own procedure and their own data stores. (Sometimes even the hardware is included.) The DBMS, whether extended or integrated, must handle the incorporation of these application packages into the organization scheme for on-line information management. It must be a requirement that the primary DBMS allow us to make this happen.

Also shown in Figure 14-6 is the requirement that both the common data *and*

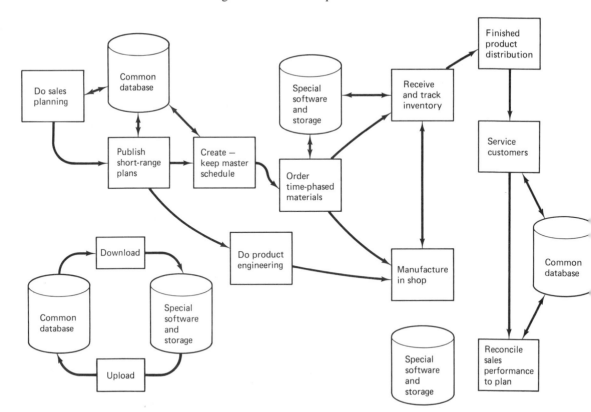

Figure 14-6   Logical Integration of Application Software and Data Stores.

the application-software-owned data be capable of being freely exchanged between all parts of the system. This is a must. Only on the assumption that this is somehow possible can we even consider purchasing application software products.

GETTING THE SOFTWARE: NARROWING DOWN THE SEARCH

The discussion on which we now embark applies equally to the selection of application software and DBMS components. Several logical steps are involved in getting software:

1. Writing down a set of detailed requirements for the software.

2. Researching the available software products and preparing a list of products that might meet the requirements.

3. Analyzing product descriptions using such general research sources as *Datapro*[3] and eliminating products that clearly wouldn't be appropriate. Causes for elimination might be such reasons as: not available for the organization's mainframe computer, serious doubts about vendor acceptance, or available only through time-sharing services.

4. Reducing the product list to the most likely candidates. In this step, of course, it would be convenient to end up with a list of one likely candidate and save a lot of time, energy, and money, but this doesn't often happen. Try narrowing down the list of word-processing programs available for a particular personal computer! With mainframe software there are not as many choices, but still we are likely to end up with several possibilities. We really do not want more than four candidates left from this selection step, and two would be better.

The way the list is reduced is mainly by telephone interviews with users. The other main strategy is to interview the vendor salespeople. Most of this can be done over the telephone. We ask the salespeople not only about their own product but also about the competition. We find out the most amazing negative detail this way. Salespeople are only too willing to expose the competition's dirty linen.

Concerning the vendor's own product we will ask for a copy of the operations manual and the general information manual—if we are really considering the product as still in the contest. Most vendors will gladly loan this material (and seldom ask for it back).

When interviewing, it is best to work from a simple form such as the following:

COMPANY:

CONTACT:

DATABASE:

FORMER PRODUCT USED:

---

[3] *Datapro—The EDP Buyer's Guide* (Delron, N.J.: Datapro Research Corp.).

REASON FOR CHANGE:
PRODUCTIVITY IMPROVEMENT RATIO:
TELEPHONE #:
REFERRED BY:
COMMENTS:

This is just an example. New systems analysts should get into the habit of not taking forms too seriously. It takes just a few minutes to design a new form that might be more relevant to the project at hand.

5. Scoring the remaining products against the requirements list. At this point the software evaluation becomes nearly a full-time concern for the analyst, who may be staff for a software architecture committee, may be on the committee, or may be the leader of a software selection team as part of an application development effort. (It is best, however, to select software as a tactical effort under the strategic plan rather than wait until the application demands purchase of the software.)

Scoring techniques vary but follow a similar basic pattern. We will examine here a typical scoring effort. Involved are four products—in this instance, DBMS component products. The same method, with different requirements, would be used to evaluate application software.

Case: Report/Query Package for an Extended DBMS

Figure 14-7 shows the requirements analysis study for a four-product comparison. The requirements are specified on the rows, and the products and the scoring results are shown in the columns. The "Note" column points to remarks of interest, which are not shown here. These contain such information as data-item name size, which can vary anywhere from seven to 32 characters. This is of interest when writing procedures using these names.

A requirement *weight* is usually assigned for such an analysis. In this case the weights are:

1 = not too important
2 = moderately important
3 = important
4 = very important
5 = vital

The particular requirement row is scored for each product. In this case a simple yes or no is given concerning whether the product supports the requirement at all. The *score* markers in Figure 14-7 are as follows:

0 = fail
1 = marginal

The software / The criterion	Note	Weight	Product A Yes/no	Product A Score	Product B Yes/no	Product B Score	Product C Yes/no	Product C Score	Product D Yes/no	Product D Score

1. IMS interface
  a. Uses primary DB, not extract
  Etc.
	1	5	Yes
			:
			:
				4
				:
				:
					No
					:
					:
						0
						:
						:
							Yes
							:
							:
								4
								:
								:
									Yes
									:
									:
										4
										:
										:

2. CICS interface
  a. Does queries under CICS
  Etc.
	4	4	Yes	1	Yes	1	Yes	3	Yes	4

3. Vendor support
  a. Quality of consultation
  Etc.
		3	Yes	2	Yes	2	Yes	4	Yes	4

4. Product efficiency
  a. Reports from multiple DBs
  Etc.
	7	5	Yes	2	Yes	2	Yes	3	Yes	4

5. Cost factors
  a. Cost of necessary package
  Etc.
	9	4	150K	2	75K	4	130K	2	200K	2

6. Reporting requirements
  a. Text reporting
  Etc.
	14	5	Yes	2	Yes	2	Yes	3	Yes	4

Overall score:
  1. IMS interface
  2. CICS interface
  3. Vendor support
  4. Product efficiency
  5. Cost factors
  6. Reporting requirements
				78
				27
				58
				81
				46
				213
						78
						27
						58
						69
						58
						192
								93
								65
								73
								129
								70
								271
										103
										103
										80
										139
										54
										356

Total score				503		482		701		840
Percent to winner				0.60		0.57		0.83		1.00

Weight legend:
1 = not too important
2 = moderately important
3 = important
4 = very important
5 = vital

Score legend:
0 = fail
1 = marginal
2 = average
3 = good
4 = excellent

Figure 14-7 Software Selection Requirements Analysis.

> 2 = average
> 3 = good
> 4 = excellent

In order to do the scoring, we take two approaches. One approach is to multiply the weight by the score for each requirement row and then to total the weighted score by requirement grouping and by grand total. For ease of comparison we then also show the grand totals as percentages of the winner, which is always 100%. Then we can say, as in this example, that product A is 60% as good as product D, and so on.

The other approach is to see where a given product falls below an acceptable requirement score in single instances that render the product impossible to use successfully. The most obvious example is when the requirement is "vital" and the score is zero. We either compromise on the requirement or scratch off the contending product. In the case shown here two products (one the total-score winner) are scored as marginal, and the requirement in question—cost of package—is weighted as very important.

How do we learn enough about these products to rate them on such specific requirements? This requires considerable skill and experience on the part of the analyst. The methods used to obtain the necessary product knowledge are a continuation of telephone interviewing of users. Usually your list of users was supplied by the vendors, and a useful but biased list results. It is wise to ask each user on the original list to provide the name of another user. This way we get a list not dry-cleaned by the vendor and obtain a more critical perspective. In addition to vendor-supplied user names it is an advantage for the analyst to have other contacts, such as through professional societies or hardware vendors or colleagues at local-area users, to broaden the interview base.

Besides conducting telephone interviews, it is often necessary to select a user who might be visited so that the product can be demonstrated and discussed at first hand. It pays to live in an area where there are many such choices nearby. If not, some travel will be involved. The interview techniques discussed in Chapter 13 are quite useful when applied in this context.

The documentation research techniques already covered in Chapter 13 are also applicable. In this instance the documentation consists of the manuals loaned by the vendor. Particularly when the field is narrowed down to two choices, it becomes essential to read and understand the operating manual for the product. As can be imagined, this is highly skilled work, since the objective—understanding the product at the deep level—must be accomplished in a short space of time. Like riding a horse or eating a lobster, this is a skill that can be acquired. Blessings on the vendor that provides just enough, clear, simple documentation. Such a vendor deserves extra consideration.

During this time of research the analyst will be seeing the salesperson and speaking to him or her on the phone constantly. A game is involved wherein the vendor will try to play down the negative features of the product and the analyst will try to get the truth. Sad to say, some salespeople are under such pressure to make sales that they will dispense misleading and wrong information. The good salespeople resist this pressure and just tell it like it is, particularly if their product is a class act.

GETTING THE SOFTWARE: MAKING THE FINAL DECISION

Having narrowed down our choice of software to one or two candidates, we are now going to sign a provisional contract with the vendors involved, allowing us to test the product in-house and then either let the purchase go through or else cancel the contract, pay a small use fee, and return the product to the vendor. There are all sorts of variations on such contracts, varying from an unlimited use of the product during testing without a written agreement all the way to a contract that limits testing to 30 days and requires that testing conditions be specified.

Why are we going to such trouble? Perhaps we are spending more money on testing the product than the product itself costs. Our motivation often is not so much the cost of the software but the cost over time of using a package which sets back the whole-system development effort. Using the wrong software is a main reason for system-development failure—along with entering a new technology without the expert knowledge to handle the change. Both go together. Marshalling resources to struggle seriously with the process of getting and testing software is always a wise move. We learn as much about the new technology as we learn about the specific product.

Take the case of the organization that is going from a batch tape system to an on-line oriented DBMS. That is not easy to do! Testing specific products is one path to force us to confront the whole of a new technology.

Testing

We will be testing those requirements that the previous phases of product analysis already discussed have not resolved to our satisfaction. Can *we* do what we are told can be done?

The other type of testing we will have to do is performance analysis. If we are testing two software products that are to do the same job, we will provide procedures to demonstrate job performance, provide monitoring software to show the performance statistics, and let the horse race proceed head-to-head, may the best product win (unless the race is fixed, which has been known to happen). If only one product is under test, it is good to have a control product against which to compare it. This is not always possible.

The quantitative analysis we are usually looking for here includes such factors as core-resident usage, CPU (central processing unit) elapsed and functional time, and number of physical I/Os and logical database calls. Every computer shop has or should have such software monitor packages. An especially good analytic tool is the logical call trace. It goes without needing much comment that, when we are dealing with packages that are involved with or affect databases, a good DBA is worth plenty.

The object of the testing is to subject the software to sufficiently stressful conditions so that if breakdowns in efficiency are going to occur later, they will occur now. It is also wise to be prepared for surprises which were not even conceived of in the test planning. We should be prepared to pick up on these heuristic opportunities and learn something about the software and the technology.

Regarding the testing procedures—it is not usually necessary to write complicated programs to achieve the desired test. Usually extremely simple programs will suffice.

One problem with constructing a good test involves the need to convert data from a production data store to a test data store. If the test involves the database itself, it will be necessary to learn enough to perform a simple data conversion. This is not so simple in practice. It is worthwhile. The DBA will learn a significant amount about the product during the testing phase—and so will the analyst.

Presenting the Analysis Results

We now write it all down in a project report. First we tell them what we will tell them, then we tell them, and then we tell them what we told them. This little dictum is not a bad summary of the methodology involved in making such reports and presentations. (Usually it is necessary to make a report *and* a presentation, therefore it is good to include good graphics along with the report text. This is a subject we have discussed before concerning the system specification.)

The content of the report consists of:

1. the original selection and elimination results
2. the interviews, both in person and on the telephone, by which the product selection was further narrowed
3. the results of the document analysis that went on at the same time
4. the final scoring report, indicating the products recommended for in-house testing
5. the results of the in-house testing
6. the recommendation concerning a course of action
7. appendices covering the technical material

It is important to retain sufficient detail, at least in the appendices, so as not to lose the knowledge gained about the product and the technology during the testing. This knowledge by itself is worth the costs involved in doing this type of testing.

Let us repeat for emphasis that the process of getting the software has told us much about how to use the software.

USING THE SOFTWARE

The analyst, who has now gone to the considerable trouble of getting to know the software and the software vendor, will usually have to make some arrangements to devolve from being the in-house expert. No good analyst wants to be stuck with being the software expert during implementation. To some extent we strayed out of our basic territory to do this task (not an unusual happening—who else is better qualified to do software analysis?).

In order to devolve (get out from under), the analyst will have to introduce the product to such functions as the training department, systems programming, and the application teams who will be using the product to teach, design, and implement software solutions to satisfy new system specifications. This is best done by writing an operations manual, which will serve as a companion piece to the vendor's own manual and training program. This task can be very simple and brief or it can be time- and energy-consuming; it all depends on the quality of vendor support. At the very least the original report which recommended the product, along with further addenda, should be widely distributed. The analyst may also need to give presentations and demonstrations.

Customizing Application Software

The special problem with application software, which had to come up during the testing, is how to best fit the package into the system development architecture that includes the organization's primary databases. Data must flow to and from the application. With some packages this is easy; in other cases it is a major problem. On the level of the personal computer (which many people can now relate to) it could be a necessity that outputs from spreadsheet and graphics software be available for word-processing tasks. Especially in the case of the extended rather than the integrated DBMS software solution this may turn out to be impossible because of a lack of standards. We will reluctantly leave this problem hanging. Sometimes good answers are not to be found at the present time.

The other equally thorny problem with application software is the need to customize the software internals. If we insist on changing the vendor's source program logic, the vendor cannot support the package. If we do not change the vendor's software internals, we must be sure the vendor has given us the type of exits and entry points to the vendor package that we require. In these exits we will call those subroutines that will customize the vendor software and change it from the inappropriate solution to our problem to the appropriate solution. The anecdotal wisdom in the trade has it that for every dollar of software delivered price from the vendor the user organization will spend an equivalent dollar before the package is truly satisfactory. By this rule of thumb the $150,000 inventory tracking system is really going to cost $300,000 (which still beats spending $1,000,000 for in-house development, particularly if the product has been around for a while and is suitably housebroken and debugged).

Whatever the software involved, a pleasant happening awaits the wandering analyst. In no time at all the people who are going to do development work with the software product will know considerably more about it than said analyst, who by that time is searching other horizons.

15 *ESTIMATING BENEFITS, COSTS, AND TIME*

DEFINITION AND PURPOSE OF COST-BENEFIT ANALYSIS

An organization usually has more contending investment options than it has resources to cover. The most important senior management decision making involves the acquisition, development, and committing of resources to achieve those benefits which advance the organization's objectives. Selecting the best projects to realize these objectives is often extraordinarily difficult.

Information systems development, the *raison d'être* of the systems analyst, is only one of the possible investment choices. It is always vexing to the managing directors to have to deal with the cost-benefit analysis associated with developing information systems because of the difficulty getting a handle on a proposed system's real costs and benefits as they will actually be in production. The cost-benefit analysis that came out of the systems analysis process is often far from the actual costs and benefits. Typically, the costs (and time-in-process) are underestimated— and the ultimate benefits, surprisingly, also are understated.

Often the senior managers will give up in disgust on a good information-system project because of the uncertainties involved. It is far easier, and the risk is far lower, to invest in something like the construction of a new plant facility, where the costs and development time may be much more reliable. The facility may cost $50 per square foot in one place and $10 a square foot in another place.

Other major costs of the facility, such as electricity, local taxes, and telephone, are also relatively solid numbers. Contrast this to the difficulties usually involved with the so-called intangibles in information-system development.

Tangible cost-benefit means that a monetary value can be assigned to the cost or the benefit. Conversely, *intangible* describes something to which a monetary value cannot be assigned. Usually we try to develop some index easily converted to a monetary value, such as the number of lines of delivered source code or computer CPU time. If we cannot develop such an index, we cannot develop a tangible cost or benefit. We will devote some time in this chapter to understanding some simple methods by which an intangible cost or benefit can become tangible. Of course, there will always be intangible benefits, which is one of the nice things about well-supported natural systems.

From the point of view of the general systems theorist (and at this point let us confess that it is impossible to do first-class systems analysis without being a general systems theorist) the problem that vexes the senior managers is to some extent the generic problem of the difference between a natural system and a device. A plant facility, no matter how fancy, is a *device* with a finite number of parts and part relationships in the small-number range and therefore subject to not unreasonable quantitative analysis. A human social organization of even moderate size has a much more complex arrangement of "moving parts." In particular, the relationships between people are significantly more complex. Consider the limited relationships in a device with three moving parts as compared to a household with a mother, father, and one child. Consider the potential complexity of relationships in this human family organization.

This is not to say we cannot arrive at very usable estimates for the development and use of a system within a social organization. We can. But we will have to use somewhat different estimating methods than those employed to evaluate the investment in a corporate jet airplane.

Our definition of cost-benefit analysis will be:

That method by which we find and estimate the value of the gross benefits of a new system specification.

That method by which we find and determine the increased operating costs associated with the above-mentioned gross benefits.

The subtraction of these operating costs from the associated gross benefits to arrive at net benefits.

That method by which we find and estimate the monetary value of the development costs that produce the above-mentioned benefits.

Those methods by which we show the time-phased relationship between net benefits and development costs as they relate to cash flow, payback on investment, and time-in-process taking (or not taking) into consideration factors such as inflation and the tax code. In short, the calculation of actual net benefit as cash flowback over time.

THE TWO LEVELS AT WHICH COST-BENEFIT ANALYSIS IS CONDUCTED

This book has dealt with systems analysis as the specification of current and new systems on two levels. One level is the specification of the whole system as the organization strategic plan, which was the subject matter for Chapter 11. Several earlier chapters covered the methodology for specifying systems at the application level. The strategic plan was shown to produce as an output a phased schedule for application development. The case was made that the same methodology used to develop application-level new system specifications was usable for the development of the strategic plan. The difference was that the charting for applications went down to the base process level and the charting upon which the strategic plan matrix charts were based went down only to the second or third level of system decomposition.

This approach also defines the two levels of cost-benefit analysis. Cost-benefit must be done at these two levels. Certain investments must be made prior to the time for application development. Such investments as the physical support system, particularly mainframe computers and their operating software and data transmission systems, must be planned for, and installed, well in advance of applications development.

Also certain costs involved with the management of common data cannot properly be assigned to the earliest applications. These costs must be separated out and assigned to corporate accounts, whether as expensed (assigned) items or long-term capital investments. They are both one-time development costs and operating costs. They may occur at the time of strategic, long-range planning or during applications development. They will occur at these two different points in time, no matter which data management approach (ranging from application data files to common database) is adopted by the organization.

As might be expected, the cost-benefit methodology will not differ when applied to these two levels of investment planning. What will differ is the risk or confidence factor, since the finer we can break down the requirements that come out of the specification the closer we can estimate the costs. This also makes it easier to conceptualize the benefits.

The final introductory words we need say about cost-benefit analysis over time is that at the application level it is an iterative procedure, to be redone at a minimum at design and implementation time. Development costs increase rapidly with changes in benefit requirements as we get into design and implementation. This is not surprising. In device cost-benefit, there would certainly be a radical effect on costs if, in the middle of the construction of a residential house, a requirement was added to put fireplaces in every bedroom. This emphasizes again the need for good enough strategic planning so that reasonable cost-benefit analysis can be put into place at the earliest possible moment. Meanwhile, the actual investment can sometimes be postponed until the latest possible moment.

THE QUALITY OF INFORMATION

We need to develop a feeling for the *quality* of information before getting into a detailed discussion of identifying and computing benefits and costs. We are indebted to James Emery, who uses this term in his classic paper "Cost/Benefit Analysis of Information Systems,"[1] for giving us a simple terminology that covers many factors in benefit analysis. Emery suggests that, instead of thinking of information as having many attributes—such as accuracy, timeliness, clarity, relevancy, succinctness, and so on—we identify all these attributes as one term: quality—the quality of information. This helps simplify the discussion by giving us one index.

Much more important and radical, it identifies in one name the best reason for developing information systems in the future. Some organizational managements still think of information-system development only in terms of cost avoidance (getting rid of troops through automation). This is not a well-thought-out point of view.

We will be discussing information quality throughout this chapter.

SOURCES FOR BENEFITS—SOURCES FOR COSTS

Models

The availability of decision models greatly enhances the search for quality benefits. If we are analyzing an operation involving user-terminal dialogues as integral to its process flow and can see what a quality improvement in response time would do to the rest of the operation, we can be reasonably assured of being able to calculate a tangible benefit. (Remember, we have said that the difference between a tangible and an intangible benefit is whether a monetary value can be assigned to the benefit.)

One of the important uses of the decision model is to test the *sensitivity* of that which we wish to predict to variance in the predictors. Some benefits are highly sensitive to changes in the quality of a predictor, and some predictors have little if any effect upon the benefits. It may be that itemizing a bill has little to do with

[1] James J. Emery, *Cost/Benefit Analysis of Information Systems* (The Society for Management Information Systems, 1971), pp. 16–46. As reprinted in J. D. Couger, Mel A. Colter, and Robert W. Knapp, *Advanced System Development/Feasibility Techniques* (New York: John Wiley & Sons, 1982), pp. 459–486.

customer satisfaction with product but that quality of product is a very sensitive predictor of satisfaction. If this is true, we might even throw bill itemization out of the model and concentrate on tuning product quality to the maximum tradeoff point. Such is the potential benefit of the decision model.

Quick Nonparametric Modeling

When a formal model is not in place, it is still possible to construct an informal model based on assumptions that have a certain probability of usefulness. These quick models are very important, owing to the lack of available formal models. An example of such a model will be given later in this chapter.

The Experienced User

No one knows better than the user the benefits of a particular requirement. That is what users are paid to know. When actually assigning benefits, we must rely mainly on the expertise of such an operations or planning manager. Our research can help organize and clarify their calculations, and our experience with like systems can act as a check, but, no matter what other support information is available, they must assign the benefit.

Historical Information About Other Costs and Benefits

It is vital for information-systems specialists to keep accurate P&Ls on past projects—what was forecast and what was the actual cost-benefit. This must be done in detail. It is possible to add up several costs but not possible to break down an actual cost that covers several items. Historical information cannot be used in raw form, but it is one of the best inputs into cost-benefit analysis.

The Experienced Technical Expert

As the user was the best source for the monetary value of a benefit, so the technical expert is the best source for the costs involved. Much of the time the senior systems analyst working at either the strategic-plan level or the application-development level is also the expert, but it is extremely unlikely that this is always true, even for a team of senior analysts. Particularly in the area of physical-system support planning the analyst must rely on such experts as the data communications specialist. Calculating all the costs of a new local-area network is not a job for a logical system specifier.

TOPICS TO BE COVERED

The rest of the chapter covers the following topics:

- identifying benefits
- estimating gross benefits
- risk assessment
- identifying operating costs
- arriving at net benefits
- identifying development costs
- arriving at the cost-benefit relationship
- methodology in support of the investment decision

IDENTIFYING BENEFITS

The main problem with doing systems analysis in this area is not in applying such a methodology as cost-benefit payback analysis or calculating the time-in-process to develop the system; rather, it is identifying and specifying benefits in such a way as to make the benefit come alive as reasonable.

Finding benefits, like most everything else in information systems analysis, is based on the system specification. So, for the nth time: We specify the current logical system at whatever scope is necessary. We develop process and data requirements for a new logical system. And we use the net requirements to specify the new logical system as it has evolved from the old current system. These net requirements (net requirements are those included in the new system alternative selected by the user out of the gross requirements first identified) each have an associated benefit. The more succinct the requirement, the more precise the benefit.

Regarding costs—it is difficult to assign an operating cost to a requirement, and it may be necessary to work at the application-system level into which the requirements are folded. Development costs can always be calculated on the system level and often at the requirements level.

TYPICAL SPIRACIS BENEFITS

We introduced the acronym SPIRACIS in an earlier chapter to describe categories of benefits. Let us use this term now to decompose "benefits" to a lower level of detail. SPIRACIS stands for:

SP—strengthen position
IR—increase revenue

AC—avoid costs
IS—improve service

Some people use the term "profit improvement" in the same context as SPIRACIS, but this is confusing, because profit improvement applies to only one type of organization and is an *effect* of SPIRACIS, not an equivalent cause.

Figure 15-1 shows typical benefits. Benefits must be identified at this level of detail or lower to be made tangible. Let us cover a few of these benefits so that we can get a feel for the different categories and the possible quality levels involved.

Strengthen Position

We covered some *strengthen position* (*SP*) benefits in the chapter on strategic planning. The tellerless banking system discussed there is an example of an automated customer-terminal dialog. A requirement like this would include such more detailed requirements as (1) the ability to automatically update employee payroll deposits from the employer's on-line payroll database and (2) the availability of money for withdrawal the same day as the normal payroll day.

This type of benefit has the primary thrust of providing a market advantage over competitors. It can far outweigh *avoiding costs* benefits in investment payback. As an example, military weapons systems are increasingly dependent on computer software and are the ultimate expression of *strengthen position* in its most threatening form.

For any organization an improvement in the *quality* of the information about a customer or a client or the information about the demographics of its locale greatly increases the chances of selling a customer a product or providing a client with the proper services. For instance, knowing that the potential customer owns a personal computer makes it possible to target that customer as a sales prospect for PC software or a subscription to a magazine about PCs. The requirement to establish a prospect file which includes the update of known purchase history both *strengthens position* and *increases revenue*.

The question then becomes: At what quality level do we need this information about prospects? For instance do we need to update the file with information that increases the deliverability of a promotion? (Which is to say, does the prospect really still live at this address or have this telephone number? Indeed, is the prospect still a live one?) What is the creditworthiness of the prospect? Along with his history of product purchasing, what is his history of payment? These requirements can go on and on, each improving the quality of the information, until we reach a point where the improvement costs too much for what we get.

Second to onboard military-system software, which allows us to blow up our competitors (hoping all the time their computers don't blow us up in response), perhaps the most interesting SP enhancement is the simulation of competition scenarios. At least it is the most fun. The use of computers for such model building

SP	IR	AC	IS
Automated customer-terminal dialogues	Faster sales-planning reaction time to meet opportunities and situations	Avoid, reduce or reroute clerical operations	Document accuracy
Customer on-site terminal installation	Accurate product testing	Avoid, reduce or reroute lower management operations	File accuracy
Customer choices expanded by inventory-control quality improvement	Accurate market selection	Reduce need for facilities	Faster transaction time-in-process (orders, payments, etc.)
Elimination of product shortages	Accurate, past financial forecasting and reforecasting	Avoid and reduce harmfully redundant data	Faster output time-in-process (promotions, bills)
Maximize price-change	Accurate, fast reconciliation of forecast vs actual product P&L's	Reduce or avoid work duplication	Faster complaint handling
Detailed, current customer profiles		Early error detection	Improved customer service
Detailed, current area demographics	Accurate, fast analysis of manufacturing and other operating costs pegged to sales plans and products	Automate routine information handling	Improved statistics on service operation
Sensitivity testing through financial/marketing models		Automate complex information handling	Better customer/client communications
Ability to store and analyze competitor information	Comprehensive on-line information on department operations showing workload vs. capacities	Reduce paper use	Better problem checking and warning systems
Simulation of competition scenarios		Automate organizational communications	Better information quality
Integration of onboard microcomputers with product line	Availability of external information such as government and stock market information — related to internal information	Avoid inventory overstocks	Improved validation of source data at data entry
Ability to add new products without delays		Avoid inventory shortages	On-board micro-computer diagnosis of product performance
Ability to change sales strategies without delays	Ability to relate all internal common, shared and private data for extended reporting	Avoid crisis ordering	Long-range, short-range capacity planning
		Know more about employee skills	
	Ability to simulate market strategies	Provide accurate information about employee benefits	More control over material and manufacturing requirements
	Better field support information and data transmission for sales/service staff support	Optimize vendor liaison	More control over schedule slippage
		Get more truthful information about operations	Improved communication between planning and operations management
	More complex scheduling of sales plans; increased promotions and customer group fragmentation		

SP = Strengthen Position
IR = Increase Revenue
AC = Avoid Costs
IS = Improve Service

Figure 15-1 Typical SPIRACIS Benefits.

will surely increase in the future. Hopefully, it will be a reasonably friendly competition.

Increase Revenue

Such sales-oriented subjects as accurate product, market, and advertising testing are good examples of *increase revenue* (*IR*) benefits. The selection of samples to be tested in advance of the full commitment of major expenses to support full scale sales promotions is a good computer application. So is the quality of the statistics involved with comparing the potential gain to the actual orders received. A difference of 3% between a control group and a test group may be an opportunity for substantial revenue increases, or, if the quality of the statistical information is not viewed with a high level of confidence, it may be inconsequential. A large market-oriented organization usually has many sample tests going on every year. The profitability of the organization often depends on the quality of the sales support information system. A benefit in *increase revenue* is usually not too hard to forecast in terms of a tangible monetary value.

Other tangible benefits that can be added to a sales support system include improving the quality of information regarding the latest manufacturing and operating costs relative to a product or product line which aids in computing a better P&L. Also, faster reconciliation of P&L forecasts versus actual can improve the profitability of a promotion while still in progress—first, by picking early winners and concentrating the rest of the effort on the winners, and second, by adjusting operating and manufacturing costs according to the early sales results. These costs, such as ordering raw materials, must be expensed in advance of the actual sales promotion of finished goods. Early trends can mean a cutback or increase of costs in real time if spotted by a high-quality information system.

Other important examples of *increase revenue* as shown in Figure 15-1 include market-strategy simulation through the use of automated decision models, better field support information and transaction transmission in support of the sales and service staff, and more complex scheduling of sales plans (and more complex market fragmentation).

We see again that the big benefits of new requirements built into new information systems fall in the area of better-quality information as well as that of cost avoidance. This does not detract from the importance of cost avoidance—especially since a megabuck saving in costs has more tangible benefit than a megabuck of increased revenue, from which we must subtract a host of manufacturing and other operating costs as well as taxes. Which leads us to examples of *avoid costs* (*AC*) benefits.

Avoid Costs

We have just described why avoiding costs is so important, since a dollar of avoided cost is worth so much more than a dollar of revenue (unless we are so good at avoiding costs that we have no revenue). Avoiding costs (by which we mean reducing, eliminating, and not incurring costs) was usually the first benefit that planning and operations management thought of in the past when contemplating a new information system. Historically, the development of many of the early computerized information systems was based on the benefits of avoiding and reducing the employment of clerical personnel and the use of the facilities and equipment needed to support this personnel. As it turned out, most of these benefits were in the area of avoiding additional hiring.

Other examples of *avoid costs* as shown in Figure 15-1 include avoiding redundant data stores, avoiding duplication of work, reducing paper handling, reducing the use of professional staff to push information around rather than create or interpret information, avoiding inventory overstocks and shortages, and getting more truthful or reliable information. The benefit area of *avoid costs* is subject to good quantification into tangible cost savings expressed as a monetary value with a high confidence level.

Improve Service

Improvement of service is becoming an increasingly more important area of benefit. Consumers demand good service with a low aggravation quotient. People who see themselves as spending increasingly more for medical service, a college education, or a new car demand a proportional increase in the quality of service. *Improve service* is the most difficult benefit category to assign a tangible monetary improvement to. Yet we all know that a profit making organization cannot long remain in business without providing a high standard of service. Good service, like just about every other human activity, is based on good information.

With regard to organizations that are not for profit, such as government agencies, we know how resentful the citizenry is concerning poor service. It may be that the most damaging criticism of high government management, equal in distaste to the abuse of the rights of the individual, is the poor level of service such an organization may deliver. On the other hand, superior quality of government service would probably have the effect of making government management more attractive. Here we have raised the level of analysis to that of whole state or nation systems and confronted the fact that high quality information systems have a political value of exceeding importance.

Examples of *improve service* benefits include faster time-in-process of such customer/client transactions as orders, bills, payments, complaints, and adjustments. Other examples include

- better itemization of the transaction contents
- improved validation of source documents at data entry
- information-system real-time problem diagnosis and automatic correction or timely warning
- generally higher-quality information in all areas of the organization

• long-range capacity planning with a higher degree of confidence.

These, then, are the four categories of SPIRACIS benefits which are offered here. The reader is encouraged to add new categories, if such are to be found, or to decompose these categories to a lower, more detailed level. Decomposition to lower levels is the main game in systems analysis, as this book has tried to demonstrate in many chapters about different methodologies. Within reasonable limits, the more the benefits are particularized the better will be the results of the estimating process.

We have dwelled on the process of identifying benefits for two reasons—first, because the benefit and the benefit quality level of a new requirement determine the difference between gross and net requirements and thus determine the content of the new system, and second, because the systems analyst must be a partner with the user in determining benefit. The analyst is a limited rather than a general partner, since it is the user's responsibility to quantify the benefit of a specific requirement. But the analyst can make all the difference in the world, particularly in setting quality levels for the benefit—especially if the user lacks, and the analyst has, experience and vision.

Identifying benefits is, to many analysts, the most creative and rewarding step in cost/benefit analysis.

ESTIMATING GROSS BENEFIT

Figure 15-2 shows a benefits spreadsheet for a system based on the SPIRACIS categories. A new system contains a detailed set of new requirements. This information is available from the systems analysis that produced the new system specification. As an output of that analysis, it would be wise to decompose the requirement to SPIRACUS-category benefits. In Figure 15-2 we list each requirement by benefit group and give two pieces of information: value and probability. Then, by simple spreadsheet arithmetic, we can show a total gross benefit for each category, adjusted by probability of occurrence. This can be cross-footed to arrive at a total gross benefit for the requirement. The sum of total gross requirement benefits is the basis for arriving at the total gross benefit of the system.

What is somewhat complicated is the assignment of value in each of the requirement SPIRACIS category boxes. Figure 15-3 shows the role played by the probability factor in determining benefit value. This method of determining the expected results of a benefit calculation involving multiple probabilities is based on the work of the eighteenth-century mathematician Thomas Bayes and is known as Bayesian analysis.[2]

---

[2] J. Daniel Couger, *The Benefit Side of Cost Benefit Analysis,* Part II, *Data Processing Management Portfolio 1-01-08* (Auerbach Publishers, Inc., 1975), pp. 1–13. As reprinted in Couger, Colter, and Knapp, *Advanced System Development/Feasibility Techniques,* pp. 489–499.

SYSTEM NAME: SALES MARKETING

SPIRACIS type / System new requirements	SP Strengthen Position	IR Increase Revenue	AC Avoid Costs	IS Improve Service	Total gross benefit
Description of requirement 1	Value / Probability	Value / Probability	Value / Probability	Value / Probability	Value
Store and analyze information about competition	$2 million / 0.20			$50,000 / 0.30	$415,000
Product testing for winners	$2 million / 0.20	$4 million / 0.80	$100,000 / 0.90	$200,000 / 0.50	$3,790,000
Etc.					
Etc.					
Total benefits for the: Sales marketing system					

Figure 15-2   SPIRACIS Benefits Spreadsheet for a System.

Figure 15-3 shows a more complicated box calculation than could occur in Figure 15-2. Shown here is a requirement to program an onboard information system on a microcomputer to be installed in a luxury car. This, of course, might properly be leveled down into smaller requirements, such as program subsystem performance warning information or trip statistics. In the example shown, the analyst might have asked the user what rate of sales improvement was most reasonable to expect and gotten the answer .5% on projected car revenues of $1 billion, yielding an expected return on this benefit of $5 million. By assigning probabilities as shown in Figure 15-3, we come up with four rates, giving four dollar amounts subject to adjustment by four probability rates. These four expected returns can be added together, giving, in this case, an expected return or value of $9,350,000. If the user's hunches are right, this will give a truer picture of benefit than the estimate of $5 million. This example points out the truth, already men-

Increase Revenue: Based on $1 billion sales revenue

Requirement: Install onboard information system in automobile using microcomputer

Estimated gross sales increase	Dollar amount	Probability of occurrence	Expected dollar return
0.5	5,000,000	0.75	3,750,000
1.0	10,000,000	0.30	3,000,000
2.0	20,000,000	0.10	2,000,000
3.0	30,000,000	0.02	600,000
Total			9,350,000

Figure 15-3   Multiple Benefit Probability Calculation.

tioned, that benefits are often underestimated. It also points out how important the systems analyst's role can be.

After the gross benefit has been worked out at the requirements level, the next step is to synthesize the requirements to the systems level. Often one must work out with the user several alternative systems from less expensive to most expensive—which is to say, from fewer packaged requirements to the optimum available requirements that can be obtained through better-quality information. Usually two or three alternatives suffice. In the example shown in Figure 15-4 five alternatives are shown for a direct mail marketing system, ranging from the quick fix to the maximum realistic system with all the bells and whistles.

In the example in Figure 15-4 there exists a customer file for marketing pulls. A quick fix that would speed up by two days the pull of the names for promotion would provide more up-to-date names and addresses and improve the benefits by some percent. Beyond routine name selection and pull we get into such enhancements as sample-testing support—multiple-regression analysis where formulas are developed from sample testing to establish what customer attributes best predict pull for a particular product. In this way we can mail fewer names and still get equivalent results to mailing all the names. We can also get into decision modelling for long-range sales planning of product lines. All these might be considered stepped-up versions of the basic sales-information system.

A complete system benefits summary would be much more detailed than this example, which shows highlights only. In a presentation to senior managers, highlights are all we have time to show.

System: Direct mail marketing

Requirement add-ons / System version	Quick fix	Minimum effective new system	Average effective new system	Maximum "realistic" new system	All possible bells and whistles added
Version 1	Reduce "time in process" by two days				
Version 2		*Plus* Allow name selection by sales-controlled external tables — one week notice			
Version 3			*Plus* Full sample testing support		
Version 4				*Plus* Multiple regression analysis	
Version 5					*Plus* Decision model sensitivity testing
					10-Year marketing model by line and product

Figure 15-4 Investment Choices.

RISK ASSESSMENT

The three most important factors in risk assessment are the size of the system to be developed, the complexity of the system, and the skill level of the various groups that make up the development team. Many other, lesser risks might be mentioned, such as system obsolescence. On the basis of our assessment of these factors we will adjust the benefits and the costs involved in the development work.

Figure 15-5 shows a nonparametric assessment of the risk level when we are looking only at the size and complexity of system.

Figure 15-6 shows that a qualified development team has an excellent chance of developing a small, simple system and that, correspondingly, our cost-benefit analysis has an excellent chance of being accurate. On the other hand, the chances of an average qualified development team's accomplishing a large, complex system without encountering serious problems such as gross overruns are only fair, and the cost-benefit has become unreliable.

Figure 15-6 shows that the way to lessen the risks of developing large, complex information systems is to have a better-than-average development team. The team that consists of expert analysts, expert designers, and at least average programmers has at least a good chance of developing a successful system within reasonable cost-benefit forecast boundaries. This assessment of the substantial importance of team and team-member quality has been established by a number of major studies.[3]

The critical factor in team performance is, in the author's opinion, the quality of the systems analysis. Systems analysis front-loads every other development task. Excellent systems analysis avoids the big disasters that cannot be overcome by good design or programming. The worst thing that can happen to a systems effort is to solve the wrong problem. The next worst thing is not to solve the right problem. In either case, once the analyst has set the course, the designer is helpless. Tom DeMarco regards the main rule in systems analysis as avoiding such disasters:

SYSTEM RISK:

Size \ Complexity	Simple	Average	Complex
Small	Excellent	Very good	Good
Medium	Very good	Good	Fair
Large	Good	Good	Fair

Figure 15-5 Risk Assessment: Size and Complexity.

[3] Barry W. Boehm, *Software Engineering Economics* (Englewood Cliffs, N.J.: Prentice-Hall, Inc., 1981), pp. 426–428.

SYSTEM: _____

Team expertise / Project scope success rating*	Analysis			Design			Implementation		
	Expert	Average	Beginner	Expert	Average	Beginner	Expert	Average	Beginner
Excellent	Excellent	Good	Fair	Excellent	Very good	Good	Excellent	Excellent	Very good
Very good	Excellent	Good	Fair	Excellent	Good	Fair	Excellent	Very good	Good
Good	Excellent	Good	Poor	Excellent	Good	Fair	Excellent	Good	Fair
Fair	Very good	Fair	Poor	Very good	Fair	Poor	Very good	Fair	Poor
Poor	Good	Poor	Poor	Good	Poor	Poor	Good	Fair	Poor

*See Figure 15-5. Note: If project scope success rating is less than good,
do not attempt project without an expert analysis team.

Figure 15-6   Risk Assessment: Size/Complexity versus Team Performance.

"The overriding concern of analysis is not to achieve success, but to avoid failure."[4]

So, as a general rule in risk assessment we should observe the following: if the project scope success rating in Figure 15-6 is less than *good*, do not attempt the project without an expert analysis team.

IDENTIFYING OPERATING COSTS

It is hard to separate the study of a benefit from the functional dependency the benefit has to the operating cost. Take into consideration the relationship in Figure 15-7 between information quality (the essential benefit) and the cost of information.

[4] Tom DeMarco, *Structured Analysis and System Specification* (New York: Yourdon Inc., 1978), p. 9.

SYSTEM REQUIREMENT: _____

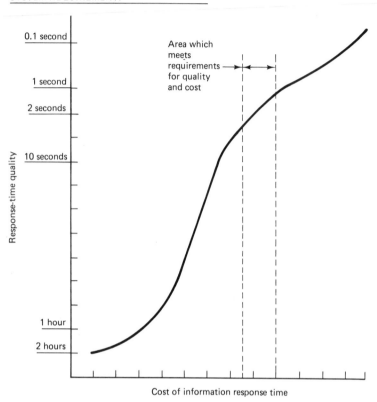

Figure 15-7   Information Quality Benefits—Taking Costs into Consideration.

In this case we are considering response time. We can see that the cost of improving response time is quite modest when response time is one hour, increasing rapidly as we go to two seconds. After two seconds response time has been achieved, the rate of increase in information quality does not change appreciably.

Two seconds is a quality level where the thought processes of the terminal user are well supported, and an increase to .1-second response time brings only a slight quality improvement. However, the cost of .1-second response is much higher than that two-second response time. The analyst needs to keep costs in mind all the time when discussing requirement benefits with the user. The first question to ask the user is: Can the department do the assigned work properly with one-hour response time to inquiries? The first question is not: What response time does the department require? The main weakness of users in assigning benefits is that they rarely know the facts concerning the consequent costs.

Another approach is to give the user a quality-level range of choices along with the consequent operating costs. After all, it is the user that will pay the operating cost (if the accounting system is sophisticated enough to handle this cost distribution—all too often the user has nothing to lose and goes for all the possible bells and whistles in a system whether or not they are required).

Quick Quality/Cost Estimating

As we go through the gross requirements and decide on system alternatives consisting of net requirements, it is sometimes possible to avoid a detailed cost analysis of a system—particularly if the analysis is an in-house effort and the benefits are very tangible. Certainly a great deal of money and other resources can be saved if we can do this. It is surprising how often a benefit is so necessary that the only cost-benefit analysis required is for the user to insist on it.

Figure 15-8 illustrates the case for this informal acceptance of the costs of a particular benefit at a particular quality level. The area $A-B$ shows a situation where a cursory nonparametric review of payback shows that without question the benefit-to-development-cost ratio is clearly a GO situation, and a detailed analysis of benefits to operating costs—while perhaps necessary for control reasons—is not a factor in making an investment decision.

On the other hand, the area $C-D$ is so obviously poor in terms of benefit quality received in comparison to development costs that the project may be rejected without further ado.

The situation in area $B-C$ is not so obvious. This is the area of possible reasonable payback, but unless extensive cost-benefit analysis is carried out, the risk involved in the investment cannot be measured. At some point along the cost curve between B and C there is a hypothetical breakeven point between reasonable and unreasonable payback.

Therefore, on cursory examination, benefits X_1 and X_2 are clearly worthwhile, Y_1 and Y_2 are worthwhile, Y_3 and Y_4 may not be, and Z_1 and Z_2 are clearly not.

The real-world situation that makes this quick method of establishing investment decisions attractive is that most organizations have a huge backlog of clearly worthwhile X-type projects waiting to be done. Y-type projects in this environment are rarely considered unless there is another reason besides tangible benefit for making the investment. Such a reason might be a risk factor, such as current system obsolescence.

Operating-Cost Centers

The operating cost we are interested in is the increase (if any) in costs that will occur because of the introduction of new requirements in the new system. We are not interested here in a decrease in operating costs because we will take that as a benefit (*avoid costs*). An increase in operating costs will be subtracted from gross benefits to give net benefits, which we then relate to development costs.

An increase in operating costs is caused by adding resources. The main cost centers here are labor, physical-system hardware ongoing costs, and vendor software ongoing charges.

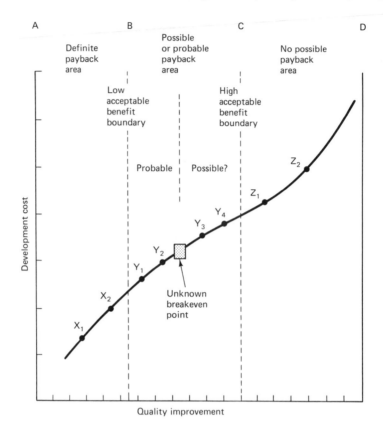

X₁ X₂ = costs less than benefits
Y₁ Y₂ = costs probably less than benefits
Y₃ Y₄ = costs possibly more than benefits
Z₁ Z₂ = costs exceed benefits

Figure 15-8 Quick Quality-Cost Estimating.

At the applications level we are usually talking about adding people and different types of terminals (telephones and printers are as much terminals as are video display tubes).

At the strategic planning level the costs involve such items as the maintenance, lease and rental charges, and tariffs of data transmission systems and staff hiring and training.

One does not apply the cost of a local-area network to the first application in the new plan. We must not apply the cost of designing a common organization database or the entire cost of the data administration function to the first application.

Each organization must work out its own cost distributions. When this is made a standard, the analyst can work out new system costs at the application and strategic planning level, and everyone will be much happier with the projected costs of developing and operating a new system.

The basic rules for working out operating costs are similar to the calculations that we will cover in regard to development costs. But first we must reinforce the discussion of net benefit.

Arriving at Net Benefit

We have said that net benefit is gross benefit minus the increase in operating costs. Let us see why net benefit is so hard to calculate:

The current system and the new system may have overlapping or reorganized domains. Maybe in the old system five function centers performed the purchasing process. Perhaps in the new system one function center performs all purchasing, including certain items that formerly were purchased by bypassing all formal purchasing centers. For instance, the training department used to buy books directly from the vendor, but now a requisition is sent to the purchasing department and a purchase order is cut.

This makes net benefit hard to do. In addition to fragmenting new systems to the individual requirement level and assigning a monetary value within a certain probability, we must do the same for increases in operating costs. We must fragment operating costs of new resource usage to those units for which the accounting department has established standard costs. In cases of overlapping domains we must spread the costs between systems according to the user's best judgement. There is, for instance, usually a standard cost for labor plus fringe according to job classification. If the yearly rental for the new DBMS benefits several new information systems, this cost must be either expensed to a corporate account or spread between applications.

IDENTIFYING DEVELOPMENT COSTS

Figure 15-9 shows a model for forecasting the development costs of an information system. In some cases these same cost factors can be used at the requirements level. Certainly the calculation must be done at the system or application level in all cases. Even if the investment decision is obvious, there is the factor of capacities management to consider. Deploying all the troops on one front could leave us with a critical problem on another front.

The biggest question in identifying development costs is: *how long will it take?* If we know how long it will take, we are halfway home and can do effective capacities management. The systems analyst is not charged with capacities management, but the new system specification developed by the analyst and accepted as a project by the user does affect capacities. In fact the project will never get done if the resources are not available when needed for as long as needed.

Labor in-house	Rate/hr	# Hours	Cost
Management	$50	50	$ 2,500
Analysis and design	37	600	22,200
Programmers-senior	28	200	5,600
Programmers-junior	22	200	4,400
Secretaries	12	100	1,200
Clerical	9	50	450
Systems programmers + DBA	34	10	340
Data administration	40	110	4,400
Office systems specialist	37	10	370
Communications specialist	37	40	1,480
Technical writers	28	0	0
Librarians	15	0	0
Total			$43,240

Contract labor	Rate/day	# Days	Gross cost	Discount	Net cost
Consultants	350	5	1750	0	1,750
Expert programmers	250	30	7500	0	7,500
Average programmers	200	30	6000	0	6,000
Total					$15,250

Vendor package	Purchase price	Customizing cost (also in labor costs)	Total cost
PPLUS-XA	45,000	35,000	$45,000

Physical system costs	Standard rate	Time	Costs	Extra costs	Total costs
Computer			5000		5000
Transmission			3000		3000
Terminal			500		500
Timesharing			—		
Other facilities + materials			1000		1000
Total			9500		$9500

Note: User costs not included. System development total cost = $112,990.

Figure 15-9 Costing a System for Development Work.

We will discuss developing time estimations at some length, but first let us identify the development cost centers. We will divide one-time development cost centers into

- resident labor
- contract labor
- vendor software
- physical system

The example shown in Figure 15-9 leaves out such corporate level indirect costs as facilities improvement. The analyst is cautioned to be sure these costs are accounted for. If a new office building will result from the implementation of the new system, somebody had better know it. We will take into consideration sudden gross increases when we discuss physical systems.

A Possible Semantics Problem Related to Physical-System Costs

We have lumped together in this presentation of development costs two different types of one-time physical-system costs. One is the operating cost of the physical system necessary to support the new application when it goes into system testing and then goes on the air as a system in production. The second, a one-time development cost, relates to the system developer's use of physical resources such as mainframe computer time in support of the software development effort.

Labor Costs

The top table in Figure 15-19 shows resident labor costs. Once we have estimated how long the work will take (and the different levels of complexity involved), we can translate this to labor needs. The table shows the different job classifications of information-system development workers. Each of these classifications has a standard labor and fringe rate. If desired, each individual has a personal labor rate, but this is rarely used except in quite modest enterprises. Standard rate times hours gives labor development cost.

(For capacities management, which is a digression not otherwise to be discussed in this book, start time and end time for each individual in the form of something like a Gannt chart of project implementation will probably be a complementary task that could often involve the systems analyst.)

Many organizations employ outside labor, usually hired by the day or week. The purpose is usually to maintain the stability of the permanent staff. Hiring permanent staff for a project and then laying these people off is not a very desirable practice from the viewpoint of either the organization or the community. However, there are peaks of development work which is best handled by the large available labor force of contract workers. The calculation is: cost = (day rate × number of days) − discount.

Regarding the type of development work, it is usually best for the large organization with in-house expertise to job out some programming and do analysis, design, and systems support in-house. However, for the small organization or the large organization lacking some particular expertise in a new technology it is often critical to contract for an expert consultant, who will not only avoid the critical pitfalls but also act as a trainer to develop in-house expertise. These are the reasons why a very large part of the development cost could involve the contract-labor category shown in the second table in Figure 15-9.

Vendor Software Purchases

Vendor software purchases usually involve a one-time license cost and an ongoing maintenance cost, which is often 10%–12% of the license cost. Whether such a one-time cost is taken as a long-term capital investment, is expensed to a corporate-level organization account, or is spread as an expense among developing information systems is a matter of accounting policy for each organization to decide. This policy will certainly have a palpable effect on the development cost/net benefit analysis.

The third table in Figure 15-9 shows the total cost of vendor software as consisting of the purchase or license price plus the customizing cost (for applications software). In the case of applications software the customizing cost is often under-estimated. As mentioned in the last chapter, the anecdotal wisdom in the industry seems to be to assume one dollar of in-house customizing cost for each dollar of purchase cost. Please accept this as an unsupported norm. The author is not aware of any quantitative analysis to develop a perspective on this increasingly more important cost-benefit factor. It probably depends on the development area.

Physical-System Costs

This, as we have touched on, can be the trickiest cost of all. The physical system consists of application host computers, personal computers, terminals, storage devices for data, and data transmission systems. The trick with physical systems costing is understanding when the add-on of a terminal or personal computer in support of an application will force a major expenditure in the area of mainframe computers and their storage devices and controllers or of a major data transmission unit.

Take the case of the 2001st telephone. If you have a switch which supports 2000 telephone hookups, the 2001st telephone is going to cost a bundle. Harder to estimate is when the outgoing trunk lines will become too congested, resulting in erosion of service on calls. Since there is also considerable service necessary to maintain all these telephones, many organizations establish a standard operating cost per telephone which includes all known factors—so the 2000th telephone may cost $10 to $15 or more per month before a call is made.

This example holds 'true for all physical-system costs. The systems analyst working on the typical application will usually find that standard operating costs per terminal (such as phone, VDT, or printer) plus immediate control equipment, such as terminal cluster controllers, will suffice. In this case there will be no development costs in the area of physical systems besides those needed to support the software development work itself. This is not true for the very high level (VHL) systems analyst working on the strategic plan. In that case large physical system purchases are development costs at the corporate level.

The fourth table in Figure 15-9 shows the one-time physical system costs normally associated with systems development. These costs could include

- computer costs
- data transmission costs (where applicable)
- terminal usage costs (possibly—but not often)
- outside time-sharing costs (a major cost consideration)
- other facilities costs

For many analysts the computer in-house standard costs for computer use and/or time-sharing costs are the only physical-system cost concerns. In-house computer usage costs, if standardized, might involve the standard rate for the configuration of mainframe plus peripherals × elapsed time. Outside time-sharing costs usually involve a formula that considers connect time, storage use, CPU time, and I/O quantity.

Costing a System by Development Work Type

Figure 15-10 shows another set of cost-category identifiers. In this example, instead of using type of resource to show costs, we use type of systems development or development work type.

Sometimes systems alternatives can be seen more clearly when we follow the development work cycle. At any rate, this method is vital to planning the development work, and it shows in development terms where the investment cash is flowing. It is in these terms that we examine *how long will it take?*

We have arbitrarily used the columns to show corporate versus application level expensing. Another view might show rate times time as in Figure 15-9.

HOW LONG WILL IT TAKE?

We have divided development cost centers into resident and contract labor, vendor software in support of applications development, such as we discussed in the last chapter, and the physical system. In this discussion the physical-system timing

Development category	Expensed to entire organization	Expensed to application, product, or user area	Total
Management			
Analysis			
Design			
Programming			
Unit testing			
Systems testing			
Development team training			
User training			
Operations manual			
Technical documentation			
Conversion			
Parallel testing			
Postinstallation tune-up			
Presentations			
Subtotal			
Mainframe computer			
Other computer			
Terminal			
Data transmission			
Time-sharing			
Other facilities			
Materials			
Subtotal			
Total			

Figure 15-10   Costing a System Application by Development Work Type.

problem refers only to the physical-system necessary to test the software development.

The problem of physical-system availability for software testing can be vexing but does not usually call for a great deal of quantitative analysis. The analyst might be responsible for assuring that physical-system components necessary to implement the software development are to be found when needed. This could be a vendor or other user installation as well as in-house.

It is in the area of time-in-process for software development based on the productivity of human labor that we have our main problem doing cost analysis. Together with the essentially subjective nature of defining the monetary value of net benefits, the crux of the matter is whether any cost-benefit analysis of such a complex thing as forecasting the value to an organization over time of a new information system is a black art masquerading as a science. Many books and many, many articles have addressed the need to limit the area of black art and expand the area of reliable quantitative analysis concerning development costs. Managers with responsibility for investment decisions demand tangible forecasted costs as well as realistic benefits. If we know how long it will take, it is easy enough to determine the cost.

We will present here two different approaches to estimating development time which have received considerable attention (and even some acceptance):

- Barry Boehm's COCOMO model as developed in his book *Software Engineering Economics*[5]
- IBM Corporation's AD/M (Application Development/Maintenance) Improvement Program with particular regard to the Function Point method of estimating (referred to hereafter as "the Function Point method"[6])

To forecast software development costs we must first estimate time. To estimate time we need a decision model containing a time predictor sufficiently sensitive so that small variations in the predictor will reflect in time-estimate changes.

Before we discuss these two approaches to estimating software development time, we should make some general comments concerning rule-of-thumb measurements of time which are in more common use than these microanalytic approaches.

The best way to estimate time is do yet another job similar to those already in production. It is important in this case to keep good documentation so that elements of the previous job that effected the estimate are retained as useful history.

Let's say we are estimating an application involving an on-line system with ten transactions, 35 screens, routine validation of adds and updates, a query capability, no unusual complexity of behind-the-scenes operational processing, and five output batch reports. And let's say we recently did a similar job involving 20% more transactions and batch reports.

If we are using the same software development tools (and last time was not the first time we used these tools) and the quality of the programming team is about the same, then we can reasonably expect to estimate this job by making some simple adjustments to the last job's actual time. This is to say that if we can

---

[5] Boehm, *Software Engineering Economics*, various references to COCOMO model. See general Index.

[6] Allen Albrecht, IBM Corporation, "Measuring and Estimating Application Development and Maintenance," Guide 57.0 Conference Report, Session No. DP-7234A, 10/10/83.

accurately estimate team performance level, job complexity, and software tool potential and have kept good records on similar jobs done before, we can predict time in process with sufficient accuracy to please anyone.

The problem is that we (unlike the folks who build office buildings) probably have not done the job before, probably find ourselves confronted with new tools to learn, probably find ourselves constantly working with teams with unpredictable skill levels, and are probably surprised by unexpected developments in job complexity, no matter how good the analysis was. Such is the current situation concerning the development of information systems for complex social organizations using a rapidly changing technology that has few product standards and encourages a relatively unstable workforce in a fiercely competitive marketplace where experts are hard to find.

It is amazing, given this environment, how good some analysts are at time estimating. It is still largely a black art. Two cunning tricks employed by the less-than-straightforward analyst or manager to handle estimating errors are:

- always overestimate the time and then, if the real estimate proves true, use up the extra time polishing the brassworks
- do not estimate overtime for the team, who in their job classification probably do not get overtime; then cultivate an atmosphere where a vast amount of hidden overtime is used to meet the schedule[7]

These are examples of the nonartistic aspects of the so-called "black art."

Another way, not as accurate, of getting a sense of time in process (mentioned in the last chapter on getting and using software) is to test a new software product in a realistic situation where some estimates can be made regarding team learning time. This can be extended as a concept to include testing a large system development effort by doing a similar effort of much smaller scope. Then extrapolate time estimates on the big project from the experience with the small project.

Another approach is to use someone else's historical experience with a similar set of cost components. Some jobs are close to generic in detail—such as developing materials management software in-house. Maybe the shop down the road just did this and is willing to share acquired knowledge.

Let us turn now to the modelling techniques which attempt to provide a computational solution to this problem of resolving how long it will take.

The COCOMO Model

Barry Boehm's book *Software Engineering Economics* is a far-ranging study of "software engineering," which is another hot phrase for performing the system development cycle.

---

[7] Tracy Kidder, *The Soul of a New Machine* (New York: Avon Books, 1981), p. 257. "He was astonished. The technicians were taking home more than twice as much as he was, on account of all the overtime."

The goal of software engineering is to realize a successful software product by conducting a successful software development and maintenance effort.[8] There is the fervent hope that the software development process in support of whole natural systems is amenable to reductionist techniques of precise measurement. This idea, which comes out of the economist's bag of tricks (where it has not worked too well at predicting the national economy[9]), as much as out of the industrial engineer's plant management bag, has now descended upon the information development industry. The government, in particular, which is becoming more and more dependent on computer-based information systems, would like to have precise estimates for complex development tasks involving whole systems.

"How long will it take? Build us a huge thing, the exact nature of which is only dimly realized by us the buyers, consisting of about 162,500 moving parts interacting with 1210 people, using 122 developers, very few of whom are experts, using new tools that are not yet mastered, which do not have standardized interfaces with other tools, and carry the answer out to six decimal places."

Author's sarcasm aside, there is value to the development decision model when properly interpreted and supported by the old rule-of-thumb estimating skills acquired by the experienced practitioner. The COCOMO is a good example of a quantifiable model which has value in the right context. Certainly, the detail in support of the model in Boehm's book along with the extensive corroborating reference material is invaluable scholarship and represents a major contribution to the study of how software gets developed. In addition the book is quite readable.

Still, we must beware of the pitfalls involved in applying mechanistic answers to natural-system problems. If that is done, Boehm cannot be blamed; it is the fault of the analyst. And let's please remember that insofar as the reason for the cost-benefit analysis is to make a go–no go investment decision and get some rough parameters on resources required and time needed, which can be constantly readjusted during the development process, our efforts can be very successful. Let's just not confuse the users about the confidence level of our analysis.

COCOMO stands for COnstructive COst MOdel. There is Basic COCOMO, Intermediate COCOMO, and Advanced COCOMO. Basic COCOMO is a very reduced and compacted formula, as we shall see.[10] Intermediate COCOMO accounts for a variety of modifying "cost drivers," which we see illustrated in Figure 15-11.[11] These cost drivers have a major impact on the basic model, and an understanding of their nature is vital to estimating time and cost. Advanced COCOMO not only shows the cost drivers but accounts for them by project phase.

---

[8] Boehm, *Software Engineering Economics*, p. 23.

[9] Fritjof Capra, *The Turning Point* (New York: Bantam Books, 1983), p. 192. Quotes by Arthur Burns and Milton Friedman.

[10] Boehm, *Software Engineering Economics*, pp. 61–62.

[11] *Ibid.*, p. 642.

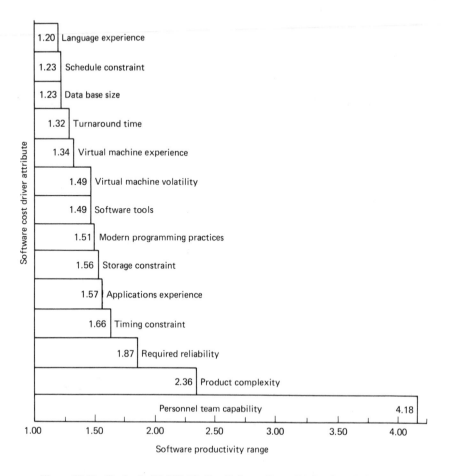

Figure 15-11 Boehm's COCOMO Cost Drivers. Barry W. Boehm, *Software Engineering Economics,* © 1981, p. 642. Reprinted by permission of Prentice-Hall, Inc., Englewood Cliffs, N.J.

The Basic COCOMO

Basic COCOMO consists of two equations:

$$MM = 2.4(KDSI)^{1.05} \quad \text{and} \quad TDEV = 2.5(MM)^{0.38}$$

where   MM   = man-months

KDSI   = one thousand delivered source instructions

TDEV = the number of months estimated for software development[12]

[12] *Ibid.*, p. 62.

Boehm gives an example[13] where, given that the initial study has determined that "the size of the program will be roughly 32,000 delivered source instructions (32 KDSI)" and the in-house programming team has been developing similar programs for several years, the following characteristics of the product can be developed using Basic COCOMO.

Effort:	MM = 91 man-months
Productivity:	32,000 KDSI/91 MM = 352 DSI/MM
Schedule:	TDEV = 14 months
Average Staffing:	91 MM/14 months = 6.5 FSP

where FSP stands for "full-time equivalent software personnel." Quite of lot of interesting information of major consequence!

In the opinion of the author the COCOMO is most useful when used for development work that will result in implementation largely in procedural languages. The introduction of nonprocedural language tools and the outright purchase of application software tends to reduce KDSI to a less-than-meaningful index. COCOMO is based on KDSI.

In addition, we are in a period when the programming costs of procedural language development are lessening when compared to the other costs of analysis and design, which are expanding. There are systems being developed where the analysis and design effort takes well over 50% of the time. COCOMO includes analysis and design time in the model and does a best possible job of quantifying this development phase. Nevertheless, the author is uneasy with the level of estimating accuracy possible in this area. Every little mistake up front echoes louder and louder down the development life cycle, causing the greatest delays in development. Can COCOMO hope to be sensitive enough to the vagaries involved?

COCOMO seems most relevant when implementing systems in Ada, BAL, COBOL, and similar procedural languages. These systems would include development of software where economy of space and execution time is paramount. Such systems include software in support of hardware devices and software for sale as a general development tool where speed and space efficiency *strengthens position*. More important to the general user organization, complex operational systems which are still not amenable to development with nonprocedural language tools and the huge current inventory of systems done in procedural languages are candidates for COCOMO or COCOMO-type analysis.

We should mention at this point that Boehm's book is highly recommended reading concerning the entire subject of cost-driver analysis far beyond the use of models.

The *cost drivers* shown in Figure 15-11 are not dissimilar to those we have

---

[13] *Ibid.*, p. 63.

already covered. As we have said, it all comes down to people quality and experience, job complexity and quality level, and tool quality. We will see these same or very similar cost drivers used in any good discussion of information-system development-cost estimating.

The last comment we need make about Figure 15-11 is to repeat how important it is to get and keep top-quality people. Most sophisticated information-system managers know this, and the competition for top graduates of top schools is fierce. For the less-than-huge organization the secret to success here is to develop a quality in-house training program, where the best entry-level people obtainable can be upgraded in a few months. This is, as Figure 15-11 shows clearly, the best way to cut development costs: get the best people, train them well, and hold onto them.

The Function Point Method

The Function Point method of estimating application development and maintenance (AD/M) is very satisfying to the author because it fits in with one of the basic premises of this book.

Figure 1-1, the first illustration in this book, shows the author's view of the generic systems model as a set of processes using resources such as stored data to convert inputs into outputs. It is at this high level that the Function Point model is constructed. Allen Albrecht, in presenting the Function Point model to the Guide 57 conference (Guide is the IBM users organization which meets twice yearly) on November 10, 1983, said regarding lower-level predictors such as KDSI (especially when going from a third- to a fourth-generation language): "It is as illogical as measuring the output of coal miners in 'pick strokes' instead of 'tons of coal'."[14] Not a bad simile. Of course, on the other hand, a lot of "black art" can creep into the less-than-precise predictors involved with the Function Point method.

Figure 3-11(a), the baseball team system specification, is an expanded example of Figure 1-1, which comes close to the information needed for the Function Point model.

There are five elements to Function Points:[15]

1. Inputs, such as on-line screen or batch transactions or recycles from another system.
2. Outputs, such as paper or screen reports or transactions to another system.
3. Inquiries, which are input screen transactions that produce an output.
4. System data stores.
5. Common "interface" data, which are data stores shared with other systems.

Albrecht says "We list, count, and weight the elements to get Unadjusted

---

[14] Albrecht, "Measuring and Estimating AD/M," p. 665.
[15] *Ibid.*

Function Points. Then we adjust this result for general internal processing complexity with . . . other characteristics . . . ." Here is an almost complete list of these adjusting characteristics according to another paper at the Guide 57 conference:[16]

- use of communications' facilities
- use of distributed processing
- performance constraints
- system environment load
- expected transaction load
- data-entry method
- on-line updates of master files
- application logical complexity
- reusable code
- conversion ease
- installation ease
- operational ease
- multisite/multiorganizational use
- built-in flexibility for change

We see here a great deal to do with product size and complexity but no adjustments for team and individual ability—which we, along with others such as Boehm, consider to be the major cost modifier.

We also see here a more subjective model. The judgements concerning the complexity of the five elements, the importance placed on each of the product cost drivers, all greatly affect the model. The author is reminded of the time when his late-model car, for which the book value was high, was judged by the unimpressed salesman to be worth very little money even though in good running condition and with a "clean body," because it had 95,000 miles on the odometer. The modifiers are everything in life.

No matter how impressive the equations are, the author sees these many skillful judgements as very much like the old "seat-of-the-pants" method by which the experienced analyst comes near the target of answering the question: how long will it take?

One advantage of such a volatile model is the list of checkpoints that must be considered. It is good to have to consider Boehm's cost drivers or the estimation inputs for the Function Point method or any of the other lists of cost drivers presented by other sources.[17] A problem that occurs is: everyone comes up with

---

[16] Howard A. Rubin, Hunter College, "Art and Science of Software Estimation: Where Are We Heading?" (Guide 57.0 Conference Report, Session No. DP-7234B, 10/10/83), p. 688.

[17] Rubin, "Art and Science of Software Estimation," pp. 679–681.

a different list. There seems to be no agreement on the adjusting factors that temper the estimation. The author comes down on the side of the COCOMO cost drivers as a good place to start when constructing this necessary list.

According to Albrecht's presentation at Guide 57, IBM is still using the Function Point method to retrospectively examine AD/M at each of its internal information-system development sites. The author does not get the impression that IBM is yet forecasting work to be done with Function Points. Results have apparently been satisfying, and some correlations have been made: "an analysis of 21 PL/1 projects has shown that about 66 lines of PL/1 source code approximates one Function Point."[18]

Figure 15-12 shows an illustration of Function Points from Albrecht's presentation.[19] The relatedness to the system models presented throughout this book is obvious. Somewhere there will be a model developed with the precision of the COCOMO relative to systems built on procedural languages and the good universal fit of the Function Point method which will advance the art of estimating somewhat closer to the desired goal: a method for all system development work.

This business of estimating will always be a high-skill job, not something a dutiful clerk can load into a model machine without knowing all the fine gradations of the modifiers. Perhaps building a better model is a long-term exercise for some erstwhile reader of this book.

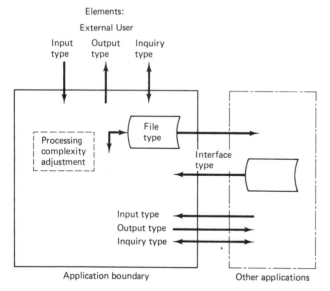

FUNCTION POINTS

Elements:

External User

Input type   Output type   Inquiry type

Processing complexity adjustment

File type

Interface type

Input type
Output type
Inquiry type

Application boundary   Other applications

Figure 15-12 Estimating Development Costs with the Function Point Method. Reprinted with permission of GUIDE INTERNATIONAL, Inc. Chicago, Ill.

[18] Albrecht, *Measuring and Estimating AD/M*, p. 666.
[19] *Ibid.*, p. 671.

Doing The Cost-Benefit Analysis

Having with great difficulty and with the use of much subjective judgement and the best advice of many expert users and technical staff arrived at net benefit and development cost, we can now present this information to those people in the organization responsible for investment—and also to those responsible for the capacities management of all the resources, particularly the professional information systems staff.

First we should notice again that, although the forecasting of development time and benefit as a monetary value may now seem to the reader to be done with less than a satisfactory methodology, it is also true that we can come close enough to meet the requirements of the investing group. In terms of benefits, most jobs that are done just darn well *have to be done* in order for the organization to remain viable. This is not to say we do not reap great gain from excellent analysis.

In terms of development costs it is more important to have good numbers, but even here the estimate will be reforecast several times before the project is shipped out the door. If the organization changes key cost drivers (like taking away the superstars in exchange for beginners), then by all means reforecast cost and time-in-process estimates. The closer we get to actually purchasing software and specifying base-level modules, the easier it becomes to estimate. In many parts of the United States it is now possible to hire programmers on quick notice to code up those base modules and those reports in nonprocedural languages.

What is usually necessary in any cost-benefit analysis is to show cash flow over time. It is seldom enough to say system A will cost $1 million and will pay back $2 million over a five-year period. Cash-flow questions, stated in the vernacular, are of the following sort: When and how many bucks do we have to lay out and when and how many bucks will be coming back? Show us over, say, five years, the cash flow out and the cash flow in. At what point in time do we have a positive net cash flow wherein the money coming in starts to exceed the money going out? How much will we make over five years?

These are the minimum questions, and usually not much more will be asked of the systems analyst. Other questions either can be answered by the information systems analyst who is adept at return-on-investment (ROI) analysis or, perhaps better yet, turned over to the professional financial analysts in the organization, using the systems analyst's cost-benefit as a base of information. It is one thing to do an ROI on a single piece of equipment. It is quite another thing to attempt an ROI on a complex development project involving many pieces of equipment and many kinds of software development.

Reasons for wanting to do a complete ROI workup include the special factors that need to be considered, beyond showing cost-benefit over time in the present value of cash. Here are some of the complicating factors:

1. The inflation factor (the discount rate of money) is important. Money which is laid out in year 1 is different from money coming back in year 5 because of the

inflation factor. Steep inflation can mean a major adjustment to the actual value of money. A big problem here is that the rate of inflation is often quite different for an individual organization than what is reflected in the government inflation index. The individual who invests money at 15% annual yield and has a 5% mortgage is in a quite different situation from the individual who has no money to invest but has a 14% mortgage. Yet for both the government inflation index may be currently 12%.

The organization that lays out cash to finance an information-system project is in a different situation than the organization that has to borrow at current inflationary rates. The organization that can keep costs even and raise prices at the inflationary level is not going to worry all that much about inflation. These complicating factors may require a financial analyst (another kind of systems analyst). A mechanical plugging into the ROI model of a constant inflation rate may not shed much new light on the investment process.

2. Also important is the effect of the tax code. The net benefit after taxes would seem to follow a national standard of about 49%. Such is not really the case. Some of the largest commercial organizations in the United States pay little or no taxes in any particular year. Also the tax code is constantly being changed by interpretations in the tax courts and by new legislation at the federal, state and local levels. As this chapter is being written, the federal tax credit for increased investment in research and development which excludes from taxation 25% of the increased annual R&D spending of a business over the average of the three previous years has a 50–50 chance of approval, and then it will probably be much more difficult to qualify for than in the past.[20] Upcoming tax legislation will likely have an effect on the investments by business organizations in new information systems. This, too, is hardly a factor for the systems analyst to deal with without professional advice.

3. Other modifiers in ROI analysis are more easy to apply to individual pieces of equipment than to entire information systems. These include the resale value of disposed capital, and the book value of an abandoned resource. Also there is the consideration of lease/rent versus buy.

4. Some other interesting benefits over time are not included in classical ROI methodology. These include the guaranteed increase in benefits over time as the developers, users, and maintainers "tune" the system to top anticipated performance and then beyond. As we have said, the analyst usually underestimates such benefits.

5. Among the complicating factors also are the heuristic benefits of the "value-added" system. An elegant system done by a team of true professionals, including the type of analysts who can "feel" the whole system (which is always different from the sum of the parts), will always produce benefits which are unexpected,

---

[20] Edward Warner, "Permanent Tax Credit for R & D Given 50:50 Chance," *Computerworld*, 5/14/84, p. 19.

given sufficient system complexity. Here we enter the area of the true intangibles, where the hard-nosed investor recoils in uneasiness and even contempt. Yet the benefits are there! The architect of some repute never knows how enhancing to the fortunate indwellers is the beautifully designed house. The same is true of the beautifully made information system. The users find that the system does not let them down as they ask more of it. The next designers find the system responsive to change.

Some ROI Methods

We will not show any of the sophisticated ROI methods but instead list some that the reader may want to investigate:

1. *Payout* or *payback*. A simple method frequently used. Does not take the project life into account.
2. *Annual return*. Shows annual percentage return. Does not take the project life into account.
3. *Average book*. Does account for the life of the project but cannot evaluate different patterns of income or savings.
4. *True rate of return*. Considers the time value of money and deals in net, after-tax figures. This method is known variously as:

   - the discounted cash-flow method
   - the interest rate-of-return method
   - the internal rate-of-return method

5. *The profitability index*. A true rate-of-return method showing ROI based on present-worth values. We describe the investments and estimate the capital requirements, do a depreciation worksheet, do a time schedule of expenditures and savings or income, and show graphically the profitability worthiness index and the time to *payout* of the investment.

The Basic Cost-Benefit Analysis

Figure 15-13 shows the minimum required from the systems analyst without getting into a lengthy ROI covering the special points mentioned. Usually this is more than satisfactory.

Figure 15-13 shows a total accumulated net benefit of $2,930,000 for a total cost of $1,120,000 (cross-footing the total-cost row). This covers the years 1985 (when the costs begin) through 1991. We have a negative cash flow until 1988, when the benefits start rolling in. From 1988 on we start to make money, but in this particular system we get no payback for over three years from project approval time.

	1985 Actual	1986 Est/act	1987 Est	1988 Est	1989 Est	1990 Est	1991 Est
Development costs	27	55	82				
Operating costs			94	195	200	200	200
Maintenance costs			12	25	25	10	
Total costs	27	55	188	220	220	210	200
Gross benefits			50	750	750	750	750
Net benefits	(27)	(55)	(138)	530	530	540	550
Net benefit accumulation	(27)	(82)	(220)	310	840	1380	1930
Intangible benefits accumulation				250	500	750	1000
Total accumulated net benefits				560	1340	2130	2930

Figure 15-13   Cost-Benefit Analysis: Project Payback.

Development costs begin in 1985 and end in 1987. At that point (1987) changes to the system are shown under maintenance costs. 1987 was the year the user accepted the system as a production system. Operating costs hit a peak in 1988 and then slowly decline as the system is more finely tuned. Benefits are divided into tangible benefits of $1,930,000 through the end of the analysis and "soft" or intangible benefits of $1,000,000 through the same period.

Perhaps we should be skeptical about this quantity of soft benefits. In this situation it would be better to give supporting detail, showing a benefit probability analysis such as we covered in Figure 15-3. It might well turn out that much of this soft or intangible benefit would prove to be very solid indeed.

At any rate, much work went into this simple page of figures. Many people made many judgements, striving to find that perfect point between too little and too much. To the extent that the whole organization was able to contribute to the hoped-for accuracy, this document is one in which we can place confidence and use as a basis for good investment decisions.

16

INFORMATION SYSTEMS DESIGN

This chapter will introduce design techniques. The next chapter will demonstrate these techniques in a case study. Systems analysis, the specifying of logical system models, cannot be done without an understanding of the requirements of systems design. The primary output of the designer of computer information systems is the design specification from which the actual system is built and/or bought. The systems designer (who may be the systems analyst wearing another hat) is the most important recipient of the system specification. The orderly passing of the baton in the system-development-cycle relay race requires the systems analyst to be aware of the designer's input requirements.

MANUAL VERSUS COMPUTER SYSTEM DESIGN

From the point of view of the systems analyst doing a logical system specification for a new system there is a continuous flow of information handling to be analyzed. Some of this information processing is done without computers being involved, and some is done, unseen by human eyes, at electronic speed under the control of computer programs.

The analyst seldom recommends the use of a computer unless the nature of the task is such that processing speed or the handling of complexity or the need for information quality makes such a solution a requirement. The practice of systems analysis does not require the use of a computer in the solution.

Often the so-called manual operations precede the computer processing, preparing the input for machining. Then outputs of the computerized system emerge as reports and queries for human handling, interpretation, and decision making.

Sometimes the process involves an interactive dialogue between human and computer, with either the computer programs leading the dance or the user in control or a mixture of both. This dialogue can be very satisfying, given a good system design, since an intimate relationship is formed between the human user and the software via the terminal and the keyboard. Good on-line software has that effect.

The last task of the systems analyst before tackling the cost-benefit analysis we discussed in the last chapter is to draw boundaries around those transform boxes on the diagram of the new system. The activities included inside the boundaries are to be included in the computer design effort. Usually several versions, ranging from plain (less inclusions) to fancy (more inclusions), are drawn up and prepared for presentation to the investing users. The users select the version desired and perhaps redraw the boundaries. This is the operative specification. Part of this specification is the list of requirements. Transactions and data stores are indicated and usually made resident on the data dictionary. This application takes its place as part of the overall strategic plan.

Now we have two kinds of design going on—the new system outside the boundary to be accomplished without the computer directly involved, and the included processes to be in the domain of the computer solution. The system outside the boundaries, by the way, may include a considerable amount of electromechanical machine automation. For instance, in an operation that does much of its business through the mail, electronic envelope-slitting machines and count-by-weight scales could be involved in upgrading to a new system. The systems analyst, working with the users, is often involved in this kind of design. Systems analysts are, after all, concerned with the functioning of whole systems, not the computerized component alone.

The computer-system design is still the major design concern of the analyst most of the time. The computer design is the realization of a detailed model of the analyst's partly realized conceptual view. It is only from the design model that the actual system can be implemented. In fact much of what used to be called *implementation* is now, rightly, called *design*. Design is the critical interface between concept and reality.

This chapter and the next will be concerned with computer-system design. Analysts are almost always called upon to do system design (as well as manage implementation projects). Besides, for the purposes of this book, it is easier to get a perspective on systems analysis, the specification of current and new systems, by knowing the task requirements of the major recipient of the analytical effort: the designer.

The main output from the design phase will be a detailed set of flow or structure charts along with the accompanying definition of the data, both moving and at rest, with which a program can be coded, a software package can be selected and customized, a nonprocedural reporting or data-entry specification can be produced, and a physical support system upgrade can be acquired. In addition, there will be a cost-benefit analysis available to be updated. Further along in this chapter we will detail the design phase outputs.

REASONS FOR DOING COMPUTER-SYSTEM DESIGN

Among the reasons for doing system design are the following:

1. Assure delivery of information at the required quality level (response time, accuracy, and conciseness, for instance).
2. Meet the requirements of the specification through the designed program logic.
3. Map the logical specification as the system/program/module flow.
4. Reconcile data analysis to process analysis. What are the data needs for each process? What are the processes that transform the data?
5. Create a software product that is amenable to constant change and evolutionary development.
6. Narrow the options available to the implementation team so that the system is delivered as bought.
7. Maintain standards (security, uniformity where desired, error detection, internal audits, etc.).
8. Make a product that is delivered top down, left to right, instead of all at the conclusion of the effort.

THE "MYSTERIOUS" TRANSFORMATION FROM LOGICAL SPECIFICATION TO COMPUTER-SYSTEM DESIGN

This really could have been the title of this chapter and the next. There exists considerable confusion, which is reflected in the literature on the subject, about the nature of this transformation. This confusion does not have a basis in actual complexity. The transformation, in most instances, is straightforward. It's somewhat like sorting a deck of cards from suits within numbers to numbers within suits.

The key to the "mystery" is knowing we are dealing with systems that process inputs to outputs using resources. The system specification, whether we are using a process-oriented or a data-oriented methodology, is essentially tracking transactions one after the other until we have exhausted all the transaction inputs and the transformative effect each input has had on all the data stores. Along with this we track (as sort of an output transaction) all the reporting required from the system that flows from the continually refreshed sum of all data stores. This, as we have just said, is done one transaction at a time.

On the other hand, the computer system for an application is designed by collecting all the reporting needed from the system, designing these reports, deciding on the total data needed for these reports, designing an application view of the database system or an application data store, and so on. Following this, the transactions, and the transformative effect they will have on the data store, are considered as a group. New data elements which have transaction-control functions (setting triggers for deferred actions such as reporting, for instance) will be added to the common data store. The transaction group will go through the following three or four sequential steps:

- transaction validation
- transaction editing
- transaction sorting (for batch sequential systems only)
- transaction access to and transformation of the data stores

In the case of an on-line transaction system the transactions are entered one at a time from each terminal, and each goes through the steps mentioned above. Any given transaction can be entered many times at once, usually each to a different data record. However, the system that receives the sum of all these transactions is one whole—a transaction-processing network of panels and supporting programs which is designed as one thing: the application input system.

In the case of the batch system it is even more clean cut. According to some trigger, manual processing of input is cut off and all the input is batched, machined, validated, edited, sorted, and applied to the data store, where the transformation is made to the data store, to update it and set output triggers. Output reporting, a "line" at a time, is often done as each record is fully updated by the transactions.

Therefore, we can restate the transform from analysis to design as follows:

The various logical transactions, each charted one at a time, that make up the new system specification are converted into a design that first handles all transactions together and then handles all reports together.

We will see this illustrated as the design systems chart.

Not all designers work backward from reporting to data stores to transactions. Some work out from the data stores; some work down from the transactions. The author feels it is better to start work backward from the reporting. It is more efficient in terms of the iterations that will occur. Also there will be a tendency to avoid making an inquiry transaction out of a need for information that is best handled as a batched report. Much of the information is not needed within two seconds.

In addition, in terms of information seen by the user, the output should define the data stores (database system). There should be no user information on the database system that is not part of a current or planned user view. Remember that when we discussed database design in Chapter 10 we used a data modelling methodology that converted user documents into user-view charts which were integrated into a logical database model that was the basis for the physical database.

THE GENERIC COMPUTER DESIGN SOLUTION

Figure 16-1 (which was previously introduced in Chapter 6 as Figure 6-1) shows the overall computer design solution to either batch or on-line or mixed systems. As we shall see, almost all design solutions of any complexity involve a mix of on-line and batch processing.

Figure 16-1 reflects the discussion in Chapter 6, where input was defined as either transactions consisting mostly of precise fields, messages of not precisely defined but limited text framed by some precise identifying fields such as one might encounter in an electronic mail system, and data streams such as one might find while browsing the news wire services through a database network. We will confine ourselves in this discussion to the design of systems with transaction input.

We see in Figure 16-1 that once the transactions enter the system, there is no point where the moving data, the transactions, do not rely on the availability of the data stores. The best and most common input validation concerns the lookup to the data store, such as a database or a special validation table, to see if the input data item makes a "hit" or is a "not found." True, the data store is seldom transformed during input validation. During input editing a minimum machined transaction with valid data items may, on making a "hit," pick up off a table several data items to add to the transaction.

When we get to the main data store involved in the transaction, the application or common database, the transaction will transform the database and lose itself in the process, except for inquiry or lookup type transactions. Even the output recycling and reporting will make some light use of data-stores, and some outputs will go as transactions to other data stores. A typical example is the statistical transaction which keeps a running report on the transforms taking place in the main data store. Statistical reporting is made from this system output data store. It is particularly necessary to show this running account of data store changes when the latest updated database does not contain enough history to satisfy the system requirements.

Before continuing, we might briefly comment on several illustrations from previous chapters that prefigured our current discussion of systems design.

Chapter 7, "Data-In-Motion," had a good deal to say about the nature of the transaction set, which is the front end of the design system. Figure 7-3 shows three variations of a shop workload transaction. In the first example (a) to (b) a job list and an employee list are displayed. The supervisor selects a job # and

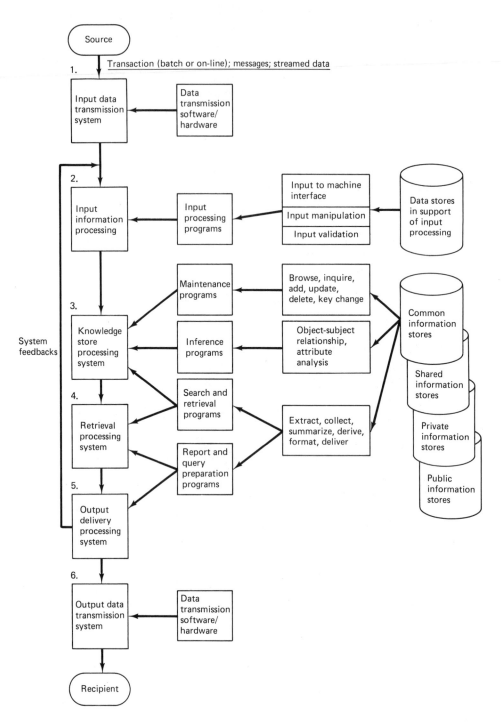

Figure 16-1 The Overall Computer Solution to Information Automation, Batch or On-line.

employee initials with a light pen or a mouse or by touch-sensitive screen; the system already knows the machine involved, since it is on-line like a terminal. After this setup the job can be run by the operator, and, when the operator indicates to the system the job is complete, a transaction is produced automatically by the system from which can be computed the elapsed hours and work quantity accomplished. From this can be computed the operator and machine rate per hour. The job code relates to a lot of other information. So, we can show work accomplished many different ways, and no data was keyed into the system.

Figure 7-3(c) to (d) and (e) to (f) are less sophisticated design solutions to the problem of capturing the same input transaction.

Figure 7-4 shows the sequence of editing steps by which this simple external transaction is converted into four more complex internal transactions to be applied to the database.

This business of transaction validation, editing, and database updating is the last frontier in the automation of procedural programming. Reporting and querying are already, for all but the most complex reporting, solved problems.

THE COMPUTER-SYSTEM DESIGN OUTPUT PACKAGE

The primary recipient of the design package output is the implementation team, but the user and various management groups such as data processing management are also interested in reviewing the design specs.

The implementation team (should they accept the design) will convert the design documents into an executable, tested and production-ready system backed by the required physical system. The implementation team is not expected to have to do any redesign work or get involved in any aspect of design except that of criticism. The implementer is, after all, the first user of the executable system and will offer valuable feedback to the designer concerning whether the system is really going to be able to be implemented as conceived. The designer will no doubt be going back to the systems analyst, who will be going back to the user in case some requirements cannot be mapped onto the automated system.

In addition to putting up a new system from the design specs, the implementing team, which is often the maintenance team, has another concern regarding the design documentation—that is, the maintenance of the production system after going "on the air." There are three reasons for maintenance:

- bugs in the implemented system
- changes in user requirements, meaning redesign of pieces of the system
- opportunities for improving the performance of the system through postinstallation tuning

In this maintenance environment, good design documentation which is kept up to date by the maintenance team is critical.

Implementing team members do not stay with a system in a well run DP shop. There is a need to move and promote these staff members. Any experienced programmer should be able to take over the responsibility for the system. The clues to doing this are the design specification and the system specification in support of it.

Going back to the role of the implementer executing a design: it's somewhat like actors in a play telling the director that certain lines won't "play." The director then goes back to the author and suggests a rewrite. Sometimes the actors just change the lines "in production," but this may change the logic of the play, which is less than desirable as a method.

The Design Package

We need now to look at the various components of the design package, and then we will look into these components in some detail. The reader is cautioned in advance that not all components need be covered in any detail in all systems. In the real work situation we need to be appropriate. We should not produce $200,000 worth of design documentation for a low-risk system that produces only a much smaller benefit. We will cover some possible shortcuts as we look at the design elements. For now we can say that buying application software and developing systems in nonprocedural languages will greatly cut down the need for design documentation.

Here are the design deliverables:

1. *The system description.* This is text to cover what the system does (processes) and what the inputs, outputs, triggers, and resources are (especially data store resources).

2. *The system chart.* This shows graphically the inputs, outputs, and programs or packages in the system.

3. *The program description.* This explains for each program the processing, inputs, triggers, outputs, and resources needed.

4. *The program chart.* This shows the same information in graphic form.

5. *The program transactions.* These transactions could be shown once for the whole system (which is good for on-line processing) or by program (which is good for many batch systems). In this batch system case the transaction processing is shown for the validation and edit program and again for the file-maintenance program. The sum of both these descriptions is the full text for each transaction.

6. *The program modules.* A modern computer program is a set of partitioned and leveled modules starting with the executive module and descending the module hierarchy to the base-level "worker" modules. We need to show the program decomposed to the module schema or chart, and we need to see the workings of each module—that is, module input, output, process, triggers, and resources. Programs are written from module definitions.

7. *The panel flow diagram and panel "paintings."* For an on-line system this "PFD" is the hat rack onto which we can hang all our displays and support program modules. Panel flow in a complex on-line system is most complicated, and this document is crucial. Along with the PFD there is a requirement for a panel layout or "painting" for every panel in the system—whether a menu, a data, or a help panel.

8. *The users manual.* Written in plain language with high-level charting (such as the system chart), this document tells the users what they need to know in order to use the system. This includes documentation we will not be covering in this chapter, such as job control language, back-up and recovery procedures, security procedures, whom to call when something goes wrong, and so on. It also includes pointers to the documentation mentioned above and to the data dictionary (for data store resources as a minimum).

9. *The data dictionary.* The design team, which includes data administration and database administration support, has a lot to do with updating the data dictionary. Whatever was not done in the systems analysis phase of development, particularly regarding the contents of the logical and physical databases and the external database views for the application, must be done now. This includes the data modelling which was described in some detail in Chapter 10.

10. *The cost-benefit reanalysis.* As pointed out in the last chapter, cost-benefit is done iteratively. Particularly regarding costs of development, the designer is in a better position to forecast than was possible during analysis.

11. *The physical system.* This is the requirement for computing, data transmission, and terminal use, which is now updated by the designer from the analysis specification.

We will now go into these design elements in more detail. The sum of these elements is the story of the design methodology.

SYSTEM-CHART AND PROGRAM-CHART SYMBOLS

The system chart shows all the transactions, data stores, and reporting and the network of computer programs that make up the three-part classic design: Make transaction. Apply to database. Make reports.

Figure 16-2 shows the symbols most often used. The symbols displayed in the various parts of the figure are listed below:

(a) Various symbols used to depict manually processed input documents.
(b) The tape and disk symbols used to show the temporary data-in-motion storage devices and the more permanent data store holding devices.
(c) Two symbols used for a video terminal display.

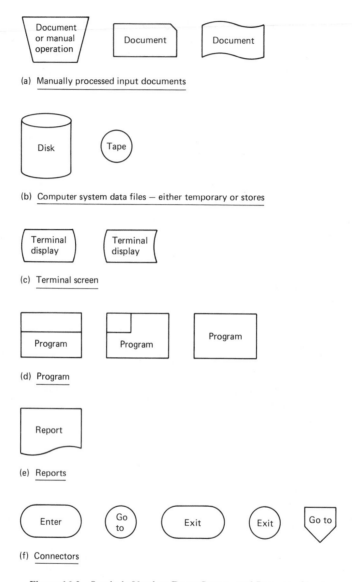

Figure 16-2   Symbols Used to Draw System and Program Charts.

(d) The program rectangle.

(e) Reports and report files.

(f) The entrance and exit and page-break connectors.

The ubiquitous pale green template used for flow charts, data flow diagrams, and (as we shall see) structure charts is also used to draw all the symbols used in

Figure 16-2. Most data processing professionals get such a template when they start out as entry-level programmers and it stays with them in their desk drawer or briefcase all their working existence, no matter what important job they finally end up doing. This little piece of clear plastic is the personal tool chest and pick-up truck of the analyst—one of the few personal tools of the trade.

SYSTEM CHARTS

We will be looking at illustrations of system charts in both the batch and on-line format. They are quite different, even though our discussion about a generic computer solution covering both situations is still valid. Almost all on-line systems contain a batch component. Indeed, past a certain level of complexity, it is hard to imagine a purely on-line system. We will look at an example of such a mixed system.

Remember that in a batch system the transactions are accumulated (batched) through a preassigned time period. When the trigger goes off (time is up), the collected transactions are run into the system and processed to the point where the transaction is completely absorbed into the system. These transactions either become new master record adds, act as deletes, signal the output of retrievals, or change existing master records. Those are the choices. Having performed one or more of these functions, the transactions are gone.

In an on-line system, transactions perform exactly the same functions and need the same attendant validation, but in this case they are entered, usually by people, sometimes by machines, from terminals. They are entered one at a time per terminal (although many terminals could be entering transactions simultaneously to the same data storage files). As each transaction is entered, it is validated, edited, applied to the database, and often creates a new or refreshed screen to view or use. It goes without much saying that the program modules used by a number of terminals have to be available for simultaneous module use—so the support program module must not contain any conditions dedicated to a particular terminal task.

These on-line systems present the main source of contention problems to the database administrator, particularly in the case of a primary database (as opposed to an extracted database) where updating is occurring at the same time as inquiries.

Figures 16-3 and 16-4, taken together, show the generic system chart for a batch system. We say "generic" because all such systems are composed of the same essentials; just the names of the transactions, data stores, and reports change. Often, these batch systems have on-line query/report systems hooked onto the back end.

Figure 16-3 shows the input processing of a batch system. Here we see some of the many kinds of batched input that usually go into such a system. Manually prepared batches of transactions are entered into an on-line data-entry system by data-entry clerks, each having a terminal for entering the data from documents

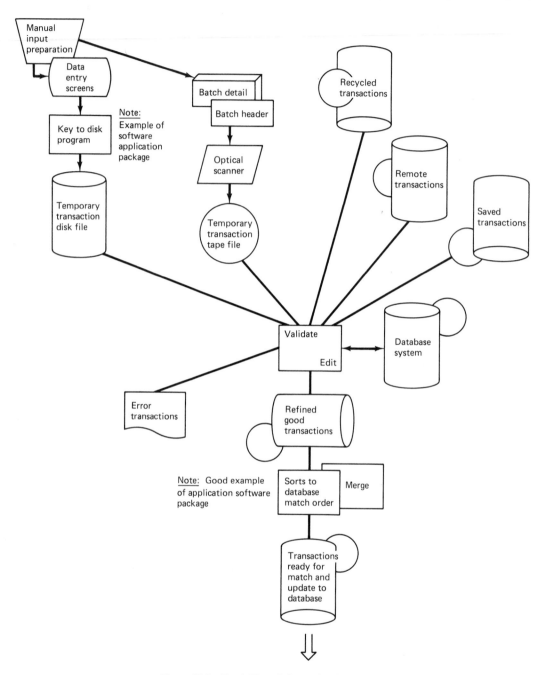

Figure 16-3 Batch-Type Information System—Input Flow.

that are stacked alongside their terminal. There is a key-to-disk program which is doing some simple, advance validation of the input and updating a temporary transaction data store. This program was probably purchased from the vendor that provided the physical system of terminals, minicomputer, and storage units, so here we have, right off, an example of a system program that is acquired as an application package rather than developed in-house.

We also see, as input, documents batched for entry into an optical scanner. This is a very cost-effective way of entering turnaround documents, with the only variables being contained in the header document that precedes each batch of detail documents. Perhaps there are three prices in this input. Then we have one header for each price and batches of otherwise like documents in back of each header. By "turnaround" we mean that the document was preprinted from information already known. Perhaps what is involved is a mailing from a customer list where we know in advance the customer's identification number and the code of the product being offered. This document is perhaps part of a promotion mailed to a customer. The customer checks the "yes" box and returns the document, which becomes a candidate for quick processing through the optical scanner.

Other inputs, already machined, include transactions that have recycled out of this information system (perhaps it takes two "passes" to accomplish a complicated multirecord inquiry transaction in a tape storage system), cycled out of other possibly remote systems and transactions which have been saved in electronic form for the assigned time of processing. These latter transactions, already machined, which are shown in Figure 16-3 going into the validation run or program might just as rightly go into the next run, which is the sort.

Errors out of the validation run eliminate those transactions from this update. The errors are processed by pulling and attaching the original source documents for human review, and new batch transactions are reentered. In an on-line system most of these errors would be fixed on the spot.

There is usually a transaction sort/merge in these batch systems. It depends on the storage device and the volume of master records that need processing. If we have the master records on a DASD (a direct access storage device such as a disk pack) and if the transaction volume is low and if we have no other requirement to look at master records in this run except to apply transactions, then we could apply unsorted transactions to the master file. Usually this is not the case, and we sort the transactions to the order of the sequential master file and apply them serially by a transaction-to-master-file match. Often master files for which there are no transaction hits need to be used anyway because of output reporting requirements.

Figure 16-4 shows the rest of the system processing in this type of batch system. The transactions are matched and applied to the master file, producing an updated master file which becomes the input master file next time the system is run. The updated master is now the current, the input master is now saved as current minus one, and the master file that was used as input in the last running is saved as current minus two. If the master file is on disk, it is periodically dumped

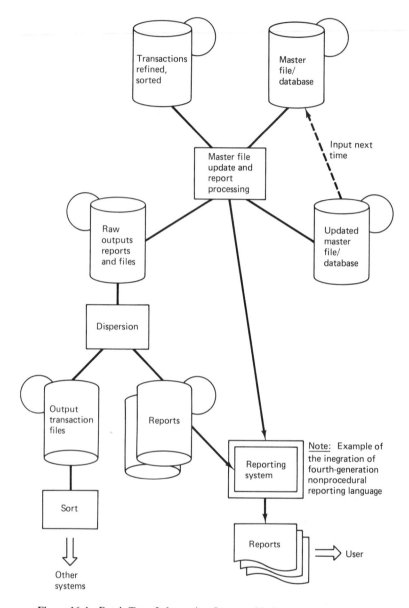

Figure 16-4   Batch-Type Information System—Update and Output Flow.

to tape and stored in a bank vault or under a very big rock somewhere, according to some kind of a disaster plan perhaps devised during the formation of the strategic plan.

The back end of the system covers system outputs. These outputs are, as we have discussed, the reason for the system and are a map of the user requirements. Data for outputs is collected as each master is fully updated and written out.

An example of why we write outputs from fully transaction-updated master files is the case where a change of address might exist as a transaction to a file that contains a name and address for customers whom we wish to promote if they meet our selection criteria. We certainly don't want to promote anyone without the latest information concerning where he or she resides. We certainly don't want to bill a customer when a payment transaction is in the input stream.

Further, we are concerned about the sort order of the transactions applied to any one master record. We don't want to write out an order acknowledgement during an order update and then afterward apply a change of address. First we want to apply the change-address transaction and then the order transaction. Therefore, we will design our transaction codes with this in mind and also sort on transaction code, making sure address changes have a lower code value than order transactions.

Back to the back end of the system: We will usually have a dispersion run to deal out to different output files outputs that have very different kinds of processing of raw to finished outputs information. After that we will have sort/merge runs to get the outputs in the right order for reporting or file delivery.

This sort/merge is another good example of an application package bought from an outside vendor. No one, in their right mind, would attempt to write a home-grown sort/merge. This would be like inventing and putting together our own typewriter because our letters are personalized. This example tells the whole story of how system development is evolving. It is becoming less and less likely that anyone would want to write a report program in-house rather than specify a report to a nonprocedural language system in user terminology or something near to it. Following a little further behind in evolution, it is also becoming less likely that we would want to develop our own on-line input system in procedural code.

In the example shown as Figure 16-4 we have something called "Reporting System" where no details of programs are mentioned. To the extent that we do not have to develop our own report programs in such languages as COBOL we need not specify these programs in any way. These programs are the proprietary internals of our fourth-generation nonprocedural reporting application package.

There are in every system some reports that are so complex that they must be programmed in a procedural language. But who made these reports so complex? Was it the systems analyst? Could the reports have been simplified so that a nonprocedural language could have handled them? Often this could have been done.

System Chart for an On-line Information System

We now turn to another view of the system chart. Figure 16-5 shows an on-line user-terminal interaction for accepting transactions and updating an application data store.

There is less conformity concerning the way an on-line system is shown on an overview system chart. With user-terminal dialogues any number of variations are possible for the same basic problem. Here we have chosen to show major panels (screens) and panel support programs. In this mythical system we have only three types of transactions coming off a main menu: order entry, payment entry, and inquiry. Concerning orders, the system allows us to browse and select from the existing logical records and allows a change, delete, or inquiry after the browse. Also allowed for orders is an add transaction to create a new master record. The payment, according to the vagaries of this system, allows only an update or a delete. An add would be invalid.

In any case when we would transform the database by a change or add, we have a validation program to insure insofar as possible that the data entered is good. The validation/edit program for the add of the order is almost the same as the similar program for the change of the order. The validation logic, as a standard in this system, proceeds as follows: The fields of the transactions are entered all at once by the terminal operator for either an add or field changes. The entry is a matter of filling in the blanks as directed by the screen. When the record is presented for validation, only the first error encountered by the program logic is returned to the screen, highlighted, and with an error message. There is room on this particular screen only for one error message.

The operator corrects this error and reenters the transaction. If there is another error further down the program logic, it is detected and also sent back with an error message. This looping continues until all errors have been corrected. At this point the transaction is edited and passed against the database. This is one common form of on-line validation.

This truncated example of an on-line system is only one of many approaches to on-line entry of transactions to a database. What is held in common is the need to communicate effectively through the screen flow, not to leave the user stranded in a dead end situation, validate all transforms to the database, edit transactions as necessary, and perform database accesses.

Database access for the purpose of file transformation can be a simple change to a one-level database or it can be a very complicated affair involving a multilevel, multipath group of "segments" in a hierarchical database or many tables in a relational database. In either case the file access can be quite a headache. What happens, for instance, if half the segments or tables have been updated and the transaction cannot be fully applied for logical or mechanical reasons of failure? Backing out this transaction can be messy.

Fortunately for us, such stand-alone on-line application development products

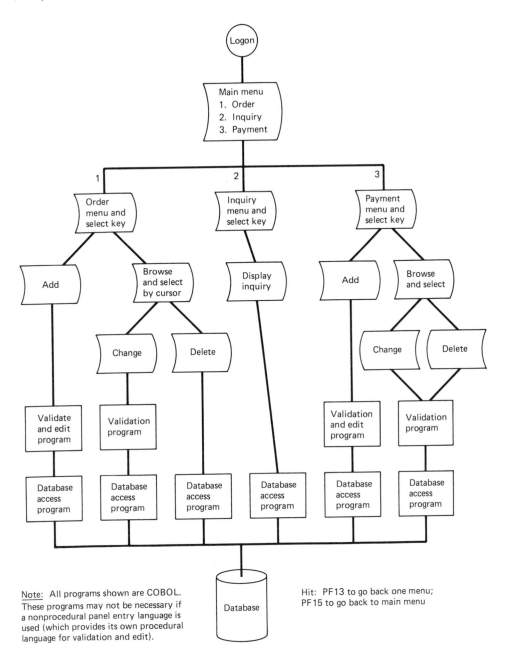

Figure 16-5   On-line Information-System Chart.

as IBM's DMS[1] and Oxford's UFO,[2] to name only two of several available, are capable of handling file access automatically. The same is true with the integrated DBMS packages which we discussed in Chapter 14. These packages also handle panel-to-panel flow such as going from a menu to a browse-and-select panel to a change-data panel and back to the menu again.

We should think the design problem over carefully before we continue to develop such on-line applications using procedural languages. The complexity of the application should be such that this development must be done.

Many fine shops still develop designs that must be programmed in COBOL. Their experts at command-level code have gotten very good at using this approach, particularly since so many systems are essentially the same (as we have gone to some lengths to point out). The author's experience has been that using a good application development tool (such as UFO) will develop most systems faster and less expensively than doing the same systems in COBOL. Every year that goes by makes this judgement more sound. These nonprocedural products are constantly expanding their area of effectiveness.

To complete the story on system-charting the on-line information system we have Figure 16-6, which shows the output handling that accompanies the system charted in Figure 16-5. Figure 16-6 shows an output system driven by a fourth-generation DBMS. The database may or may not be the one that belongs to the DBMS. A DMS may be used instead of a DBMS. We will recall from previous chapters (such as Chapter 14) that a DBMS is a database management system that has an owned database but can also interface with several of the major databases such as IMS. We will also recall that a DMS is a data management system that does not own a database but can interface with and report from the major commercial databases.

In Figure 16-6 we see the typical functioning of the DBMS. Nonprocedural statements called "procs" or "programs" are developed on-line and stored for batch production whereby reports are produced. The job control language (JCL), which is a de facto industry standard, runs these production jobs in an IBM environment. Reports are produced according to a predefined schedule of triggers or are dynamically scheduled by a scheduler. Output transaction files are just another form of reporting but to a different end device.

This system also has a query system, which is shown on the system chart. The database system being queried is hopefully an extracted rather than a primary database, but this depends on the immediacy requirements of the user as interpreted by the systems analyst.

Other combinations are possible in drawing a system chart for an application. Those just shown are currently the most common.

---

[1] Development Management System (DMS)/Customer Information Control System (CICS), IBM Corporation, Armonk, N.Y.

[2] User Files Online (UFO), Oxford Software Corporation, Hasbrouck Heights, N.J.

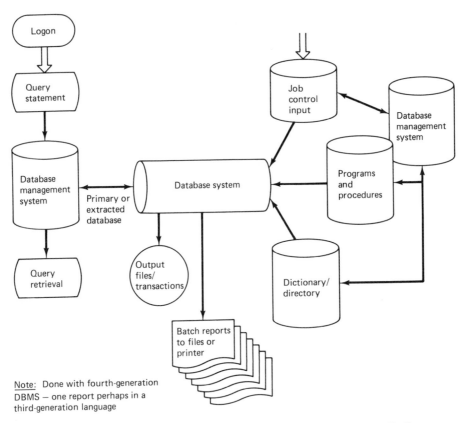

Figure 16-6   Batch Reporting System and On-line Query to Accompany On-line Information System.

Program Charts

It is not necessary to show an illustration of a program chart, since it is exactly the same as the system chart and uses the same symbols. The difference is that the program chart shows only one program in the system. Also it shows the inputs and outputs in more detail. For instance, a table of values read into the program would be shown on the program chart but may not appear on the system chart, which shows only the major inputs and outputs.

SYSTEM AND PROGRAM DESCRIPTIONS

Along with the system and program charts the designer provides some brief text covering at a high level what the system or the program does in the way of processes. This text also goes into a little detail concerning the inputs and outputs and triggers of the system/program and discusses the contents of the stored data resource. This information allows the users of the system design detail specification to understand the big picture, so that they may properly place the little detailed picture. The detailed picture is the modules that make up the programs (the program should be comprised only of these modules) and the specific serial processing steps in each module.

An example of a program test description might be:

Program AA02: The Post Validation and Edit Sort

Program AA02 passes as input all the validated transactions that are entered into the system through Program AA01, the validation and edit program. These transactions consist of two separate transaction data sets. One data set is the customer file transactions. The other data set is the product file transaction. This sort must collate each data set into the following algebraic ascending order of fields:

Customer output transactions:
   fieldA
   fieldB
   fieldC

Product output transactions:
   fieldD
   fieldE
   fieldF
   fieldG

The AA02 sort has a generalized sort criteria which can accommodate both output sort criteria in the following way:

Sort Control Field	Customer Output	Product Output
fieldV	fieldA	fieldD
fieldW	–	fieldE
fieldX	fieldB	–
fieldY	fieldC	fieldF
fieldZ	–	fieldG

The sort control fields shown above are built in front of the existing transaction by taking an exit in the first pass of the sort at E15.

In the last merge pass of the sort the sort control field is stripped off the output transactions, the sort output is deleted, and two special data sets are used to write the output—one the customer data set, the other the product data set. This is done in exit E35.

This type of information is not found on the program chart. It is necessary information to allow the implementation team to understand the workings of the modules that must by programmed to customize the sort application package to meet the requirements of this particular use.

The system description is similar but not as detailed. As with the program description, the overall mission of the system is stated briefly. Each program in the system is described without getting down to the module level. Sort AA02 might be described in the system description this way:

> Program AA02 is a sort that takes the validated, edited transactions out of AA01, sorts them, and splits them into two transaction files. One transaction file is to be applied to the Customer database and the other transaction file is to be applied to the Product database.

THE PROGRAM TRANSACTIONS DEFINITION

This design output is in text form. It could be written as paragraphs or it could be made in table form. What must be given is:

- the transaction code
- the name of the transaction
- what the transaction represents in user terms (Payment, Cancel Deceased)
- the purpose of the transaction (for instance, indicate on the customer file that the customer is deceased)
- the effect on the database system
- outputs produced during the process of applying the transaction to the database (perhaps a paid order generates an acknowledgement letter to the customer that includes another promotion or a premium gift)

Here is a situation where the documentation is going to duplicate documentation in the form of detail module charts which will show in another form the processing shown above as text. The author advises writing this text, even though we will be charting the same processes. This type of documentation can go in the users manual where the module charts are not appropriate.

THE PANEL (SCREEN) FLOW DIAGRAM
AND THE PANEL "PAINTINGS"

This design output is done for on-line systems. When we talk about the "user-terminal dialogue" in general terms, what we are discussing in specific design terms is the panel flow diagram. No design element is more important in the development of on-line "dialogue" systems than the panel flow. What we are discussing here is the system of menus, submenus, data panels, and help panels that are, to the user, the reality of the system.

Program modules are directly related to these panels. It is possible to design and implement an entire on-line system with no other documentation than the panel flow diagram showing in-line modules for such functions as validating input data and actual panels depicted as the user sees them. These data panels are the transactions. When the user sees the payment panel or screen to be filled in and entered, the user knows all about the transaction contents. Add to this design output the charting for the support modules, if present, and it is quite possible to do the implementation.

However thorough the designer needs to get to produce the necessary documentation, it is advised that the panel flow and the actual panel views be done as early as possible and used for consultation with the user. This implementation must be done during the design phase, not later. It is not necessary to be able to actually work the panels so as to really update a database or retrieve data to show the panel flow to the user.

What is done is to put up the panels without anything in back of them except the ability to branch to other panels using the instructions on each panel: "enter 1 to add—2 to delete—3 to browse—4 to change—5 to return to menu." Those branches should work for the demonstration. Using the appropriate application development productivity aid, this panel flow can be put up very quickly from the panel flow diagram.

Of course each panel must be "painted" (show the panel literals as they will actually appear in a live system). The panel flow diagram should always be accompanied in the design package by a "painting " of each panel. This can be shown on panel description layout forms or, better yet, as a printout of the developed panel right off the tube.

When the users see their system actually on the terminal, they know something is happening. More important, the user can point out mistakes and make changes early in the design phase, thus significantly reducing costs and preserving the good morale of the design and implementation teams and the users.

Figure 16-7 shows an example of a section of a panel flow diagram. The figure shows a typical panel flow where a function is selected from a menu. The value associated with the function, such as

1 – add a new record

2 – browse and change

3 – browse and inquire

4 – browse and delete

5 – return to main menu

6 – help

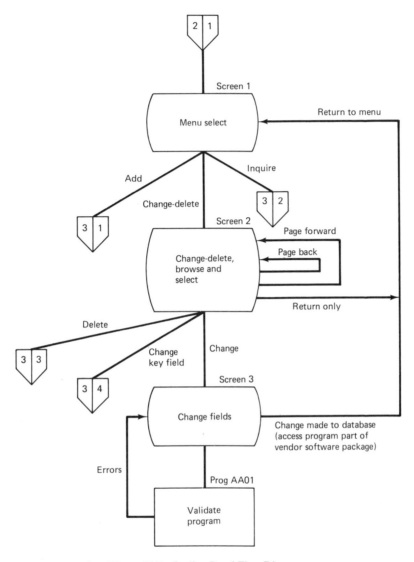

Figure 16-7 On-line Panel Flow Diagram.

is placed in a standard location, such as over the cursor in row 1, column 1, and upon pressing the enter key we go to the panel associated with the function. In this instance we go to a browse panel.

The browse panel usually shows just enough file information, one line to a record, to identify the record we are interested in. Perhaps we see the record key and a description field. In the example shown we can do several operations from the browse screen:

1. select a record to delete
2. change a key field
3. select a record to change non-key fields
4. page forward
5. page back
6. return to menu

To select, we chose a function and move the cursor to the record line that we wish to select. There are many ways to devise these panel flows, which makes it all the more important that we show this panel flow to the user early on, when we can make many changes in a matter of minutes or hours.

This panel flow which we have touched on is, of course, under control of a program. A majority of panel developers now use some kind of application development software to automate this laborious task of panel flow control. What is often not automated is the panel data validation, and what may or may not be automated, depending on the package used and the complexity of the problem, is the movement of data to and from the panel and the database system. We say "system" because more than one database may be accessed from a single panel.

To change data on a panel as in the Figure 16-7 example we must first access the data and bring it to the panel for viewing. This may be as simple as accessing a flat table from a relatively simple one-level database structure, or it may be as complicated as accessing table (segment) data across multiple, multilevel, multipath databases in a database system. (We discussed such database systems in Chapter 10, "Modern Database Design.")

In Figure 16-7 we are assuming good panel flow development software and a database access problem that is not beyond the capacity of the package to handle automatically. What we are not assuming is that the data validation is already defined to an in-line data dictionary. This program module must therefore be designed and coded as part of the application. Therefore we are showing as part of the panel flow PROGAA01, the validation program which takes the changed data from SCRN3 and checks it before updating the databases involved.

This program, upon encountering invalid data, brightens (or converts to reverse video or blinks) the offending data field and sends the panel back to the terminal user with an error explanation message, which is usually located at the bottom of the panel (a designer option, just like everything else on the panel).

A designer option? Hopefully the organization has a book of standards. The layout of the panel should be subject to one set of standards across the organization. It is very confusing to users to see many different formats, depending on who developed the panels.

Panel Painting

Unless the user wishes otherwise (and does not ask for a panel view that violates the published standards), the panels should be kept as simple as possible. Panels should not look cluttered. The choices available to the terminal operator should be clear. Anything the operator might have to do (such as quit in the middle) must be covered by the function choices, and the operator must never, never be left painted into a corner from which there are no logical choices except to turn off the tube.

The problem arises, in designing panel content (and panel flow), whether the design should be slanted toward the first-time user or toward the experienced user. Two weeks after beginning the use of a new on-line system the operator, who was at first delighted with the simplicity and clarity of the panels, now becomes frustrated with the menus and submenus that must be traversed to reach the data. Even such data panels as the browse and select may become a "pain in the neck" to the experienced operator. Two design solutions come to mind regarding this problem:

- Design a two-tier solution to this problem, with one path for the novice and one for the expert.
- Design a solution for the experienced operator and put up with user gripes about confusing screens for a couple of weeks.

The choice taken will vary according to the problem and the funding. For instance, an on-line system for public use where there are always newcomers to the system is quite different from the situation in a data-entry department with low employee turnover.

Whatever level of panel complexity is desired, the dictum still holds: stay as simple and as unambiguous as possible, given the circumstances.

THE DATA DICTIONARY AND THE DATABASE SYSTEM

In Chapters 9 through 11 we devoted much space to a consideration of how an organization's moving data becomes resting data. We described the strategic plan for the organization's total information system, where a high-level review is made of the organization's need for information entities (groups or tables of information). This was done by looking at the total documents and other types of user views of the needed information. This entity information was stored on the data dictionary. Following that effort, and whether the strategic plan was done or was not done, we described the system specification for a particular new information system, which is the main job of the systems analyst.

The system specification not only described the processes supported by the new information system but also defined what data was needed for the system. Through a process of data modelling, with or without automatic support software, the user views, mostly output requirements, were synthesized into a logical or subject database system. This subject database system was then used as a model by the database administrator to build the physical database schema, the conceptual schema, and the external application schema.

All these schemata are retained on the data dictionary, replacing the entity definitions, which are much more fluid and speculative. There is really no getting away from a data dictionary. If there is no formal software called the data dictionary, there must be an informal manually maintained collection of data descriptions which is the de facto data dictionary.

With these database system views in place, what remains to be done before implementation can be tested is to load the databases with actual data.

It is important to realize that there is no clear demarcation between the systems analysis phase and the systems design phase, especially when it comes to building the new databases. This is an evolutionary situation. Also, whatever is not done by the systems analyst must be done by the system designer. A computer system cannot be implemeted without a live database resident on a physical device.

Incidentally, this gradualist demarcation between analysis and design also applies to the cutoff between design and implementation. Design specifications are tending to become much more detailed, leaving very little for the program coder to do besides map very explicit charting and/or text into the language of choice. The extent of design development varies from shop to shop. The larger the shop, the more likely it is that design leaves very little development creativity to the programmer. This may or may not be good. The quantitative information that says it is good is offset by the fact that so much of our best software is developed by small shops run by "super" programmers who do the design also. Probably it is true that the more esoteric and challenging the software, the better it is to have the super programmer analyze, design, and act as lead programmer in implementation. This individual wears three hats, and sometimes these hats go on and off and back on again iteratively. Given a really challenging computer problem of limited scope, it is better to trust one individual wearing three hats than three individuals each with a hat.

There is no doubt that in the case of data analysis and development and the storage of results on the data dictionary, the designer must make adds and changes to the work of the systems analyst.

THE COST-BENEFIT REFORECAST

In the last chapter, which was about cost-benefit analysis, we illustrated the minimum cost-benefit analysis needed to complete the new system specification package. Figure 15-13 was presented as an example. The system was funded by the investors based on that presentation. We said at that time that cost-benefit was an iterative process.

By the time the design package which we are now presenting is complete, there should be very little mystery left except for knowing the effectiveness and problems associated with vendor application packages. The main question left after the accomplishment of a thorough design is how long it will take the implementation team to put the design model into a functioning reality. In many cases this is not difficult to figure closely. We can look at the programs and the program interfaces and the program modules (which we have not discussed yet) and the module interfaces and come very close in our estimate.

Therefore, the reforecast of cost-benefit by the designer is a particularly necessary operation. This reforecast may be quite different from that done by the system analyst. Maybe the analyst figured the system as costing one million and the designer says, no, now we know the system will cost two million. It may well be that the investor will scrub the project at this juncture, fire the analyst, and take a loss on the very considerable investment already made. Maybe the project will be folded into another project or deferred for a while.

It is just as likely that the designer will indicate that the rest of the costs to implement the system and make it production-ready are below those estimated by the analyst. There is a tendency to think that an unknown endeavor will take more time than it actually turns out to take.

At any rate, the same exact cost-benefit exercise done by the analyst is now done by the designer. This should always be done. The user should be educated to ask for it. Probably this education will not be necessary, since the user is usually both very familiar with cost-benefit and very concerned about the costs and time-tables involved.

THE PHYSICAL-SYSTEM REQUIREMENT

We will include in this discussion the *application software requirement*. Sometimes application software is purchased not so much for what is provided by the present version as for what is offered in the next version which is still in development. There is some risk here that the software will not be properly debugged and installed in time to phase in with the in-house development. This can be most frustrating. Vendor software delivery is always more difficult to predict than vendor hardware delivery. Also hardware tends to be more bug free, and what bugs there are can usually be fixed in short time if enough fuss is made. Particularly in the early days of computing it was a matter of course that the hardware would be delivered with a wretched excuse for software or no software at all, and maybe a year later the software, having gone through multiple revisions, began to be tolerable. This was and continues to be most stressful to the almost helpless development team, the disappointed users, and the disenchanted management.

The Physical System

Again we have an imposing problem of phasing. The physical system consists of increased need for computers, storage devices and their control units, data transmission systems consisting of hardware and vendor software, and terminals along with their cluster control units. All this hardware, to the extent that it is not in place already as a result of long-range strategic planning, must be in place according to what can be a complicated schedule.

The first item on this schedule concerns the resources necessary to show the on-line panel design to the user.

The next item to be scheduled is adequate hardware to test the implemented system.

Third, we must schedule the delivery of the hardware needed by the users to begin working the production-ready system.

Last, we have an iterative need to add hardware at those points in time when the user has a production requirement to expand use of the system. This includes exporting the system to remote locations, either by extension or by replication. By *extension* we mean putting more points on the existing network. By *replication* we mean setting up a similar system in another part of the organization. Perhaps the organization has a system installed in one organization entity in New York City and wishes to replicate the successful system in another owned organization entity in Paris. Besides the software problems involved, there will certainly be a need for more hardware.

PROGRAM MODULE DESIGN

Program module design is generally considered the "'name of the game'" in developing in-house information systems. All the parts of the total design package we have been covering in this chapter are support documents for the essential document of design, which is the program decomposition into levels of modules and the detailed specification of the processing of each module.

It is from this "spec" that the implementation team will create the code that makes the program. This is done one module at a time, although many modules may be concurrently under development.

We will show two methodologies for developing modular programs:

- the flow chart method
- the structure chart method

Developing modular programs is really the same essential exercise as developing the system specification. We decompose a large, complex problem through leveling and partitioning until we reach a base level where it becomes futile to decompose further. We try, insofar as possible, to defer process to the base level modules. Modules above the base-level are mostly decision-making in nature and act as transport managers that send data and switches (triggers) up and down the module hierarchy. The top-level module acts as the program executive and super dispatcher.

Modular design is best learned by looking at examples, and that is what we will do here. But first we will introduce four terms made part of the lexicon of modern computer terminology by the Yourdon "group,"[3] past and present, who have had such a profound influence on computer-system design. These four terms all refer to module design. They are:

1. *Cohesion.* Cohesion refers to the degree of association within a module. It is good for a module to have a high degree of cohesion on a "functional" basis, which means that all the contents of the module contribute to the performance of only one task. There are other types of cohesion besides functional and the reader is advised to read one of the references cited in footnote 3. *Structured Design* by Ed Yourdon and Larry L. Constantine is good for a comprehensive look into this and the other three subjects. For a concise and satisfactory discussion Gane and Sarson do quite well; so do DeMarco or Weinberg.

2. *Coupling.* Coupling is the term for the degree of dependence between modules. Whereas we want internally cohesive modules, we want loosely coupled, independent modules. This makes a lot of sense. We want to be able to maintain these modules as contained units without worrying too much about the effect a change to a module will have on other modules. What we want to avoid, for instance, is introducing a bug in some random module while fixing a bug in another module. We also want to avoid a situation where we are afraid to enhance a module because we do not know the extent of the risk that other modules will be adversely effected. This module independence implies that we will restrict the data available to a given module to only that data that the module must have to perform its task.

3. *Span of control.* Span of control refers to the number of modules that are managed by a control module. There is a upper effective limit, which some say is between five and nine.

4. *Scope of control and effect.* Scope of effect refers to decisions in the program and what modules are affected by that decision. Scope of control refers to a command module and all modules subordinate to it. Decisions should be placed no higher in the hierarchy than necessary to place the scope of effect within the scope of control. Again, for a good discussion of this problem refer to footnote 3(b).

The author has felt it right to include here a brief mention of this theoretical approach to program module construction. However, out on the firing line of design, these considerations must fit within the framework of good common sense. It is good to have cohesive modules on a functional basis, but the experienced designer will know when it is safe to package several tasks into a module that shares something in common, such as a subordinate module or a piece of common

---

[3] Structured design module characteristics: (a) Ed Yourdon and Larry L. Constantine, *Structured Design* (New York: Yourdon Inc., 1975); (b) Chris Gane and Trish Sarson, *Structured Systems Analysis: Tools and Techniques* (Englewood Cliffs, N.J.: Prentice-Hall, Inc., 1979), pp. 176–222; (c) Victor Weinberg, *Structured Analysis* (New York: Yourdon Inc., 1978), pp. 83–205; (d) Tom DeMarco, *Structured Analysis and System Specification* (New York: Yourdon Inc., 1978), pp. 283–331.

data. Remember, too, that most of the problems with systems are between modules not inside modules. Interfaces are the weak link in all designs, and the more modules the more interfaces.

Regarding what the Yourdon group calls "coupling,"—that is, module independence or the lack of it—this independence has been the goal of information-system builders since the beginning of commercial programmed computing over thirty years ago. We certainly do require modules that can be changed without worrying about other than subordinate modules. We also want to be able to reference a module before we get around to building the module internals as we implement top down.

Regarding module-to-module hierarchical relationships, these must to a great extent follow the logic of the systems analysis. There is no convincing reason to follow any pattern that does not unfold naturally. A payment or order process module probably will have more complexity than a module that controls simple lookups. The module hierarchies will differ here from more to less complex. The work modules at the base level will break up naturally; some of these base modules will be quite large and some will be only a few instructions. That's all right. We will give the complex module structures to the experienced programmers and the easy module structures to the beginners.

Figure 16-4 showed a system chart for batch sequential information processing. The center of that system was the master file update and report processing program. Figure 16-8 shows that program decomposed into its major modules. We will first use the flow chart method to show this program, since we have already covered the flow chart methodology in Chapter 3. We called this method neoclassical flowcharting, since it is something of a revival in some circles. Also we show the data involved with each module.

Figure 16-8 shows "the mainframe"—the level-one control module to which all the rest of the modules are subordinate. This is a "classic," because it has not changed since the first days of commercial computing, and most of the programs of this type are exactly like this example without much deviation. The model shown here can be copied as-is to be the mainframe for any like program. The only exceptions that we can think of to the batch sequential master file update, as shown, are the situations where update and retrieval are done in separate programs or where a group of records on the master file are processed on two levels, the group level and the logical record level.

Terminology: Modules and Subroutines

Many designers and programmers use the words module and subroutine interchangeably as if they meant the same thing. This is quite true. They do mean the same thing. There are two kinds of modules/subroutines:

- accessed by more than one controlling module/subroutine
- accessed by only one controlling module/subroutine for convenience of coding and because it is a cohesive package performing a well-defined task.

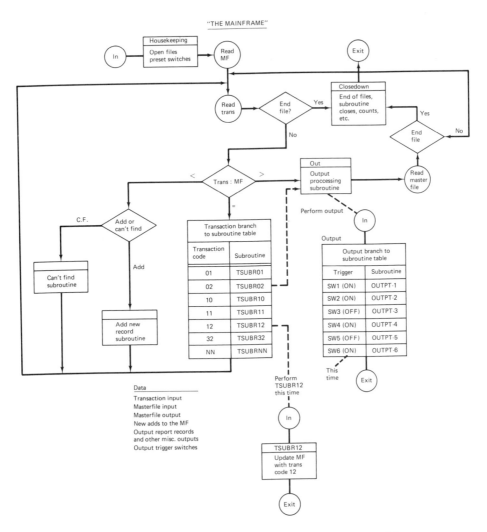

Figure 16-8 Classical Batch File Maintenance Transform of Sequenced Input to Output.

Modules or subroutines can be either called or performed by the controlling module. A called routine is outside the formal boundaries of the source program and is usually linked at object time to make an executable program.

We will feel free here to use the word "subroutine" henceforth, particularly when describing a flowcharted program. When we get to structure charting, we will use the word "module," since that is the accepted convention.

Figure 16-8 shows the charting at the top level of the program. All other charts will be subroutines which are subordinate to this "'mainframe.'" This mainframe routine as shown does all the reading and writing except for the report and miscellaneous output, which is done at a lower level as needed.

This chart shows there are six subroutines performed at level one. Two of these are performed only once in the program. These do housekeeping at the start and end-of-files closedown at the end.

When the program is first entered, all the files must be opened so that reads and writes can occur. There may be certain switches to be set to certain conditions. There will be tables of program variables to be read in and stored, and in this case the master file is read once only, so that proper matching can occur. And there may be messages to display to signify that the program is now ready for the great loop of processing that will come to an end only when both input files are at end-of-file.

When both the transaction file and the master file inputs are at end-of-file, there is a one-time EOF or closedown subroutine which will close all files, force out all pending work, and write end-of-job reports. Finally there will be a message saying the job is done in good order.

The other four subroutines, which together make up the great program processing loop, are the

- add new record subroutine
- can't find subroutine
- transaction update subroutine
- output processing subroutine

By the time the program is fully charted down all levels, there will probably be several hundred subroutines in a key program like this.

The key to the logic of the batch sequential program, matching fully refined transactions to a master file, is the three-way decision which is made after comparing the transaction key to the master file key.

If the transaction key is less than the master file key, there exists either a new master file to be made from the transaction or a can't find, depending on the code of the transaction. Some transactions are meant to set up new masters in this condition; others are meant only to match and change or report on a master.

If the transaction key is equal to the master key, we have a "hit" and a record to update with a transaction (or perhaps a transaction to reject if an insert is a must for this transaction). A hit means we update the master and produce whatever reporting or miscellaneous outputs are meant to occur. The transaction = master hit path subroutine at the second level is shown as a large branching table that performs one subroutine at the next level for each transaction. This transaction subroutine will very likely call other subroutines down several levels. Perhaps there will be fifty or a hundred different kinds of transactions.

In the data flow diagram of the system specification, each of these transactions was very likely a different flow path. They all come together in the design to do their designated function, which is to transform the master file and then disappear (unless they are recycled or rejected). If the reader grasps this difference between analysis and design, a major breakthrough in understanding has occurred.

In an on-line system the same process has occurred even though the master (the application view of the database) has been updated in direct random order rather than batch sequentially and the updating has occurred one at a time from many terminals. This doesn't change the fact that from the designer's perspective the process is the same: a lot of transactions to pass against a lot of logical masters.

The analysis has not been mapped into design by taking round bubbles and mapping them one-for-one into square module boxes! What has happened is what we have just shown.

The final subroutine at the second level is the branching routine that occurs when the transaction is greater than the master. This means that the master is fully updated. For instance, if a change-of-address transaction has come in to the program, it has already been applied to the master, and we have the freshest possible delivery address.

In this system as shown there is probably an equally compelling function to be carried out along with the transaction update function. That function is to examine every record in the master file to see if a given record is a candidate for a writeout as an internal or external report. An external report might be a promotion to a customer at the best possible address. An internal report would give a performance snapshot of the master file. These are the user views.

It may be that the transaction file updates only 5% of the masters during any given run. The reason to run through the file sequentially is that 100% of the records may need to be looked at for output reporting, or file purging, or a number of other possibilities, according to various triggers that are probably read into the program during housekeeping. After all, why would we go to the bother of maintaining a master file except for these types of reasons?

In Figure 16-8 we have a subroutine called OUT which is performed to all fully updated records if at least one switch is on. All the switches are tested at once in OUT or OUTPT. OUT performs OUTPT, which could look like this when mapped into COBOL:

```
OUTPT.
    IF ALL-SWITCHES OFF GO TO OUTPT-EXIT.
    IF SW1 ON PERFORM OUTPT-1 THROUGH OUTPT-1X.
    IF SW2 ON PERFORM OUTPT-2 THROUGH OUTPT-2X.
    IF SW3 ON PERFORM OUTPT-3 THROUGH OUTPT-3X.
        .
        .
        .
    IF SWn ON PERFORM OUTPT-n THROUGH OUTPT-nX.
OUTPT-EXIT. EXIT.
```

There may be dozens of these subroutines at the next level. Each will perform subroutines which will perform subroutines down to the base level, which will perform system subroutines that are not of interest, such as date format or square root. These may not be base-level; they may be regarded as sub-base-level.

It is quite easy to establish a subroutine-naming standard which identifies the subroutine flow so that we don't lose track of this giant pyramid of leveled subroutines. This must be done, and doing it does not require much creative effort.

The main difference between the flow chart method as just presented and the structure chart method which we will soon discuss is that

- the flow chart method shows modules as boxes within boxes within boxes. . .
- the structure chart method shows boxes under boxes under boxes. . .

The other important difference is that structure charts show the data and switches on the chart between the boxes, and the flow charts show the data and switches along with and inside each module process flow chart. Note the data and switches in the bottom left of Figure 16-8 which are associated with MAINFRAME.

TOP-DOWN IMPLEMENTATION OF A DESIGNED SYSTEM

Still using Figure 16-8 as our example, we can see how a system can be delivered to the user long before the design is fully implemented. In fact we can implement before finishing the design if we wish. In this example let us say that processing orders and payments is 60% of the benefit and the other 45 transactions account for 40% of the benefit.

Having gone in some detail into MAINFRAME, which is level one, and controlled the modules in level two, we can see clearly that, using a language like COBOL, it would not take one programmer too many hours to code (or copy the code) for these levels. Having done that, we must make sure that just the orders and the payments pass through the validation/edit/sort input system processing. In the main program shown in Figure 16-8, following the coding of the top levels, we code the order and the payment and then use a fourth-generation language to make the finished reports from the raw output that is produced. Then we have a deliverable system that is something more than a prototype. Much to the delight of the users and the management, the system is running in at least the systems test mode long before the last transaction and the last report are accomplished. This is top-down design and implementation. This is modular design and programming.

Not only have we produced the orders and the payments but, just as important, we have debugged and made field ready the main structure upon which all modules will hang. Now the problems in implementation will be circumscribed low-level module problems if we keep our modules cohesive and loosely coupled. These modules can be distributed to coders for mapping to a language and unit testing. The role played by each coder is very circumscribed, since the module/subroutine includes the data and process needed for just that task.

Module Internal Design

We have already given the various methodologies used by designers. This was done in Chapter 4. A review of the discussion associated with Figures 4-7 through 4-11 will reintroduce the reader to the options available. It boils down to whether we show the coder, and other concerned people, a detailed text or pseudotext description or whether we define module processing internals with a flow chart. Chapter 4 covers the pros and cons involved with each method.

The flow charts or text may be oriented toward a problem view of the processing internals or may be more technically oriented. Example One below shows problem-oriented text; Example Two shows text oriented to the coder and probably not readable by any larger audience.

Ex. One:

Test to see if Quarterly Net Orders Report is to be made this running of the file. If yes, write out a report detail line.

Ex. Two:

IF QUARTNOR-SW EQUAL TO ONE PERFORM QUARTNOR THROUGH QUARTNORX.

Example Two shows a situation where the designer is really doing coding as a definition of module internals. This is not uncommon. Often modules internals such as this example are given to junior designers. The coder (programmer) is left with very little to decide except how to take this "design" and put it into machine form in a production environment.

RESPONSIBILITY FOR DATA STORE DEFINITION

While process logic and transaction data definitions go from analyst to designer to implementor, the data store development flow is somewhat different. In the larger, well-run shops the analyst and/or the designer model the data requirement from the user views as defined in Chapter 10, creating a database logical schema which is then translated into physical, conceptual, and application external database schemata by the data administration (DA) and database administration (DBA) teams. The DA defines these databases to the data dictionary and extracts an application view of the data (by loadable quanta such as segment or table). This loadable segment or segments definition can be written to a library data-set, which can be loaded directly into the procedural language program or program module.

The implementer gets the names of moveable loads, such as segments, along with the module definition. From these names the connection can be made to the data dictionary through the DA group.

Data can be loaded into this database through batch data conversion programs or through entry screens or other techniques of data entry. Essentially a series of ADDS will load the data, so a partially implemented system where adds are coded can be used to load the data onto the database/data store.

Loading data into a sophisticated fourth-generation DBMS is the same, but much of the action is off-stage. The whole concept of program modules is out the window unless exits to procedural processing must be taken owing to processing complexity, in which case we are back in the program module business.

In relatively simple decision support systems which capture and retrieve information in straightforward ways we would still model the data from user views and arrive at a database concept. This concept would then be defined to the inboard data dictionary along with the validation rules. Then we would use the DBMS capacity to build panels for data entry and retrieval, and we would also use the automatic reporting facility to extract both predefined and query information. We would extract queries by saying to the DBMS:

Dear DBMS,

From the PERSONNEL file please print me a report showing Employees and Employees-Department and Length-of-service and Wage-band if they are fluent in Chinese.

Sincerely yours,

This facetious example actually is wordier than a more realistic example would be.

In the case of the DBMS just mentioned, the DA would probably be responsible for defining the concept to the DBMS data dictionary. This doesn't leave too much for the programmer or even the designer to do, and thereby hangs many a tale of conflicted interests.

In this situation the analyst would assume the remaining design responsibilities. What will not go away in the brave new world of future information systems development are the analyst and DA functions—the analyst to understand the current system and invent the better system, and the DA to manage the increasingly more complex information and metainformation (information about the information).

GLOBAL AND LOCAL DATA

This concept carries the modular concept for process into the area of application data. Since it is possible to store data on an applications library as separate loads, we can classify the data load as being systemwide, programwide, or module only.

It is then possible to configure the production system to always load global data by program and load local data by module. For what it is worth, this makes it possible to store the values ONE (1) and TWO (2) once only rather than in every program as local literals. More importantly, tables which apply to many programs across an information system can be maintained "globally."

This brings up two important standards for systems written in procedural languages. Never use literals within a language statement (SWITCH = 'YES'). Never put constants into a program that can be read in as a table of constants. We want "table-driven" systems, so that if the system changes constantly in a predefined manner we change tables and not program code. Many think the combination of these two situations is one of the main reasons for high maintenance costs in procedural-language information systems.

STRUCTURE CHARTS

We have shown flowcharting as a design methodology in the example of Figure 16-8. There is another methodology which will accomplish the same design tasks. This is the structure chart method. At first look structure charts seem quite different from flow charts, when looking at a top-down modular program. They are really much more alike than different. Here, again, are the differences:

1. Boxes within boxes versus boxes below boxes. Flow charts, as we have seen, have command modules which call modules which in turn call modules—down as many levels as needed to reach the base level. Module AA performs modules AA-1 and AA-2 and so forth. AA-1 and AA-2 are likely to be on different pages than AA but are quite easily referenced, as we can see from this paragraph.

    Structure charts, on the other hand, show the performed modules right below the calling module on the same page (if possible). This is the boxes-below-boxes concept. The code produced from either technique may be the same.

2. Structure charts show the data names and switches or flags on the same page as the module flow on the lines between the modules. Flow charts using the technique described in this book show this data within the boundaries (on the same page) as the target module process flow.

    Sometimes, when doing structure charts of some complexity, the traffic flow on the lines between the structure chart modules creates serious traffic jams, and it is necessary to further abstract the data, eliminate the data-switch flow altogether, or significantly lessen the number of module boxes which can appear on any one page. Since what we are about is drawing a nonlinear picture of a linear process, this is a serious limitation, because we want to see as much on a page as is possible so as to grasp the module relationships and flow.

    However, for many problems, and when done by a dedicated advocate, structure charting can give a view of process *and* data flow that is most satisfying.

3. Structure charts do not handle well the processing which is on the interface path between modules. This is good! We don't want processing that is not in a module. However, we often do want to see key decisions that affect module flow. Structure charts have a clumsy way of showing this decisioning. On the other hand, it is all too easy, when using the flow chart method, to sneak in nonstandard processing.

The feeling for wanting to do this is quite compelling to the designer who is trying to make an elegant design. The cumulative result of this practice is bad news, since we know that interface problems are a main cause of system failure.

EXAMPLES OF STRUCTURE CHARTING

Figure 16-9 shows the symbols and symbol combinations used in constructing structure charts. As in all the other charting techniques we have covered in this book, there are very few symbols needed to construct the most complex design solutions.

Figure 16-9(a) shows a module which does some process. It is used in this one flow only and is not available for access in other paths. Part (b) shows a similar module. However, the two vertical lines within the box indicate that the module is shared by other calling modules.

Part (c) shows a module-to-module relationship or flow wherein module A calls or performs module B. We are told this by the arrow going from A to B.

Part (d) shows module A calling or performing modules B, C, and D in that exact left-to-right order.

Part (e) shows the data and flags or switches flowing up and down along the connecting arrow between modules. An X on the line indicates global data such as an entire master record.

Part (f) shows an abstract example of the module components just described working together to provide design information. There is a command or executive or mainframe module which calls module A. Data is sent down to module A for processing. Shown coming out of module A to a lower level is a new symbol, a branching diamond to handle an OR condition (not the same as a process LEFT first and then RIGHT; this says process LEFT OR RIGHT).

Module A, when it receives the data back from the next lower level of modules, sends the data back to the command module along with a switch. Note that the tail of a data arrow is a blank circle and the tail of a switch arrow is a filled-in circle.

The command module then passes the data received from module A down to module B, which in turn passes the data down a level. When module B gets the processed data back from the rightmost module, it passes the data up to the command module.

The command module receives the data from B and passes it to C. C sends the data down a level for ultimate processing. Here we see a new symbol, a connector for a data flow that goes to another page—perhaps to section 2, page 1.

That, then, is the model for the structure chart model, showing the necessary components. For a more detailed presentation of this charting technique the reader is advised to read Chapter 9 of Gane and Sarson's *Structured Systems Analysis*. They give less than Yourdon-Constantine's *Structured Design*, but quite enough for the average reader. Both books are listed in footnote 3.

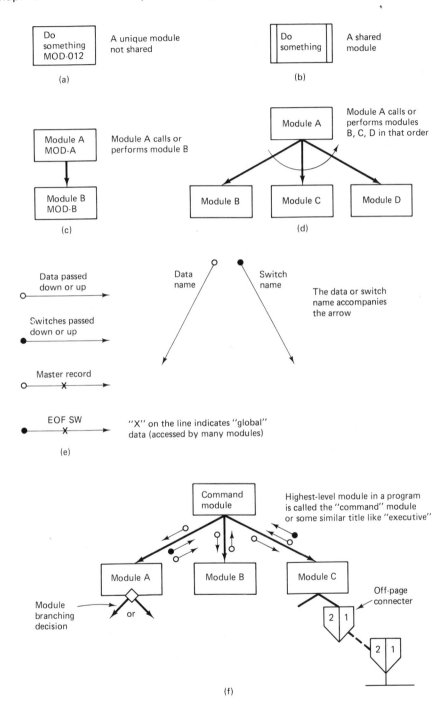

Figure 16-9 Symbols Used in Structure Charts.

Structure-Charting an On-line System

Figure 16-10 shows a structure chart for a change update screen. In this example some kind of a productivity aid for formatting screens is being used. This software has a language of its own, which can link, if necessary, the automatic screen maintenance to a set of programs which do data validation and database access calls. Module LINK004 is the link module and either branches to the submenu from which the change was selected or processes the data changes entered. In the latter case the command module for procedural processing is called from the link. This program is PRGM004. PRGM004 calls modules to lock up the database record, validate changes, do the file update, and release control of the database record. Going down one level, MOD004-3, the control module for file access, reads in the

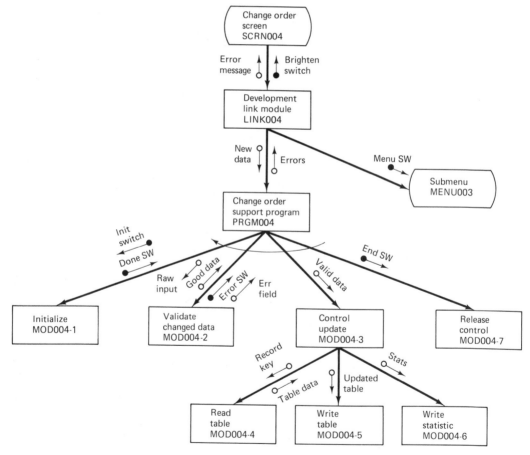

Figure 16-10 Structure Chart Example, On-line. (See Figure 16-8 for comparison with flow chart method.)

appropriate record(s), writes the updated record, and writes out some kind of a statistical control record also.

Hopefully without confusing the issue too much, it should be mentioned that on-line application development software usually goes through many phases, called *cycle points*, between the entry of screen data and the appearance of the next screen or panel. There are a minimum of four cycle points:

- post data entry from panel
- pre read of database segments or tables
- post write of database segments or tables
- pre display of next panel

Procedural program modules can hang off link modules at any one of these cycle points. Different on-line update functions are appropriate for different cycle points. Post data entry we probably want to validate changes or adds. Post database access we may want to check the status code for the results of the file update. Assuming one panel updates two segments, were both updates successful? In some cases on-line data-entry systems can get very complex indeed, and we become more dependent on software productivity tools.

Structure-Charting a Batch System

Figure 16-11 shows a structure chart for the same batch system shown in Figure 16-8. Figure 16-8 shows the design for a "mainframe" command module and indicates how we would flow chart a much-used design for a top-down modular program structure commanded by "mainframe." Observe how this is done in the structure chart method in Figure 16-11.

We see that MAINFRAME is shown simply as a box with a module name. The module internals for this box are very slight. The modules HOUSEKEEPING, GETMF, GETTRAN, MATCH, GETMF, and CLOSEDOWN are performed. The horizontal curved arrow means that those modules under the arrow are performed iteratively for each cycle of input. HOUSEKEEPING, CLOSEDOWN, and GETMF-ONCE are performed once only. As in the Figure 16-8 flow chart version, the bulk of the program—and it is usually the largest program in any system of this type—is controlled by the matching of transaction to master file. In the structure chart view the module MATCH is really the command module, sending data and switches down to the modules that control the work of inserting, rejecting can't-finds, applying transactions that find master files, and doing outputs.

It is worthwhile to take the time to see the sameness between the Figure 16-8 flow chart view and the Figure 16-11 structure chart view. Anyone who does that successfully is in a good position to make a judgement concerning what methodology is best for his or her application design.

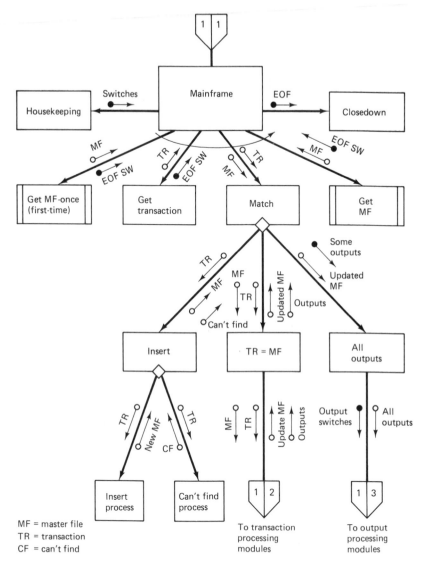

Figure 16-11 Structure Chart Example, Batch.

In the next chapter we will demonstrate through a case study most of the design elements covered in this chapter.

OTHER SYSTEMS DESIGN CONSIDERATIONS

Before closing this chapter on design elements we should briefly touch on three other methods which aid in making design solutions more effective as implementation models.

Table-Driven Systems

We have mentioned in passing the table-driven system. The designer should always consider the importance of all the semiformal data moving around in a system, mostly burnt into program code, which should be formalized as small data stores (tables) used to modify an information system at execution time.

To the extent that the designer is able to capture this critical data and find simple and reliable ways to store and retrieve it, the design will be a joy to implement and maintain. Nothing will have a more significant impact on system maintenance costs.

The nice fact about table-driven systems is that the user usually gets to maintain the tables, thus insuring that the people most concerned about the data are those responsible for it. There are a few good dedicated table-oriented database management systems now coming onto the market. A good example is TLIF,[4] a product made in West Germany which operates under the IBM CICS (Customer Information Communications System). An alternative is a modern integrated DBMS such as we have been discussing throughout the book. This would be a more expensive solution for an organization that did not already have this DBMS.

Design Standards Manual

We have mentioned analysis, design implementation, and maintenance standards. These standards must be kept in some formal document such as a DP standards manual and be seriously maintained and rigorously enforced (but don't go too far with standards and stultify the inventive process with foolish, unimportant, and unenforceable standards!).

Nothing is more destructive to the information software and hardware development process than the lack of standardization. On a national scale it means equipment and software that doesn't "talk" to each other and frustrates users that try to configure an overall system.

On the organization level, standards mean that on-line panels should have a consistent format, messages should be consistent and clear so as to clarify the user-terminal dialogue, and systems analysis and design methodologies should be similar across either the entire organization or at least the major divisional subsystems of a large organization. The choice of methodologies has been deliberately limited in this book to the two most likely alternatives in each area. That is quite enough of a choice.

Standards should apply to the way systems are implemented and maintained. Programming languages which are very versatile should be circumscribed to follow a subset useful to the organization. Maintenance procedures should be written down and adhered to.

---

[4] TLIF, Agena GmbH, Stuttgart, West Germany.

Most of these standards, but not all, are in the domain of the system designer and the data administrator.

In the last analysis the standards manual is a main concern of the management of an information systems department.

Project Scheduling and Tracking Techniques

These techniques run the gamut from simple Gannt charts to computerized PERT and critical-path-method systems. Designers must include in their long list of responsibilities the necessity to include a project tracking method in their implementation plan.

17 THE BAKERY PROBLEM—A CASE STUDY IN DESIGN

Kobler's was in deep trouble on many fronts. The official name was Kobler & Son Family Bakery Inc. By either name the business was losing money. The number of customer accounts and the gross revenue kept increasing but net revenue before taxes was exceedingly disappointing, given the apparent facts.

Kobler's was easily the top quality commercial bakery in the large northeastern United States city in which it was located. It had started off as the type of quality neighborhood bakery where on a weekend morning customers lined up out in the street waiting to get in. The smell of the fresh-baked breads, rolls, and cakes that wafted out the door was one of the keen pleasures of city living, and expensive cars brought customers from the more affluent suburbs back to the old neighborhood just to get hold of the white bags and boxes printed with the name: Kobler.

Somehow, try as they might, the shiny new shopping-center bakeries in the suburbs could not make baked products to match Kobler's.

When Old Man Kobler, as the founding genius came to be known, brought his son Arthur into the business, things started to change. Arthur had gone to Brown University and the Harvard Business School, where he had graduated with high honors. After a stint in a top accounting firm, Arthur had turned down a partnership to return to the bakery. He had a dream.

The dream was to turn his dad's bakery into a large commercial operation while retaining the quality for which Kobler's had become known.

To say that his dream was realized was an understatement. Arthur Kobler succeeded in getting the quality commercial accounts that put the bakery into the big time. These included most of the large airlines, many of the fine restaurants and hospitals, and a large share of the suburban public school and corporate cafeteria business. On this base it was an easy move into the supermarket consignment business, where aggressive route sales staff had to fight for shelf space. Stale returns from this business had to be kept to a minimum.

In the process of this fast growth the physical plant and staff that made up the bakery changed markedly. Now there was a fully equipped bake production operation with master bakers and their assistants which could handle the upward-spiralling expansion. Now there was a route sales staff of over 100 to drive the route delivery trucks, deliver the orders, handle the collection of payments, get the next day's order, and pick up the stale. Now there was a truck service operation to handle a fleet.

Through all of this growth Old Man Kobler stayed in control of the daily operations. He was not about to get his name ruined among his cronies in the old neighborhood by turning out commercial garbage. Arthur could handle the long-range planning, make the big deals, and keep the financial records, but Old Man Kobler stood looking over the shoulder of the master bakers as they converted the old recipes into mass-production bake runs. He could taste the sourdough used in the hard-crust rye bread and tell whether the acid level was all right. If it wasn't, he would spit it onto the floor of the antiseptic lab room and turn red. "'You call that junk Kobler rye?'" he would yell. The master baker would inform him that the laboratory equipment showed the sour level to be the same as the day before, which batch he had also personally tested and finally approved.

"Yesterday was Tuesday. Today is Wednesday!" the Old Man would shout. While the baker tried to figure that one out, O. M. Kobler would change the mix himself. It sounded crazy to the baker, but there were passengers who rode on certain airlines because of Kobler's baked goods.

Old Man Kobler had a personal, confrontational relationship with every route salesperson. The drivers departed from the bakery with fully loaded trucks between 3:00 and 4:00 A.M. and returned to the bakery between 1:00 and 3:00 P.M. The Old Man was always there waiting in the check-in room outside the order clerk's office when they returned. What took place was a scene of barely controlled bedlam.

All customer complaints and cancellations went to the Old Man. When the salesperson came in, there was Kobler confronting the salesperson with the problem. "What are you trying to do to me?" Kobler would ask, eyeball to eyeball. He was also concerned to know about how the customer was paying and what was the percentage of stale to delivered. Despite his seeming harassment, the route salespeople deeply respected Old Man Kobler. After all, he did know their problems, and he never did fire anyone even after threatening to do so. Sometimes he quietly helped out people, particularly the bakers and route sales staff when they got into trouble, financial and otherwise. When there was a death in the family, Old Man Kobler would show up at the chapel to pay his respects. Actually, as it turned out, despite all the irritations involved with working for him, he was not only respected, he was loved.

Then, one day, Old Man Kobler died in the check-in room after a shouting match with a route person. He fell to the floor, blood gushing out of his mouth, and at the hospital, was pronounced dead of a massive heart attack.

Arthur Kobler was stunned. He was a planning expert but had no taste for daily operations. He was a quiet man, a sort of latter-day aristocrat who could not relate to the exuberant bakers and route drivers as his father had done. He promoted the chief order clerk to order manager and the chief baker to production manager and retired back into his office to do long-range planning. Five months later the bakery was in serious financial difficulty. The bakers had autonomy and the drivers were more independent, which was seemingly good, but there were a number of ugly scenes developing and there was a sense of evasiveness in dealings between people.

The bakery had never kept very good records. The process was extremely informal and subject to serious errors. The key to the information flow in the bakery was the order receipt process, which the Old Man had always personally monitored. Here was the essential information that drove the bakery operation:

1. The route salesperson took the daily order from each customer—so many rolls, so many white bread loaves, so many cakes, etc. These were the customer-order transactions.

2. The route salesperson also summed up the total quantity by product needed to satisfy the sum of all the individual orders. This was the total-product-order transaction by route.

3. The route salesperson also filled out a form showing stale returned by product for each customer. This was the customer-stale-return transaction.

4. On many orders the route salesperson maintained the billing/payment records. It was a confusing situation, since institutional orders and telephone orders were maintained by the billing/payment clerks. Payments were turned in on a customer-payment transaction.

5. There were windows between the order clerk's and the billing/payment clerk's offices and the check-in room. The salespeople went to the order window and turned in the order and stale transactions after going to the billing/payment window to turn in the payments. Sometimes there was a conference to see if a customer still owed money. If so, the salesperson wrote this information down on a scrap of paper to tend to tomorrow.

6. The order clerks wrote down all the product bake orders onto a master bake order as each route checked in. Meanwhile phone calls were being made to get the phone-in institutional orders. This was all done under tremendous pressure, since the bake order had to be in the hands of the bakers as soon as possible so that baking could commence and finish in time for the 3:00–4:00 A.M. deadline. Every night was a new production cycle. Meanwhile late trucks were straggling in

well into late afternoon. Even without Old Man Kobler's presence the scene was one of great confusion and tension.

7. Finally the master bake order was sent to the baking operation, where it was broken up into product recipes, giving a bill of materials for each product run— so much milk, so much white flour, etc. This required a good deal of feverish paperwork on the part of the master bakers.

8. From this raw materials list the ingredients were pulled from inventory and sent to the PORTION and MIX operations to start the baking process.

9. When the bake order was done, it had to be boxed and bagged down to each customer order for each route. This route-customer bake order slip went into the front seat of each truck, while the baked goods represented therein went into the back of the truck.

10. Customer bills also were sent down by Billing and Payments to be put into each truck showing amount due to be collected.

11. The stale goods were returned to the bakery where they were converted into bread crumbs, croutons, stuffing, and cake crumbs for cake toppings. What couldn't be used was sent to local pig farms or dumped. Minimal records were kept on stale returns.

This was the essential information process by which the bakery was run, leaving out all sorts of things like purchasing and payroll/personnel records, accounts payable and receivable, inventory control, and many other pieces of information also needed to run the bakery.

It was this essential process which broke down in the months after the Old Man died. It didn't just break down. It was also sabotaged. The route salespeople, aided by the bakers and loaders, began to rob Arthur Kobler to an alarming degree. They couldn't do it to Old Man Kobler (without his knowledge), but they surely did it to the innocent son Arthur. In terms of general systems theory, various subsystems, having lost top down control and proper feedback mechanisms, begun to function as independent systems with goals that varied from the whole-system goals.

Salespeople began to set up personal accounts on their routes which were off the record. Deliberate bakery overruns of product were made available to the route sales staff, and the trucks went out on the road with significant amounts of unaccounted-for baked goods. Stale was even sold illegally to small restaurants in poor neighborhoods.

Anyone not going along with this deception became a subject for character assassination and was forced into leaving Kobler's. Arthur and the new managers generally supported the crooked employees out of fear. These, after all, were the people the bakery had to rely on to survive. But even the payments began to disappear.

The information system was so sloppy and full of genuine errors that it was hard to get substantive facts about the details of the day-to-day operation.

Finally, things got so bad that Arthur Kobler considered putting up the family bakery for sale and going off to teach management science at a little college somewhere. He even suspected that his father had known about the situation and had manipulated it to keep commissions and wages down and unions out. It had probably been going on for years under the old man's delicate control. Arthur now understood his father's dying words to him: "'Watch out for the route SOB's."

More important, perhaps, than the theft was the general confusion that reigned at the bakery, particularly at those critical moments of information and product transfer:

- when the route people checked in during the late afternoon arrivals
- when the master bakers converted the master bake order into a raw material pick and production schedule
- when the baked goods were bagged and boxed by truck, customer, and product
- when the route people picked up their loaded trucks in the early morning hours
- when the route people made their customer stops and delivered the wrong goods and incorrect bills—the airlines in particular were furious with short deliveries or inappropriate substitutions

Perhaps the worst change to occur was the general decline in the quality of the once-famous Kobler baked product.

Matters came to a head when Kobler's was mentioned on a local TV station during a debate by four professors on the "decline and fall of standards of excellence in modern life." Shortly after that, Arthur Kobler was standing in line at a supermarket checkout and overheard two women discussing Kobler's. "Kobler's rye bread just isn't as good since Old Man Kobler died," said one to the other, speaking knowingly.

That was the last straw for Arthur Kobler. The next day he called in an information systems consultant recommended to him by an old school friend who ran a successful business. In the condition the bakery was in, it couldn't even be sold at a decent price.

THE ARRIVAL OF THE SYSTEMS ANALYST

Fortunately, the consultant was an experienced systems analyst who took a broad view of systems problems. The details presented earlier in this chapter are from the findings summary which accompanied the data flow diagrams of the Current System Study which was the first thing the analyst did. Arthur Kobler wanted an immediate solution. What he got and paid for first was a statement of the problem. Without an understanding of all aspects of the problem, serious mistakes can be made while implementing premature solutions.

It took the analyst about two weeks to dig out all the grim details. Lengthy interviews were conducted with employees who had left, with old cronies of the Old Man, and with members of the current staff at all levels. The problems came out one by one. It took a trained interviewer to ask a few critical questions and wait patiently for the truth to come spilling out.

It did not take much longer to see the essential information flow that was the nervous system that ran the bakery. The steps of this process have already been related. The analyst recorded these steps into a set of current system data flow diagrams.

Figures 17-1, 17-2, and 17-3 are the key data flow diagrams necessary to follow the vital transactions through the processes that drive the system. The reader will recognize these figures as identical to Figures 4-3, 4-4, and 4-5, which were covered in Chapter 4. After taking a second look at the system specification and discussing the new requirements and the new system specification, we will proceed to examine the design by which the analyst endeavored to give Arthur Kobler a chance to save a business enterprise well worth saving.

THE CURRENT SYSTEM AT KOBLER'S BAKERY

Figure 17-1 shows ten major transforms needed to process the information in this system. Going into the first transform are all the transactions that drive this system. Actually in a very pure data flow diagram we might show each transaction going separately through the system. After all, the transform flow we are looking at applies to orders more than to payments and stale. Also, payment and stale transaction processing have their own peculiarities not shared by orders. Nevertheless, this DFD, which is appropriate enough for our discussion, will show the main data flow in this enterprise, and we will split the transactions out one level down, as shown in Figure 17-2.

The following bubbles define the high-level context of the system as shown in Figure 17-1:

1. Get orders, payments, stale, etc.
2. Turn in the transactions to the order and billing clerks
3. Convert route orders to bake orders
4. Plan baking requirements
5. Get raw materials
6. Make bake order
7. Distribute bake orders to customers by route trucks
8. Service trucks
9. Load trucks
10. Deliver orders.

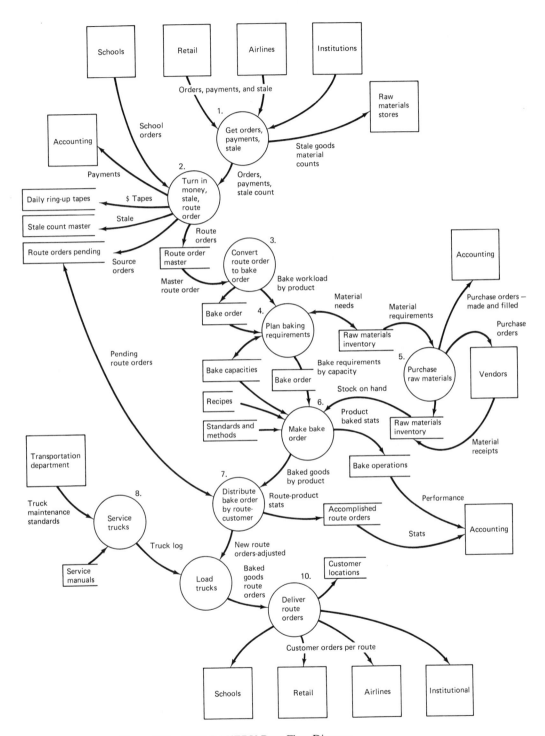

Figure 17-1 RUN BAKERY Data Flow Diagram.

We see in Figure 17-1 all the data stores needed as a resource to run this organization. We will soon see that these stores can be grouped into three databases:

- PRODUCTS
- CUSTOMERS
- ROUTE-SALES-PEOPLE

and that we will need such other data stores as Accounting, Vendor, Raw Materials Inventory, and Finished Goods Inventory, among others.

Figure 17-2 shows the second transform (2) on Figure 17-1 in lower-level detail. We will see that this process (whereby the route salespeople turn in the day's results and place the order for the next day) is the key to turning things around at Kobler's.

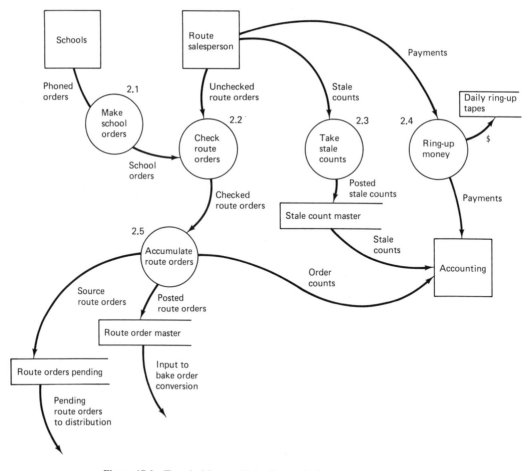

Figure 17-2 Turn in Money, Stale, Route Order.

Figure 17-3 shows what we need to accomplish after we successfully capture the best possible data regarding orders. Using the product recipes for bulk processing of each product, using the quantities specified by the master bake order, the production goes through the nine steps necessary to deliver the products to the hold area, where the product is loaded onto the route trucks just before the salespeople-drivers arrive for the next day's work.

What we do not see here is a lower level to bubble 6.9, the process by which the bulk finished goods is portioned out to the requirements of each route. With good input information, good instructions to the bakers regarding the raw material quantities, and good control over bakery destroys due to baking errors, the information system should be able to completely control the product development and distribution to each route truck. Even a "fudge factor" can be built in and controlled so that certain routes can carry extra product to conform to the necessities of making last-minute changes according to customer's needs.

This is the current system study the consulting analyst made to understand, and let the user understand, the essentials of how information controlled the process and, conversely, of how faulty information was destroying the process.

The following new requirements were developed with Arthur Kobler: These requirements could be met only with a computer-oriented information system.

1. In the future there should be no doubt regarding what product, in what quantity, is delivered to each customer on each route.

2. There will be no doubt as to the order received from the route person. A signed copy of the order received will be given to the route person, and another signed copy will be retained by the order clerk. The database for this information will be secured from unauthorized changes after the the order is printed.

3. There will be no doubt concerning the accuracy of the master bake order, which is the sum of the individual route product orders.

4. There will be no doubt concerning the product placed in the route trucks for next day's deliveries. These will be the baked route orders by customer and product, which in most instances will be identical to the original order. The rare substitutions will be highlighted. All quantities will agree. The concerned parties will receive a copy of this accomplished set of route orders. These parties will include the driver, the truck loader, the customer, and the management.

5. There will be no doubt concerning inventory levels for raw materials and such finished goods (as fruit cakes) that can be stocked. There will be no doubt as to the raw material requirement needed for the future. There will be no unaccountable shortages allowed. (This is possible since the raw materials are available in the city on quick notice—such items as milk and flour, salt and butter. Other items such as vitamins could be stored longer.)

6. There will be no doubt concerning what quantities of raw materials are needed to accomplish a particular bake order. The bakers can vary the batch size as they will and still get the correct end results in terms of quantity of finished goods. The

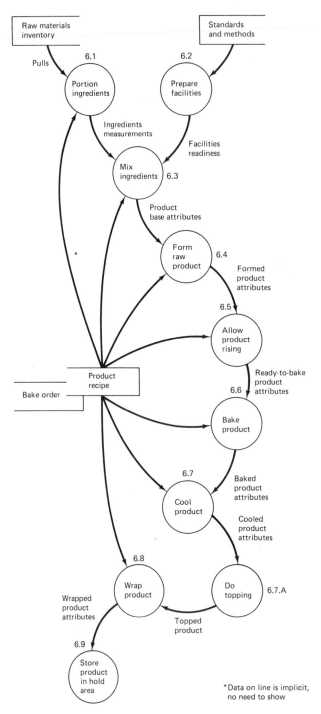

*Data on line is implicit, no need to show

Figure 17-3   Make Bake Orders.

bakers will not be allowed to produce quantities in excess of the computer-developed bake order without the written permission of the plant manager. Arthur Kobler, or a person in management designated by him, would get a copy of such an overrun.

7. There will be no doubt concerning what is delivered to each customer. The customer will sign for the delivery and be billed from the computerized information system according to rate times quantity by itemized product within order by date. Any discounts will be controlled by the computer.

8. There will be no doubt, given a customer's past stale-return history, what percentage of stale to expect from a given customer. There will be no doubt concerning a normative value for stale returns by customers (what constitutes a reasonable stale return). Route salespeople will be encouraged to lower delivery levels to customers with abnormally high stale returns.

9. There will be no doubt about what is a reasonable daily mileage total by route.

The following requirements were brought up by the analyst or by Arthur Kobler and his management team and rejected for the present. The systems analyst was not asked to do alternative studies at this time.

1. The full automation of the baking process.
2. The installation of microcomputers on board each route truck and at key customer locations to speed up the whole process of placing the order and baking the product. Route truck telephones were also ruled out for the present.
3. Any change to the current operation of the baking process itself beyond better information.
4. The installation of any monitoring devices onboard the trucks.

In the course of discussing the effect of the successful implementation of a system that met these accepted requirements, the management of the Kobler Bakeries decided on the following business strategy to deal with employees who had been engaged in illegal operations with a consequent increase in their personal income.

It was decided to try to save these employees and also their side business acquisitions. Kobler would first get evidence of illegal activity, then give amnesty, then provide incentives for acquiring new business—in effect buying out the route salespeople and their baker accomplices.

The following incentives would be established.

1. Special bonus commissions for new customers
2. Profit-sharing plan for all employees
3. No prosecution; conditional amnesty
4. Bonuses to bakers for product successes

In addition, it was decided to do the following:

1. Find a replacement for Old Man Kobler. Some feisty, tough outside, soft inside, operations manager who could speak the special language of the bakers and route people and feel comfortable doing it—the difference being that this individual would be working within the constraints of the new information system and would not carry around the future of the organization in one head.

2. Use a time-sharing approach to the implementation of any new computer system which would be developed by the consultant. Use a fourth-generation integrated DBMS for development. Time-sharing was expensive, but, given Arthur Kobler's skepticism about the future, it was less risky than immediately installing an on-site host computer system. Regarding the DBMS, it was decided that speed of development was more important than customization capability. Also Arthur Kobler did not consider the size of the business large enough to support an on-site programming staff that included such specialists as systems programmers and database administrators. Yet he had some sophisticated requirements. After installation of a successful system, he could see acquiring the necessary physical system and employing a small programming staff (unless he sold the business first).

3. Establish a public affairs program which would include such newsworthy events as sponsorship of quality baking contests.

4. Do not let up on the quality of the product.

Did the systems analyst who was brought onto the scene as an information specialist participate in decisions like the ones just mentioned? Yes. Why? Because the analyst had the experience and knew that whole systems consisted of not just information but information interacting with people and phenomema. What good would a new information system be without the other changes? No reasonable analyst would want to design an information system that could not succeed because of other system malfunctioning.

Still, we can say with some surety that the key problem with Kobler Bakeries was lack of information. None of the people in the system had the information needed to run the system according to the goals that the system (the bakery) had implicitly established over the years: make some honest money by volume selling of the best product in the field.

THE NEW SYSTEM SPECIFICATION

At the level at which we have stopped doing data flow diagrams for this case study we are not deep enough into the processing details to see the processing differences (such as the method by which product is loaded onto route trucks). Therefore, at the level we have defined, the system specification will not change for the new system. The names of the transactions will not appear to change, but the contents of the transactions will. Also the reporting will change. Instead of showing the changes in great detail as the new system specification, we will observe some of this detail as the system design, which is our primary interest in this chapter.

THE SYSTEM CHART FOR THE NEW BAKERY-COMPUTER SYSTEM DESIGN

Figure 17-4 shows the on-line input and update system chart for the bakery information system. We have added some transactions here which would be part of the transaction set in any system like this. Besides add and change panels for orders, payments, and stale, we have similar transactions for order adjustment, billing credit, and change-of-address transactions for the various databases that carry a name and address, such as the Salesperson, the Customer, and the Vendor databases. There are also inquiry panels to browse and select and then display the contents of the Salesperson, Customer, and Products databases.

These panels all flow from the main menu. In this system we will have a program associated with each file, written in a language provided by the integrated DBMS. File access, validation, and error messages will be controlled by this program.

All transaction panels flow from the main menu. This part of the system chart, like the rest of the design shown here, is partial. It can be assumed that there are more transactions and some kind of a batch load program for each database that needs to start with previous history.

This system chart of the the input-update path is similar to the on-line information system shown as Figure 16-5 in the last chapter.

An on-line approach (as opposed to a batch-type system) is necessary as a design solution for Kobler's because it is required that salespersons enter their order on-line when checking in. They can see their order as it develops on their terminal, which is available to them, respond to error messages with corrections, and receive a printed record of their order, which they sign.

They no longer need to total the sum of all customer orders by product, since this is done for them by the computer system as a second report. They leave with a copy of each report. Another copy of the customer order by route report, perhaps revised by the baker, will go to the loading area where the bagging of the finished goods will take place.

When all the route salespeople have checked in, the master bake order will be produced by combining this order information entered by the route people with the product recipes which are stored on-line. The master bake order takes into account bake-equipment capacities as well as the ingredients list and produces a detailed set of baking instructions for each product run. The volumes of raw materials are calculated to produce a precise quantity of product which, when broken down to routes, will come out correctly. There will be a controlled built-in overage.

In other words, all order entry, production, and distribution flows from the route salesperson entering a good order transaction. The same is true for payments

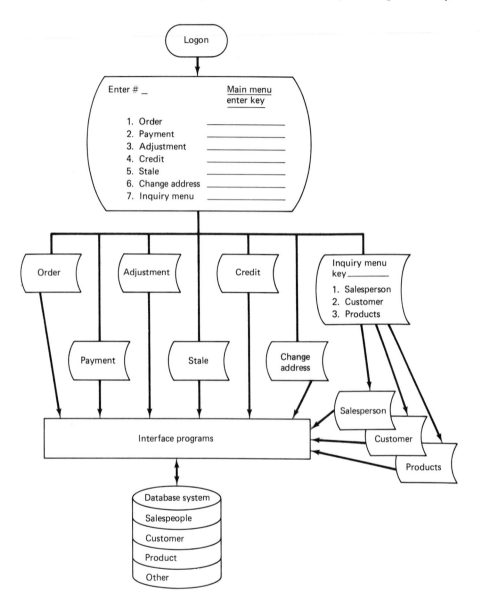

Figure 17-4   Bakery Order/Payment/Stale/etc. On-line System.

and stale. Once the transaction has been entered, the rest of the process of operating the bakery is under computer-developed information control. There are adjustment transactions, as we have pointed out, to handle "situations" that occur.

Figure 17-5 shows the output reporting and query system which completes the system chart. This system is both on-line and batch. We have already mentioned the salesperson's delivery ticket report, which is signed on the spot by the route

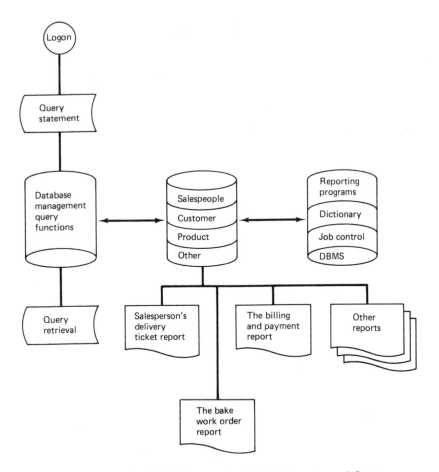

Figure 17-5   Bakery Order/Stale/Payment/etc. Output Reporting and Query System.

person. We have mentioned that a copy of this goes, via the baker, to the loading area to be used to load the route trucks. A copy is kept on file by the order clerk. The important thing is: the database system from which all reports emanate is available for other reports and queries in many different permutations.

Also shown in Figure 17-5 is the bake work order report, which we have discussed. While this is a batch-type report, it is produced immediately after all the daily orders have been received.

The basic payment report is a record of the total amount purchased daily by the customer, the payments and credits received to date, and the amount now due. It is itemized by product.

And then there are all the other reports which will make Kobler's bakery a more viable operation. When we add production capacities data and inventory data and process data (in this case the recipes), we can control the entire manufacturing operation, because what we need to run any production shop is the resource capacity related to the *workload*, both forecast and actual. If we gather history on actuals, we will have the basis for forecasting. The actual workload which is the key to all planning, control, and operations is, in the case of Kobler's, the route salesperson's daily route order. Get that right and the rest, as they say in the bakery business, is a piece of cake.

THE DATABASE SYSTEM FOR KOBLER'S BAKERY

Figure 17-6 shows two views of the database system for Kobler's, (a) the conceptual schema and (b) the external schema. The internal or physical schema can be inferred from the conceptual schema and need not be shown here. The concept shown in (a) is a predefined multipath, multilevel set of databases. A relational database at the internal, physical level would do just as well just as long as we could show dependency relationships between the relations or moveable chunks of information. This means, for instance, that for all PRODUCTS we have a PRODUCT-RECIPE and for all PRODUCT-RECIPE tables or loads we have an associated PROCESS and set of INGREDIENTS. We see these relationships in the PRODUCT database (3) shown in Figure 17-6(a). Other databases shown in Figure 17-6(a) are CUS-TOMER (1), SALESPEOPLE (4), INVENTORY (5), VENDOR (6), and two cost tables, PRODUCT-COST (2) and INGREDIENT-COST (7). Perhaps not enough has been said in this book, or any other book on this subject, about the many little table databases, usually single-level, which are necessary to run any system. Based on the philosophy that all data should be externalized (not included in program logic), we need "table-driven systems."

The various pointers that make up connections in the database system schema are shown in Figure 17-6 as broken-line arrows. These pointers are in support of the external schema shown as part (b) of the figure. The external schema, as we know, is the application view of the database system from which outputs are made and to which, in many instances, input transactions are applied.

Such a set of schemata can be built rather quickly using a modern integrated DBMS, and nontechnical users can often work from the data dictionary which is associated with such a system to build transactions and reports.

We see that in section 1 of Figure 17-6(b) we are interested in a view where SALESPEOPLE is at the first level and we are interested in CUSTOMER and PRODUCT as they relate to the route salespersons.

We see in section 2 that we are primarily interested in CUSTOMER as a view and are concerned with PRODUCT and SALESPEOPLE as they relate to customer.

In section 3 we have the PRODUCT view. We are interested in SALES-PEOPLE and CUSTOMER as they relate to PRODUCT. We also are interested in INVENTORY for PRODUCT. We have two kinds of INVENTORY for which

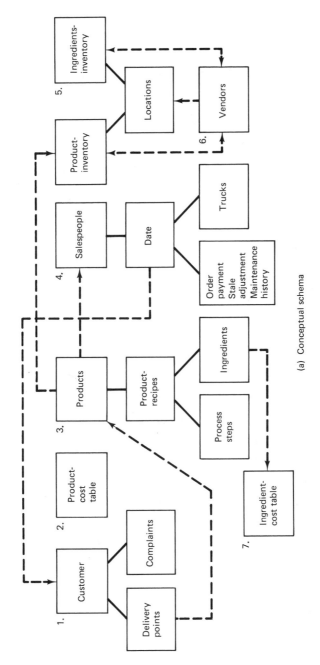

Figure 17-6 Bakery System Database System Schemata.

(a) Conceptual schema

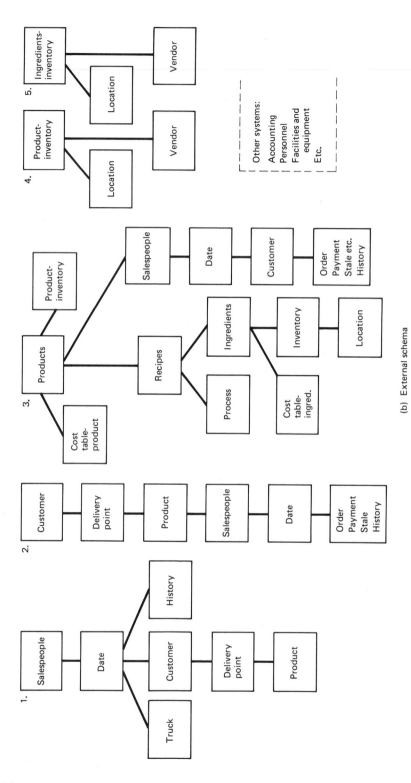

(b) External schema

Figure 17-6 (continued)

we need views: PRODUCT-INVENTORY (4) and INGREDIENTS-INVEN-TORY (5).

This external application schema is the data store as viewed by the designer, once the analyst or designer has modelled the data needed on the conceptual level and the database system is established. From these external views and the dictionary the application design can proceed, which is to say that transactions can be defined and reports made.

We can see by the side-bar in the lower right-hand corner of Figure 17-6 that there are other systems to be designed besides the main operating system, which design we are pursuing in this case study. The consulting team had reasonable hopes of being able to purchase and customize already developed software for the Accounting and Personnel functions.

TRANSACTIONS

Figure 17-7 shows the critical transactions needed to drive the new system. Other transactions, some of which we have discussed, will be needed for the full design of the inputs needed to maintain the database system shown in Figure 17-6.

The Order Transaction

The order transaction needs the following data items:

- the salesperson #, the customer #, the date for delivery, the product #, the product quantity, the truck #, the transaction code which identifies this transaction as an order and various comments as shown in Figure 17-7.

This order transaction is in a raw state going into the system through the panel dialogue, which we will show next. There will be many customers, and each customer will order many products for each salesperson order. Salesperson # and customer # will not have to be entered redundantly for each product #. That is a design requirement. Another design requirement, which is the major complicating factor in this design, is the need to update all three major databases from this one transaction entry.

If this were a batch input system we would edit three output transactions from this one input transaction to update the Salespeople, Customer, and Product databases, respectively. These three transactions would all be sorted to the order of the database and then applied in batch.

Since we have an on-line requirement, we will need random access to the three different databases, and the transaction is not consummated until the three updates (adds, since the daily date is always different) are complete. Besides the three major databases we also will have an impact on the Ingredients Inventory, but this will be done in a batch run after all orders for the day are received.

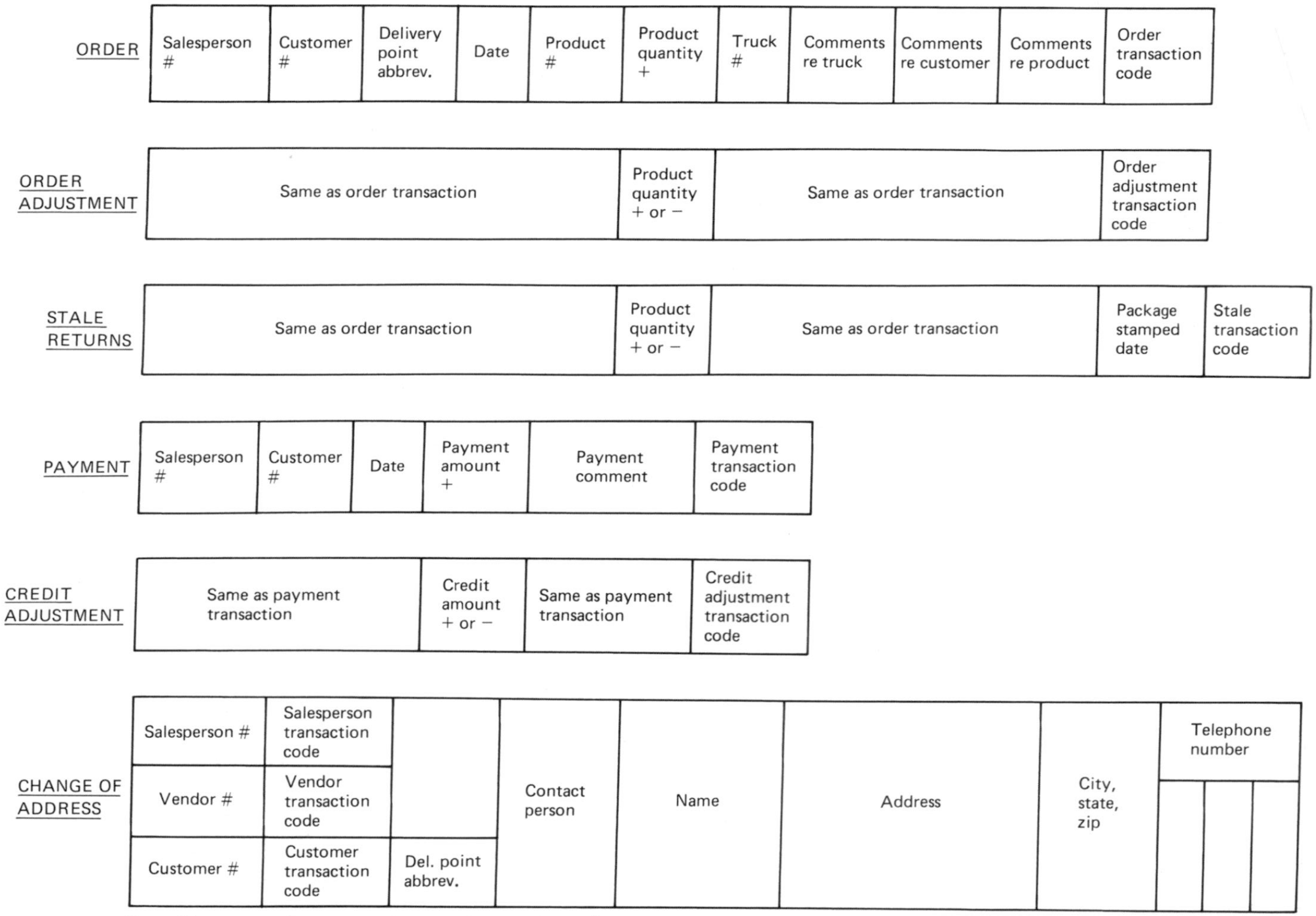

ORDER

Salesperson #	Customer #	Delivery point abbrev.	Date	Product #	Product quantity +	Truck #	Comments re truck	Comments re customer	Comments re product	Order transaction code

ORDER ADJUSTMENT

Same as order transaction	Product quantity + or −	Same as order transaction	Order adjustment transaction code

STALE RETURNS

Same as order transaction	Product quantity + or −	Same as order transaction	Package stamped date	Stale transaction code

PAYMENT

Salesperson #	Customer #	Date	Payment amount +	Payment comment	Payment transaction code

CREDIT ADJUSTMENT

Same as payment transaction	Credit amount + or −	Same as payment transaction	Credit adjustment transaction code

CHANGE OF ADDRESS

Salesperson #						Salesperson transaction code
Vendor #	Contact person	Name	Address	City, state, zip	Telephone number	Vendor transaction code
Customer #	Del. point abbrev.					Customer transaction code

Note: Many transactions are not shown, such as Add/Delete/Change of Customer, Salesperson, Product, etc.

Figure 17-7 Bakery System Transactions.

The Order Adjustment Transaction

This is exactly like the order transaction except that the transaction code is different and the quantity field is allowed to be minus as well as plus. The baker might use such a transaction in the case where an order cannot be filled owing to shortages and a substitution must be made as another order transaction. In that circumstance the unfilled quantity for the short product must be accounted for in the database system and the reporting. This is why we will need a terminal on-line in the bake area.

Last-minute changes in quantity requirements, assuming available finished product, would also be entered as an adjustment. If an airline customer schedules an additional flight and makes a last-minute phone call to increase the order—perhaps minutes before the truck leaves—there is a requirement to be able to handle that order if product exists. This implies a terminal in the truck loading area.

The "bottom line" is the requirement to have the right reporting to account for the order as baked and as loaded.

The Stale Transaction

This is another variant of the order transaction carrying a different transaction code and one more new data item. That is the stamped *use until . . .* date on the product. The shelf life of the product can be determined by the current date minus the stamped date. This shelf life can be tracked by customer and product, and shelf space can be adjusted to minimize stale returns from food markets.

Also new product testing can involve shelf-life analysis. What is the effect of refrigeration? Of removing chemical retardants? Perhaps an all-natural product can be priced to compensate for shorter shelf life. This pricing requires exact information concerning product performance. All the data is stored in this system to support not only the day-by-day operations but also the product sales analysis and sample testing.

The Payment Transaction

This transaction is complicated at Kobler's because both billing and payments occur in two different ways. One way involves the route salesperson receiving the payment; the other way involves the customer mailing the bill back with a payment to the office. The route salesperson always presents bill #1 with the delivered order and sometimes does collect money and sometimes does not, depending on the billing relationship with a particular customer. Regarding a delivery to a school, the bill must be sent to the administrative offices of the school system, whence come the payments. In this case the bill presented with the order is nothing more than a packing slip. All bills after bill #1 come from the Kobler billing and payment department.

All payments received from the route driver are entered while the driver is still in the check-in room. The route salesperson stands at the payment and billing window and hands over the money and a filled-out payment form. The billing and payment clerk enters the payment on-line, and a two-part receipt is printed out and handed to the salesperson. The salesperson signs the receipt and returns one part, retaining the other part for his or her records. Arthur Kobler really didn't want the route people to handle money at all, but the immediate positive cash flow and the reduction in billing costs were compelling reasons not to change these functional requirements. The data items needed for the payment transaction, as we can see in Figure 17-7, are:

- salesperson #, customer #, date, payment amount, payment comment, and transaction code.

The date could be automatically computed, and it would be well to display salesperson name and customer name during the clerk-terminal dialogue. Since the route person is standing at the window, the system can validate the code numbers entered by getting real-time confirmation that the names retrieved for the numbers are correct.

The Credit Adjustment Transaction

This is the same as the payment transaction except for the transaction code and the acceptance of a plus or minus payment/credit amount. This allows adjustment of a customer's account for incorrect billing, such as a "claims paid" situation where the customer has proof of a payment which, owing to error, is not on the Customer database.

Then, of course, we need change-of-address and other change-data-item transactions to properly maintain the file. We may or may not want to have an add transaction which builds an entire database logical record without receipt of an order. In this system we would want such purely file-maintenance transactions as add, change, and delete.

PANELS (SCREENS)

Figure 17-8 shows the Order, Order Adjustment, and Stale Add Entry panels. These panels are filled in by the route salesperson in the check-in room from one of the several terminals provided. The salesperson goes into the panel flow system through the main menu, as shown in Figure 17-4, and arrives at an entry panel. In the example shown in Figure 17-8 it is an order add panel, which is the key to the success of the new system. Showing this panel in advance to the user is a key step in getting design acceptance. The user will often understand the system best by looking at dummy panels.

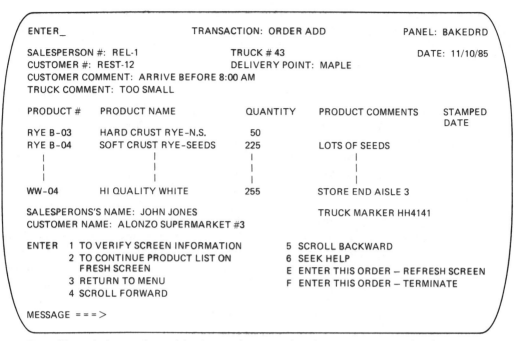

```
ENTER_                      TRANSACTION: ORDER ADD              PANEL: BAKEDRD

SALESPERSON #: REL-1              TRUCK # 43                      DATE: 11/10/85
CUSTOMER #: REST-12              DELIVERY POINT: MAPLE
CUSTOMER COMMENT: ARRIVE BEFORE 8:00 AM
TRUCK COMMENT: TOO SMALL

PRODUCT #    PRODUCT NAME          QUANTITY   PRODUCT COMMENTS      STAMPED
                                                                   DATE
RYE B-03     HARD CRUST RYE-N.S.     50
RYE B-04     SOFT CRUST RYE-SEEDS    225       LOTS OF SEEDS
   |            |                    |            |
   |            |                    |            |
   |            |                    |            |
WW-04        HI QUALITY WHITE        255       STORE END AISLE 3

SALESPERONS'S NAME: JOHN JONES                  TRUCK MARKER HH4141
CUSTOMER NAME: ALONZO SUPERMARKET #3

ENTER   1 TO VERIFY SCREEN INFORMATION     5  SCROLL BACKWARD
        2 TO CONTINUE PRODUCT LIST ON      6  SEEK HELP
          FRESH SCREEN                     E  ENTER THIS ORDER – REFRESH SCREEN
        3 RETURN TO MENU                    F  ENTER THIS ORDER – TERMINATE
        4 SCROLL FORWARD

MESSAGE  = = = >
```

Notes: When 1 is entered at top left, salesperson's name, truck marker, customer name, and product names
are returned for route person to confirm.

This panel also for order adjustments and stale entry.

Figure 17-8   The Order, Order Adjustment, and Stale Transaction Panels.

Figure 17-8 shows a real, filled-in panel. We have the salesperson, the truck, the date, the customer, and the delivery-point code identified. For that customer the product information is entered one line at a time. If the panel is filled up, we have the option of continuing product listing on a refreshed screen by entering (2) in the upper left-hand corner and pressing enter. The new screen has all of the existing information redisplayed with room to add more products. When all the products are entered for that customer, we enter the order by entering (E) and pressing enter. This action brings a new screen, but the salesperson, truck, and date information is redisplayed. Now the next customer's order is addressed.

When all customer orders are entered and the refreshed panel returns, the salesperson enters (F). This action ends the session, and the entire order has been entered. There is a return to menu, and the salesperson can go to the stale entry panel or logoff.

The product information entered includes the product #, the product name (for checking only), the quantity, and any comments (such as "wants lots of poppy seeds").

If the route people want to scroll through an order before ending the session, they can do so by entering (4) and (5) to scroll forward and backward.

There are many ways to do panel flows and panel "paintings" or logic. Many of them are equally good. This order add panel could also have served as an order adjustment and a stale panel by adding the transaction code and the stale stamp date as extra panel entries.

After order entry, the salesperson goes to the order clerk's window and signs the printed order. This printed order has the price shown for each item ordered and the total for each customer and for the truckload. This order eventually ends up, reprinted perhaps, at the truck loading area, where it is used to pick the order for each route truck from the finished, freshly baked products. It will all come out just right! A little but not much will be left over.

There is no unaccounted-for product. Every bagel is entered into the terminal, and the route person is accountable for what is loaded onto the route truck. All billing and payments must relate precisely to the order value of the loaded product. What is not sold is returned as stale where this is relevant.

This is not an impossible design solution to live with. Organizations that survive have systems like this. Any other condition is impossible to live with without an Old Man Kobler to juggle it all precariously. O. M. Kobler probably was able to do this by charging prices so high and keeping costs so low that all indiscretions and errors were covered anyway.

A PROGRAM EXAMPLE

Figure 17-9 shows the flow chart for the order panel support program. This figure shows the modular program in support of the order screen. The first module (ORD1) accepts the screen raw data into a working storage.

The next module validates those fields that do not need file access for checking. Date and quantity could be such fields, and perhaps truck #—it depends on the elegance of the validation. Perhaps the fact that the truck number is between a lower limit and an upper limit would suffice. Or perhaps the designer wants the user to maintain a Truck mini database table for absolute identification. Perhaps the salesperson # is also checked to see if the usual truck is assigned to the usual person. That would be the ultimate. Each field must be analyzed to see if the cost of the further validation is worth the risk of possible wrong data. General rule: if in doubt, validate to the limit.

After validating fields or data items not needing file access (ORD2), if necessary, a message and a switch are sent to ORD9 to process a single error message. If there are more error messages than one, this is repeated iteratively until the terminal operator has cleared up all the errors found.

Next the information received from the panel is used to do calls to the databases. When a new salesperson or product or customer comes on-line, a special entry must be made preceding an order—or, as an alternative, the first order received sets up the database. In this case the former procedure is employed.

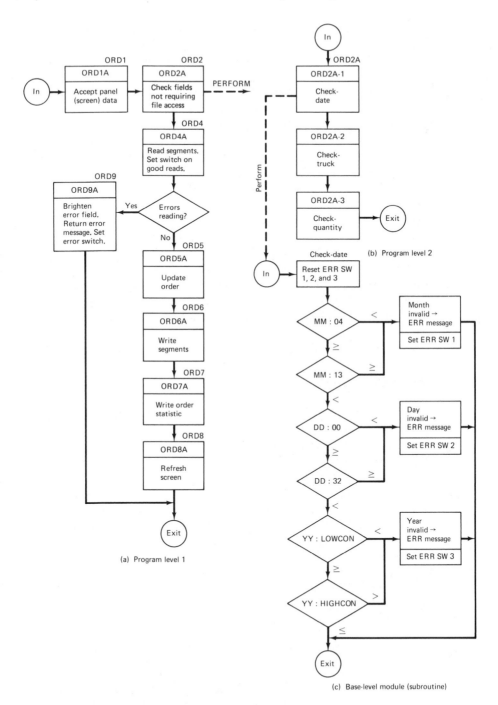

Figure 17-9   The Program for Order and Panel Support.

Therefore, when an order comes in to the system from a panel, it is uniquely an Add only because the date is unique.

This is good from a validation point of view. Given that the original setup was done right, most of the rest of the validation is based on positive hits to the database system. This series of file accesses and no-hit validation is done in module (subroutine) ORD4. If customer # on the panel is wrong, it most likely will not get a hit. If it does get a hit, salesperson must also get a hit for the panel information to pass validation. Products must get a hit on the Product database or there is an error. Just as with the other validation, ORD4 sends error switches and error messages to ORD9 one at a time (or, optionally, all at once). A panel that has cleared these two levels of validation is a fully validated panel ready for database update.

ORD5 is the module that updates the data in working storage, and ORD6 sends the data chunks back to the databases with write calls.

In this case we are updating a database (which has not been shown in any of the figures) called the Order Statistics file. This is an auditing control file, somewhat more complicated than a simple transaction log, which monitors the process and must balance to a snapshot of the database system. By a snapshot we mean passing the databases sequentially in the batch mode and deriving an activity report—which in this case would have to balance to the Order Statistics file maintained by ORD7.

ORD8 is responsible for checking the cursor selects made by the terminal operator and directing the panel flow to the next required panel view. This flow is shown on Figure 17-9 as "(a) Program Level 1."

Figure 17-9(b) shows an intermediate-level module. ORD2A is performed or called by higher-level ORD2 to check date, check truck, and check quantity. ORD2A, like any well-trained middle manager, delegates processing to the base-level worker modules ORD2A-1, ORD2A-2, and ORD2A-3.

Figure 17-9(c) shows the internals of a base-level module ORD2A-1 that checks the date. (Please don't write letters to the author complaining about month with 28, 29, 30, and 31 not being checked for separately. Please don't mention the problem with the year 2000. This is just an example of program charting presented to convey understanding of the design process. At any rate, ORD2A-1 checks for month, day, year characteristics and does things. It sends messages and sets switches.)

Again we note that top-down modular charting of a program design is not dependent on whether the flow chart or the structure chart method is used. Both methods work well or ill depending on the designer's understanding of good module development, as discussed in the last chapter.

REPORTING ORDERS

Figure 17-10 shows the delivery ticket which the salesperson finds in the truck along with the products to be delivered when the new delivery day starts at 3:00 or 4:00 A.M. It is also the order report which the salesperson received in the check-

Date: _____   RS #: _____   Route salesperson's name: _____   Truck #: _____   Marker: _____

Customer #: _____   Customer name _____   Del. point: _____   Comment: _____

Product #	Product name	Special request	Quantity	Unit price	Product price	Bakery order changes

Total order value: $ _____   Past balance due: $ _____   Total balance due: $ _____

Customer #: _____   Customer name: _____   Del. point: _____   Comment: _____

Product #	Product name	Special request	Quantity	Unit price	Product price	Bakery order changes

Total order value: $ _____   Past balance due: $ _____   Total balance due: $ _____

Total all customers this route:

Total value of products: _____   Total past balance: _____   Total balance due: _____

Bakery comments: _____

Figure 17-10   The Saleperson's Delivery Ticket.

in room the previous afternoon after entering the next-day order into the terminal following the turn-in of payments collected from the customers. It is also the ticket that went to the loading dock, or was reprinted at the loading dock in case of changes, to split the bake order to trucks. This is the key operating document in the Kobler enterprise, along with the product bake order for all routes.

We see on this series of tickets, one for a customer, followed by a summary footnote, the following information:

- on all tickets—date, route salesperson # and name, truck # and marker
- for each ticket—customer # and name, delivery-point abbreviation, and a list of products ordered by that customer

- product information on each customer ticket: product # and name, special request comment ("more seeds on rolls"), quantity, unit price (including customer discount in some cases), product price, order changes due to bakery problems, etc.
- at the bottom of each customer ticket (report) the total value of the order, the past balance due, and the total balance due including this order

In data modelling terms this is a critical USER VIEW. Working back from user views like this one, we design the database system and the transactions which make up the three parts of the information-system design. This user view should be modelled first before subsequent, less important user views are folded into it to create the logical database system. If in doubt about this discussion please review Chapter 10.

In the Figure 17-10 example we see two customer tickets and the summary footnote for the route giving the all-customers order value, past balance due, and total balance due.

Along with these delivery tickets we must have the total bake order which the system calculated for the same input: the sum of all route orders and phoned-in orders. The nice part of an information system like this is that the bake order can be ready as soon as the last route checks in, or can be forecast from a percentage of the orders received or can be forecast by looking at past history. There is no reason the bakery cannot begin the bake operations before all the routes are in, since, owing to facilities limitations, the baking is done in several batches.

Figure 17-11 shows the bake work order report. Like the recipes in any cook book the bake order for a product is comprised of a list of ingredients (a bill of materials, if you will), volumes of each ingredient, and processing recipe instructions. In the context of the bake operation the steps include portioning, mixing, forming, rising, baking, cooling, topping, and wrapping or bagging. The report shown as Figure 17-11 gives these pieces of information.

From the information in this report we could go on to automate the bakery operation itself. This is a process flow operation amenable to full automation. Remember that Arthur Kobler rejected this requirement as not justifiable.

THE PAYMENT TRANSACTION PANEL

Compared to the order entry panel, the payment panel is simple. It is shown in Figure 17-12. This payment information is entered through the terminal by the billing and payments clerk. The route salesperson can observe the transaction being entered, if a route person is involved in the transaction, and can also get a hard copy of the returned information.

The clerk enters the information above the dotted line about the salesperson, the date, and the customer and also enters the payment amount plus comments if

Date _____   Time _____

Product #	Product name	Product quantity	Ingredient list	Ingredient quantity-volume	Production comments

RECIPE PROCEDURE
Portion:
Mix:
Form:
Rise:
Bake:
Cool:
Topping:
Wrap:
Other:

Figure 17-11   The Bake Work Order Report.

any. (E) is entered to submit the panel to the system. If the payment is applied, back comes the information below the line; otherwise a message comes back.

We get back the salesperson's name and the customer's name for the numbers entered. We get back the balance due before and after the payment.

DELIVERIES AND PAYMENTS—ACCOUNT STATUS REPORTING

Figure 17-13 shows the account status report, which is a summary of charges and credits to a customer's account. This report would go to customers, route salespeople, and the credit and collections department. The problems at Kobler bakery were not limited to tracking orders and products. There was also a problem concerning delivering product without receiving payment. Cash was flowing out freely but not flowing in properly. Figure 17-13 would be one of several reports dealing with this problem.

```
ENTER_                   TRANSACTION: PAYMENT              PANEL: BAKEPAY

SALESPERSON #: GAF-1                                      DATE: 11/21/85
CUSTOMER #: AIRLINE-5          DELIVERY POINT: SAME
PAYMENT AMOUNT: 112.50         PAYMENT COMMENT: ROLLS COST TOO MUCH
— — — — — — — — — — — — — — — — — — — — — — — — — — — — —
SALESPERSON'S NAME: GARY A. FOY     CUSTOMER'S NAME: FEDERAL AIRLINES
AMOUNT DUE BEFORE PAYMENT: 112.50
PAYMENT AMOUNT: −112.50
BALANCE STILL DUE: 0.00

ENTER  E TO ENTER PAYMENT
       C TO CONFIRM PAYMENT ENTRY
       1 TO CANCEL AND RETURN TO MENU
       2 TO GET HELP


MESSAGE = = = >
```

Notes: Credit adjustment entry panel is the same but credit amount
       can be plus or minus.

       Information below dashed line comes back after entry of
       payment. Must be confirmed.

Figure 17-12   The Payment Transaction Panel.

Figure 17-13 shows order value compared to payments over any span of time desired. We see, for each customer, the delivery date for an order (perhaps to more than one delivery point) and the value of the order delivered on that day corrected by adjustments. For the same date we see payments received adjusted by credits. Then we see the running balance due, which is the net of adjusted order value minus adjusted payments. Since virtually all of Kobler's business is steady repeat orders from existing customers, we can get a good picture of how a customer pays and how a salesperson collects.

The summaries are by salesperson and entire sales staff, shown through time from a certain start date to a certain end date. This report would serve as an audit for the accounting reports, which are not shown in this case study.

This concludes the case sample of some of the most important design elements which were covered in the last chapter. It should be considered a first cut at design for the Kobler bakery, since the case is not taken from any actual design in place anywhere. The purpose is not to show a case for bakery owners to copy but to give examples of the design components in a realistic setting.

SO WHAT HAPPENED TO KOBLER'S BAKERY?

The ending of the Kobler story includes both good news and bad news. The bad news is that trying to find a replacement for Old Man Kobler didn't work out. It seems he was one of a kind—at least in the big-time bakery business.

The good news is that under Arthur Kobler's brilliant leadership, now that

Account Status Report from _____ to _____

Salesperson #: _____   Salespersons name: _____

Customer #	Customer name	Delivery date	Delivery point	Value of order	Order adjust	Payment received	Credits	Running balance due

By salesperson: Value of orders for period   _____

Value of adjustments for period _____

Minus payments for period   _____

Minus credits for period   _____

*Balance now due for period

*Also for all routes at end of report

Figure 17-13   The Account Status Report.

he had control of the necessary information, Kobler's grew rapidly into one of the great national corporations in the food business. Today Kobler's is known as Kobler's United Food Corporation.

Some additional bad news is that the quality of the product went down as the business grew. Today Kobler's United makes products not too different from those of any other mass producer of food. The hard crust rye in the open bag is gone. The Kaiser rolls are soft and there are 50% less poppy seeds (after careful sample testing it was found that this could be done without reducing sales revenue at all).

The pumpernickel is bland and the crust is not chewy. The fancy cake products are not like Grandma used to bake.

But the airlines still buy from Kobler's and so do some fine restaurants. The sandwich white, while not as good as in the old days, is still the best mass-produced white bread anywhere.

Perhaps lower standards are generally acceptable. Kobler's is still considered top-of-the-line premium.

What would the old man have said? "PPlugh. . .," he would have said, spitting the product on the floor. "That garbage is not Kobler's."

The final good news is that Kobler's, which is now public, is rated by securities analysts who specialize in the food industry as a Best Buy. Kobler's has just acquired a quality book publisher.

Arthur Kobler is on the board of a prestigious museum and is a presidential advisor on management techniques. The little consulting group that saved Kobler's from disaster is now out of business. It seems they were not charging enough for their services.

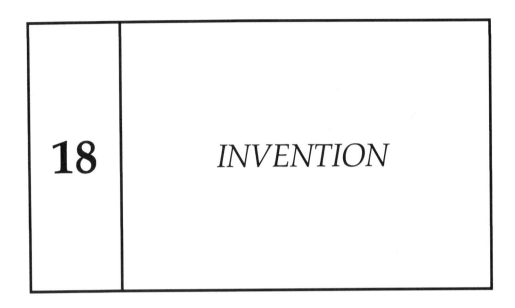

18 | *INVENTION*

There are more things in heaven and earth, Horatio, than are dreamt of in your philosophy.
—Hamlet, Act I, Scene 5

We have now covered, in all the preceding chapters, the various methodologies by which an information system of any size or complexity is specified. From these specifications a new system can be made into a working reality.

The sum of all these methods is a multifaceted tool, in support of problem solving, of no small consequence. Unfortunately, this tool is useless without a power source. This power source is invention. Without invention to interact with method there is no problem solving. The most routine of problems, well handled for the most part by good methods, requires at least some inventive ability. Without high-quality invention there is, at best, a mundane solution: costly, uninspired, short-lived, not supportive of users, often irrelevant, and hard to change. Many problems in information-system development require an extraordinary capacity for inventive skills to arrive at any decent level of system improvement. The user of the system is often better off sticking with the old system than going to a new system where method is not intimately interactive with high-quality inventiveness in the development process.

The main problem with new system development is not lack of planning, standards, and controls—as important, secondarily, as these factors are. The main problem is lack of clarity, lack of whole-system comprehension, lack of inventiveness on the part of analysts and designers.

It is possible, though not desirable, to develop solutions to systems problems with the most informal methods, given that the quality of the inventiveness is at the appropriate level (the level of excellence). It is not possible to do anything with a set of methods not subservient to the creative mind. The finest violin in the world is a useless piece of wood without a gifted player to pass the bow over the strings.

This assertion needs some explaining, since there is a school of thinking in the information-development field which believes that the elegant and powerful method in the hands of anybody trained in the use of the method is the key to development —and that creativity is a somewhat vaguely defined attribute of dubious reliability. This school of thought is cynical about the value of creative people and inclined not to trust their so-called creativity. Method, they say, vigorously applied, is the only winning strategy. Creativity is irrational, oddball, phony and inconsistent.

We will develop the case against this point of view during this chapter (as well as offer some suggestions for relating creativity and invention to the requirements of the whole system). For now, we remark that the trivialization and holding in contempt of the creative energy potential is certainly one of the deadliest of sins in human society. This attitude has caused the fall of gross empires, the latest of which was Adolph Hitler's Third Reich, where the most creative people in the nation were driven out, stifled, or killed as decadent. Himmler is supposed to have said, "'When I hear the word culture I reach for my gun."

On a more personal level, countless individuals have failed to actualize their potential to create because of an emotional attitude which says that there is something embarrassing about being thought of as a creative person.

Some of the material in this chapter may arouse negative feelings in the reader who is on the quest of "hard" techniques which are marketable. This is understandable.

The author's advice to the skeptical reader is to "hang in there" until the end of the discussion, suspending cynicism for the moment. Maybe something will catch the fancy. To those who find this discussion too irrelevant—a fond good-bye. The author has deliberately left this discussion until the end of the book, first delivering, within the limits of his capacities, the best of the working "hard" methodologies in the field.

Nothing that will be discussed here is new. In fact, the best the author has to offer is very, very old information about the creative process. Before further development of this discussion, however, we need to agree on the meaning of certain terms.

DEFINING BASIC TERMS: INSIGHT, CREATIVITY, AND INVENTION

Insight.   Insight is perception into the inner nature or real character of a thing. It results in penetrating discernment and understanding.

Creativity.   Creativity (the power and ability to create) is characterized by originality of thought and execution. Insight is the raw material of creativity. Creativity is insightful processing. Creativity is the assimilation of insight into a way of thinking. In terms familiar to our profession, creativity is inspired process grounded on a knowledge database, and insights are add and change and delete transactions to this database.

Invention.   Invention is the use of the creative imagination to make a fresh finding which can work. Invention is creativity applied to a particular problem that demands a solution. Without insight there is no fresh finding.

. . . and Invention is the product of the Creative process—and the heart of the creative process, the so-called *Aha!* or Golden Moment, is the Insight—that most wonderful and mysterious product of the complete human mind.

DEFINING TERMS INVOLVED WITH STATES OF CONSCIOUSNESS

Concentrative Consciousness

Concentrative thinking consists of processes of the brain conducted in a state of self-awareness and directed toward the analysis of retrieved information in the form of language that can be logically processed to deal with problems. The subject is aware of the ongoing process and is unaware of any hidden process. This processing is generally performed by the left hemisphere of the brain and therefore also is known as left-brain or left-hemisphere functioning.

The Superconscious

Another word for the superconscious has been the "unconscious," which lacks the aptness of expression of "superconscious." We are trying to label a mind-state which transcends the boundaries of the conscious, concentrative logic-processing mind. The superconscious is concerned with processes of the brain conducted in states of mind awareness and unawareness often beyond the control of the subject. These are either altered states of awareness or states of mind functioning of which we are completely unaware. In both cases the process is almost beyond conscious control. We say *almost* beyond control because there are still some inhibiting control systems in effect in the mentally healthy individual. For instance, dreams are usually inhibited from becoming nightmares; if not, we wake up. And the problem being dealt with by the superconscious is the same problem that the concentrative consciousness failed to solve or solved incorrectly.

This almost uncontrolled state of consciousness deals with perceptions of "things in the air," with ancient group wisdom shared by all living things, with basic nonlanguage symbols, with nonverbal patterns of relationships, and with the processing and reformulating of patterns of relationship. (We have already discussed in Chapter 6 the important fact that the difference between data and information is relationship, the relationships between subjects.) These functions just mentioned are ascribed to the right hemisphere of the brain and are also known as right-brain functioning. To make this nonverbal information available as knowledge requires the joint functioning of both sides of the brain—or, stated another way, of both levels of consciousness (ordinary and super).

Also there is a lot of "noise" (garbage?) mixed in with the essential information processed by the right hemisphere, and this information needs left-hemisphere filtering to arrive at effective knowledge.

Right-Side, Left-Side Brain Functioning

Please note that in the above discussion we have defined the left side of the brain as logical (direct verbal) and the right side as intuitive (nonverbal or indirect verbal). Having made these definitions, we are now in a better position to discuss the generic model of the creative process.

A MODEL OF THE CREATIVE PROCESS

Figure 18-1 shows the usual steps involved in creative problem solving.

1. Introduction to the problem. The problem is explained, to whatever extent is possible, to the potential problem solver.

2. The potential problem solver agrees to "sign up"[1] as the problem solver. Signing up is an unconditional commitment.

3. The necessity to solve the problem. Since the problem solver, also known as the analyst, has made a strong commitment, there is a necessity to solve the problem. The analyst is now a warrior in this cause,[2] and no matter what its objective importance, the problem is subjectively of major consequence.

4. The analyst becomes absorbed in the details of the system where the problem exists. Without absorption in detail and mastery of the detail the analyst-warrior is not a candidate open to unusual creative insight.

---

[1] Tracy Kidder, *The Soul of a New Machine* (New York: Avon Books, 1982), p. 63.

[2] Carlos Castaneda, *Journey to Ixtlan* (New York: Simon and Schuster, 1972), pp. 134–51.

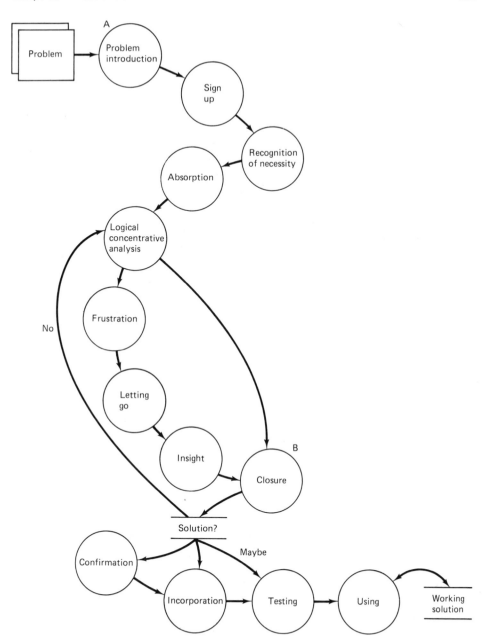

Figure 18-1 A Model of the Creative Process.

5. That part of the consciousness of the analyst that deals with concentrative thinking is called into play. This is done because many problems can be solved mainly with concentrative thinking.

If the problem is indeed now solved, or thought to be solved, we branch to (B) on the model.

6. Frustration, alas, may occur. This is not a rare event in information system development or maintenance. It is something that all systems analysts, designers, and programmers (and users) live with all the time, owing to the enormous complexity of relationships in many systems. The simple cause-and-effect relationships of mechanical systems do not apply to human systems.

7. Letting go. What follows this time of frustration for those who do not truly give up is a mock giving up, a letting go. We say to ourselves, "I guess I'll drop it and sleep on it." Or a team of problem solvers will throw their pencils in the air and someone will say, "The hell with it. Let's go for lunch."

What usually has to be let go is not the problem itself but the approach to the problem which blocks insight. What is usually involved here is giving up a tendency to try harder at seeing things in a particular way rather than turning to seeing things in a different way. Our minds are closed by our methodology, which tends to make us see something in the present in the same way a number of people have seen something similar in the past. This time, however, the backward view doesn't work, and we are forced by necessity to take the hard road to the fresh view.

Or is it the hard road? Beating our heads against the wall can be much harder than just letting go (to those capable of letting go). But we have to let go in an innocent way, not in a contrived way. A team leader who suggests "a creative-inducing tangent into playfulness" and brings out a beach ball to throw around (yet another time) may be disappointed with the results. This is not letting go, any more than counting sheep is letting go for the insomniac. Letting go is a true surrender of will.

Letting go is unclenching the left-brain control, which is very hard for people like systems analysts, who believe that if only they think hard enough in a procedural way they will knock open the enigma of the unsolved problem.

8. Insight. Given that we are truly open to the creative experience. Given that we have done our homework and know the details of the problem thoroughly. Given that we have good methods to gather and order and decompose the details of the problem. Given that we have "signed up." Given that we have tried our best at concentrative thinking and have had to give up on our assumptions—given all that, we stand a good chance of receiving as some kind of a gift from "out there" or "way inside there" the blessed insight. Insight, the "golden moment," is one of the loveliest experiences available to us. More than anything else it is what makes us human and gives the race a vast potential for growth and change. We will discuss the process of insight later. For the present let us again recognize

that insight is the heart of creativity and the ground upon which invention emerges from the creative process.

9. Closure. Closure is when the insightful moment is actualized. Without closure and use of formal methodology to record closure the moment of insight fades away like a snowflake melting. Everyone experiences moments of insight. Very few people can do anything with these insights. Most people are not subtle enough even to recognize the thirteenth wave when it flows into their consciousness and are not trained to snare the insight, magnify it, quickly evaluate it as a true insight and not just some "space junk," and record it for future examination.

Future examination is vital. Closure is vital, but premature closure puts us back in the same frustration box from which the insight was to rescue us. Not all insights solve a particular problem. Some insights are bogus. Some insights just peel off a problem layer to expose the real problem. Any programmer who has been called in at three A.M. to solve a problem that crashed the system (there are always 300 people hanging around idle until the problem is solved), and received insight into how to solve the problem and solved it, only to find another worse problem revealed, will know exactly what we are discussing here.

Therefore, closure, yes. Premature closure—no. We should always be prepared to give up the golden moment of insight as just another meaningless eructation of the brain or as only gold plate.

10. Confirmation. It is now time for what the structure people call the "walkthrough" or peer review. Our invention is documented and presented to the hard scrutiny of our colleagues. We submit our ego (creation is a function of the whole personality) to the harsh words and bruising phrases and occasional pats on the back of those whom we respect. From this point on, the invention is public property and may get revised.

11. Incorporation. The new invention is now incorporated into the entire problem-solving model—let us say a data flow diagram for a new kind of traffic control system for a city.

12. Testing. The model is tested. Engineers have nice words for different levels of testing. They speak of breadboarding and brassboarding. A breadboard is a plug board for hand-wiring electronic circuits in a laboratory environment. Brassboarding is a field-ready but only partially developed problem solution that actually works in a limited way. Unit testing and system testing are breadboarding, and the prototype is the information-system equivalent of the brassboard.

13. Using the invention. Finally we ship the invention out the door to users and wait to see what happens through time. If something goes wrong, we fix it in production.

Incorporation, testing, and using may seem like the aftermath of the creative function. A good case can be made that making a creative idea work is a lot harder than getting the original insight to the problem. There are really quite a lot of creative people around. There are far fewer people around that can take an insight and make the creation actually work in the context of the always messy real world. Taking an idea and making it into something technically, financially, and politically feasible and seeing it gain common usage may not be the essential insightful break-through which is the heart of creativity, but we prefer to pay respect to these functions as a true part of the creative process flow.

HELPING THE CREATIVE PROCESS WORK

The essence of the creative process—the sudden insight—cannot be triggered with any reliability by any planned method. What we can do is create the best possible environment, inside and outside ourselves. This is the environment that is most likely to enable the mysterious process of creation and consequent invention to take place. There is not much doubt any more that this is possible to do.

Before discussing how we can help advance creativity, we must emphasize that in order for invention to come about there is almost always an urgent problem to be solved. The passive creativity possible in a cloistered low pressure place is spiritually rewarding and contributes in the long run to the collective good but does not produce the type of utilitarian inventiveness we are addressing here. We are people with a passionate need to invent.

The Creative Society

Creativity is not available, for the most part, to the hungry, wretchedly poor, downtrodden of the earth. The inventiveness needed to just survive until the next meal is limited in scope. There are usually very few options available to this suffering and blighted mass of humanity.

The stodgy society of the complacent who are quite comfortable with the status quo and have no sense of urgency to invent is not conducive to creativity. It is the cultural radicals who stay partly outside the main culture who most support creativity. In nineteenth-century England it was the romantic poets such as Shelley, Wordsworth, Byron, Keats, and Blake who nourished the creative spark. Curiously, these are not often the people who invent things of utility. They create the ambience in which others with less clearly defined radical positions can invent such things. This was also true of the 1960s, that creative and disturbing decade which really began about 1964 and ended abruptly in 1973. 1964 was the year the shock of John Kennedy's assassination really set in. Out of this disillusionment, aided and abetted by the unfolding events of the Vietnam war, came a great outburst of creativity. Again, as in the war years of the 1940s, anything seemed possible. It all ended during the oil embargo of the autumn of 1973, when Americans competed with each other in countless gas stations across the land for a share of the dwindling supply of gasoline. Scarcity of energy brought an abrupt end to idealism.

From our point of view the interesting fact about the so-called Sixties was that it was not the peace marchers and communards who made the important inventions. It was the in-between folks, deeply affected by the cultural revolution, who decided to stay in school and get their master's degrees in electronic engineering or physics or the new biology or the liberal arts. These were the people who moved the energy of the Forties forward and caused major advancements in the information sciences. Almost all the things we are doing today—all the methodologies discussed in this book, for instance—were defined or advanced in the Sixties. The decade since then has been assimilating what was done then. These have been years of consolidation. Fourth- and fifth-generation languages, the personal microcomputer, the computer-literate end-user, AI, whole-system planning, database design for whole systems—all these ideas were invented but not yet assimilated.

Finally, we have the politically repressive society that worships conformity and looks with fear on new ideas and new patterns of thinking and doing. In general these societies greatly injure themselves by punishing creativity. Curiously, there have been many examples of great creativity existing within a repressive structure. This usually involves a tacit understanding among repressive leaders that certain elite groups must be allowed a circumscribed freedom in order to avoid the total obsolescence of the system.

Family, Friends and School in Relation to Creativity

The family is a very dangerous place in which to be brought up, particularly in relation to the nurturing of the creative individual. Our problem is we have not found anything better.

It seems that for every successful childhood (happy, creative, a natural development to adulthood) there is an example in family life of the unsuccessful childhood engendering fear, dullness, a sense of failure, and an inability to grow up. As regards nurturing creativity, it is difficult to make hard-and-fast rules, except to say that this above all is the place where the creative person is formed. The early years are the crucial years in nourishing the creative impulse which we all have.

Several things seem to stand out in the type of successful family living which nourishes creativity:

1. The child is made to feel special.
2. The parents, no matter what their position in life, display a creative, curious disposition.
3. The parents do not display to the child a fear that creativity will bring inevitable suffering.
4. The child is rewarded in the family for inventiveness.
5. The child is provided with resources that encourage creativity. This could mean making a habit of visits to the local library, reading to a child at bedtime, learning a musical instrument, creative toys such as boxes of dress-up clothing

for play-acting—anything to encourage imagination and participation and discourage passive watching.

6. A standard of discipline pervades family life. The child is meant to feel accountable and responsible. Discipline is pervasive, first outer and then inner. According to the age of the child and developmental level, discipline goes from high control–low trust to high trust–low control.

7. No one gives up on a child who is not performing at a certain level according to parental expectations. There are too many examples of late bloomers. Also, playing the fiddle poorly is O.K. Not everyone is a world class performer.

8. Sibling rivalry is discouraged. Siblings receive fair and equal treatment.

This list, though it could go on, is enough to establish the vital role of the family in nourishing the creative impulse.

Friends are more important than one might think in the development of the creative person. There are such things as bad companions of both sexes. Generally, it is a lot better to "hang around" with friends and lovers who appreciate creativity, who want to be creative themselves, and who encourage it in their friends. We all know of individuals who grew up in families that did not foster creativity and who came into their natural creative heritage through friendships formed in childhood and adulthood. Most people want the acceptance and even the respect and admiration of their fellows. Young people are particularly vulnerable regarding this need. Therefore, we need friends who will accept and validate the creative spirit inside of us.

Schools

What passes for support of the creative person in most public school systems amounts to less than a happy situation. Most creative people either endure school to gain left-hemisphere skills or pass it up as soon as possible for other types of learning, such as the mentor-learner relationship. Sinclair Lewis once told a group of students in a writing workshop to go home and write rather than talk about it. All some of them wanted was an "A". (Not all the problem is with the school resource. What good is the best resource if the input is poorly conditioned?) Generally students are taught to regurgitate the current prejudices of the system and are rewarded for accurate feedback. They are also taught to be lazy and turned off and to respond only to threat systems.

There is little time for experimentation and for a struggle to develop the real thinking process which takes place on all levels of consciousness. In the real world there are almost no true-false or multiple-choice answers except at the bit level (especially in the real world of systems analysis). The real world is full of mystery and chance events and unfathomable cause-and-effect relationships that seem to defy the so-called scientific approach.

Very few artists learn their trade by coloring neatly between the lines (not that that doesn't have a minor place in the scheme of things).

Fortunately, there seem to be on most faculties a few teachers who miraculously manage to retain and pass on the creative spark. The hope of many students who have had the inner strength to endure is to come in contact with this type of teacher at a good university and to survive until graduate school, where the survivors may finally get a chance to concentrate, experiment, dream, image, question, invent, and work with creative people. At such a sanctified place creativity is a living experience.

Personal Commitment

Leaving behind us now—with a profound sense of sadness at the cruelties of this very incomplete and unfair world—that great legion of humankind that never had a chance to live a creative life and that other vast throng that had their chance taken away from them, sometimes without their even knowing it, we proceed ahead to concentrate on the survivors—that very finite number of people who form the creative worker pool. It is out of this group that we select the systems analysts of the future.

Creative insight does not come easily. It is not likely to come to one unless he makes a personal commitment to some very particular task. As we have mentioned before, the creative person must "sign up" as an act of personal will. It won't do to show up on time in the morning and stick it out until "happy hour," shuffling a few papers, chatting on the phone, and making a few deals. In this game, personality will get us nowhere at all. It doesn't matter how handsome or pretty we are. We are given a problem, which is a set of lower-level problems. We have learned a methodology, which is a set of methodologies. What matters is that we solve our problems and make the best solutions possible. This is our commitment, and it may be fun—but it is no joke, once we sign up. Our sense of ourselves grows or shrinks based on the quality of our solutions. This is the basis for the development of the insight.

Feeling Comfortable with the Mysterious

Perhaps, our superconscious is connected to an infinite realm about which we have little direct knowledge. One of the great contributions of the Swiss psychiatrist Carl Jung was to state the case for what he called the *collective unconscious*. This is the case we are restating, here, using the term *superconscious*. Carl Jung's theories on this subject have received very substantial acceptance among the modern scientific communities.[3] We get little messages of a different kind from this realm about which we know so little, to which we are so intimately connected at the deepest

---

[3] (a) Fritzjof Capra, *The Turning Point* (New York: Simon and Schuster, 1982), pp. 361–62; (b) Rupert Sheldrake, as interviewed by Robert A. Wilson, *New Age Journal*, February 1984.

level of our being. There seems to be some vast knowledge base like a great underground river, to which we, along with thousands of others, can send roots to tap. Aldous Huxley called this the "Perennial Philosophy." For the accepting among us this is the basis for knowledge.

The seventeenth-century Quaker, Isaac Pennington, said it well in one of his collected letters: "There is That which can guide you. Oh, wait for it and mind that thee keep to it," There are many ways of "keeping to it" and we will discuss some of them. Meanwhile let us note that when theologians talk of continuing revelation or the active presence of the Holy Spirit (perhaps as opposed to a fully revealed doctrine), it is this simple concept as stated by Pennington that is involved. It is also worth noting that as we approach the Mysterious to which the personal superconscious is felt to be connected, we almost have to start talking in metaphors couched in theological language. That is the closest that language can come to what we are striving to describe.

Receiving Insightful Information Through the Superconscious

The following activities seem to be particularly conducive to opening us up to the moment of insight:

Relaxation.   Anything that relaxes the tightly controlled mind has the effect of opening a path for insight.

Sleep.   Perhaps more than any activity, sleep opens us up to the temporary control of the superconscious right-brain functions. Is this one of the biological reasons for sleep? During sleep the brain keeps functioning, more at some times than others, and the right side is in the driver's seat. (There are dual controls, and the left side has access to the braking system.) Not only does sleep provide us with dreams which can be interpreted (most of us can tell a significant dream from a mundane dream), but sleep also leads the way to the emergence of the insight about our problem. We go to sleep with the problem still running on the back roads of our mind as we deeply relax, and during this sleep the right-side brain, the superconscious, unencumbered by left-brain-fixated thinking, solves the problem and presents the insight to us. (Or is it the left side computing unencumbered?) We wake up slowly, peacefully, to the gentle entrance into our thought stream of the problem solution. What knowledge worker has not had this experience?

Fantasy.   We fantasize and daydream all the time, some of us more than others. These fantasies can be guided, and we have a subtle combination of right-brain, left-brain functioning. Some call this process of guided phantasy *imaging*. Many people have used phantasy or imaging or daydreaming to solve problems. Albert Einstein, for example, imaged himself riding a light beam into distant space, watching himself in a mirror, and received a profound insight into relativity. He proceeded to report on his space trip[4] and influenced much of modern scientific thought.

Meditation.   We will discuss meditation further in a separate section. Meditation is a practice which goes back to the beginnings of recorded history. Meditative practice is at the heart of Buddhism, Hinduism, Christianity, and in fact all major religions. It is the basis for the effective practice of the martial arts. The ancient Greeks were well acquainted with this methodology. The school of mathematics conducted by Pythagoras required an extensive period of meditation before the formal learning process was begun. Alas, meditation is now in a state of disrepute—a reflection on the hype that surrounds all good things, as bored fools move casually in and out of what is available, never lingering long enough to learn what is really going on.

The heart of meditation is one simple thing: the cessation of the internal dialogue. We just completely stop talking to ourselves. No thoughts. No guided trips. No chants. No verbal mantra. We just stop the process and watch our breathing. What happens is very interesting.

That is formal meditation. There is also the constant practice of meditation, which means keeping a quiet, aware mind in the midst of action and confusion as we conduct our daily lives. This helps keep us in the present moment, so that when we eat an orange we are fully aware of eating an orange and when we test something we are fully aware of what is going on and when we read a book we give ourselves fully to the present activity of reading the book. And when we want to think of the past or forecast the future we do just that and don't try to play tennis at the same time.

Inverse or negative thinking.   The author is grateful to the American psychologists Miriam and Otto Ehrenberg for insights into this process and for the example we will use from their book on this subject[5] to demonstrate what the Ehrenbergs call "The Power of Negative Thinking." This is the process by which a pattern of thought which is not productive in problem solving is rejected and a new pattern is sought. Whereas positive thinking is adaptive to the status quo (if indeed it is thinking at all), negative thinking does not automatically go along with the status quo. Among many examples, the Ehrenbergs give the following one, which the author particularly likes because, while it solves the problem nicely, it creates other problems. This is absolutely true for problems involved with the analysis and design of information systems. The solution often contains other problems which must be solved.

[4] R. W. Clark, *Einstein: The Life and Times* (London: Hodder and Staughton, 1973). Referenced during interview with Miriam Ehrenberg—see below.

[5] Miriam Ehrenberg and Otto Ehrenberg, *Optimum Brain Power: A Total Program For Increasing Your Intelligence* (New York: Dodd, Mead, 1984), p. 153.

To quote the example with some slight changes for brevity:

Part-Time Parents

A divorced couple with joint custody of their two children, ages six and eight, were having difficulty working out a suitable plan for child care. They tried different arrangements involving the children staying with each parent at different times. The children were angry with both parents, resentful of the continuous moving back and forth, and found it hard to keep up with schoolwork and meet with their friends. They asked to be allowed to stay put with one parent, but both parents were unwilling to give up their time with the children. What would you have done?[6]

After much discussion the parents agreed that it was better for the children to remain in one place. They reconciled this with the problem of joint custody by letting the children have a permanent place to live and having the parents alternate living at what became the children's home. In this way the parents had separate but equal visiting rights and the children could lead happier lives.

The author is more inclined to call this inverse or rearranged rather than negative thinking. As we have noted, what is fascinating about this problem is the number of new problems the solution raises. Do the divorced couple maintain three homes—his, hers, and the children's? What happens if and when they both remarry? Doesn't this tandem sharing of the children's home sustain a relationship which has failed? Is this a problem which must be dealt with at a higher level in the systems hierarchy, such as at the national level, in order to arrive at an effective solution available to all? Is the answer to discourage the divorce in the first place by rewards to those families that stay together in spite of all grievances? From the point of view of a systems analyst, dealing with problems at too low a level in the systems hierarchy is the major cause for failure to gain insight into solutions.

Then again, the children's problem was immediate and needed a "quick fix," rather than a solution that might take years to become a national consensus.

It is also interesting to speculate on what was the moment of insight that created the inventive solution given in the example. Did someone say, "Well, after all, the children's home must be near their friends and school." And then, perhaps, the concept Children's Home popped into the mind as a solution.

Peak times.   Recognizing peak times requires knowing the mind-body rhythms—your own and those of others on the creative team. Unlike the performance of a device, the performance of a human varies in peculiar ways from less than acceptable to more than satisfactory. It is only during the actual attempt at doing creative work that most people can know for sure whether they are going to be receptive to creative insights and are going to be prepared to deal effectively with them.

[6] *Ibid.*

It is possible to acquire enough knowledge about oneself and about the functioning of companions to make a good statistical guess about what will happen. Can you function up to the necessary level on three hours sleep? On too much sleep? First thing in the morning? Evenings? With a particular partner? Wise persons pay enough attention to these subtleties to gain wisdom about themselves and their relationships.

Being prepared.   Being prepared means having done the necessary left-brain homework concerning the facts of the problem. Being prepared also means being open to right-brain processes, such as seeing nonverbal patterns of whole systems and seeing how fixed ideas can be inverted or otherwise altered to achieve a new view, a new concept, a new metaphor.

Many, many other practices that are less specifically directed to aiding the creative process work quite well—such things as exercise, good times, laughter, falling in love, and on and on. The reader gets the idea.

THE ANATOMY OF INSIGHT

We have been careful to point out that insight is the heart of the creative process. In terms of total elapsed time between the onset of the creative process and the time of closure, when the process is captured, confirmed, and incorporated, insight seems unimportant. The time involved in gaining insight is a mere fraction of the total time. But as most of us know, these few milliseconds or seconds are the heart of the matter. Without these sudden insights many difficult problems are not open to satisfactory solution.

It is extremely difficult and even presumptuous to try to analyze insight. Insight is the sublime moment when we are in our most human state and realizing our best potential. What the author is prepared to do is to give an example of how certain people view the process of insight and of gaining access to insight.

Fortunately for us, we can tap on the ancient, ages-old experience of the Buddhist tradition. Nothing opens up the mysteries of insight better than that variety of the Buddha's Dharma (teaching) known as Theravada Buddhism and the associated meditative practice known as Vipassana. It is with some anxiety that the author opens this discussion and those to follow, since some of these approaches to thinking, and altered states of thinking, are not open to the kind of statistical proof so dear to the hearts of the scientific community. This discussion, however, may at least offer a useful metaphor regarding insight—something that will have to do until the "real" (verifiable) thing comes along.

Vipassana Meditation is translatable as Insight Meditation. There are now several Theravada Buddhist centers in the United States.

Life inside an Insight Meditation Center is very difficult, full of physical and psychic pain and even danger. People, mostly very young people who are at rites-of-passage times in their lives (perhaps having just graduated college), go through this difficult experience and seem to feel a lot of joy and strength if they succeed in sticking out the program. Usually attendees come for either ten days or three months.

Each attendee is assigned a room with nothing it it but a mattress on the floor. No medicine, such as headache pills, is allowed. No radio or TV. No reading or writing material. Eye contact is to be avoided. Conversation is to be avoided. People eat, walk by each other in the halls, and work (on meal preparation, in the vegetable gardens) in a state of silence. There is nothing whatsoever to do but sitting meditation, walking meditation and work meditation. There are two vegetarian meals a day. Generally only four hours of sleep is required and no more is needed. The physical and mental pain is significant.The unnatural sitting position causes great physical discomfort.

If one feels a pain in the back, one meditates on this pain, watches it arise, watches the mind attach itself to suffering the pain, and watches the pain finally go away after about the third day. If one gets an itch on the cheek, one does not scratch. The itch becomes the focus of meditation (rather than the traditional breathing exercise), and one watches the itch emerge, watches it become a source of psychological suffering, and then one watches the itch go away (it always does).

The most serious psychic pain is caused by the trauma that has been suppressed by the active conscious mind talking to itself. This inhibition to remembrance of trauma is now gone. This, along with the feeling of aloneness and sensory deprivation, can be dangerous; the seriously disturbed or disassociated individual is not a candidate for such an experience—even a ten day Buddhist retreat. There is usually a talk in the evening by a master practitioner, who may be a man or a woman, and each attendee is assigned a counselor on the staff of the center with whom they can discuss their progress.

Levels of Insight Achievement

These Buddhist teachers (whose teachings are not all that much different from the teachings of the Zen Buddhists—both carefully follow without distortions the basic Dharma or teachings of the Buddha) see the process of meditation as involving four levels of insight achievement. We achieve insight at each level by watching calmly what is going on inside our mind as we review our attitudes and behavior. This is done *only* after clearing the mind of left-brain control—that is, stopping the verbalizing, the endless conversation, with ourselves by which we confirm our "illusions" about ourselves and our relationships. This process of stopping the conversation or "centering down" involves an altered state of consciousness, more amenable to messages from the superconscious, which is palpable and easily demonstrated to any beginner. This, at least, can be called verifiable fact.

These four levels[7] are as follows:

[7] Dhiravamsa, *The Middle Path of Life* (Surrey, England: Unwin Bros, Limited; The Gresham Press, England, 1974), pp. 26–35.

Level one.   Level one deals with the conscious, waking state of existence—that state which is so much the territory of left-brain functioning. What we most watch in this level of meditative practice are what the Buddha called the five aggregates of the human mind-body system: (1) the body and its functioning, (2) the feelings we have, (3) the process of our conscious, left-brain mind, (4) our perceptions of what we sense, and (5) our habits, those "predefined subroutines" by which we function without having to think it all out each time.

We watch these aggregates, which together make up who we think we are, from the stance of the disattached observer. We do this by keeping our minds quiet, by watching our breathing and gently brushing away thought formations until something comes along that cannot be brushed away. Then, in this altered meditative state, we deal with it by watching it from "outside."

Most particularly we watch, become aware of, the constant basic conflict of life that goes on within us for as long as we live, between the negative, destructive qualities and the positive, good qualities.

Buddhists call these the Vicious Circle and the Virtuous Circle.

The Vicious Circle consists of such qualities as the tendency to do evil, to become passive, lazy, and mentally inert, to always live in the past and the future and never enjoy the present—all those qualities which create uncertainty and fear (fear being the great enemy of happiness and accomplishment).

The Virtuous Circle consists of wisdom and acts of wisdom, the tendency to do good, the avoidance of extremes in action and thought—in short, the quality of saneness.

Level two.   Level Two concerns conditionality or relationship. Relationship has come up often in the methodologies part of this book as the key to understanding and dealing with information processing. What the Buddhists are dealing with here is causal relationship: cause and effect. (In the next discussion we shall deal with noncausal relationship.)

What we are enjoined to do here, now that we have begun to deal with those inner components of ourselves at Level One, is to see how all of our best and worst thoughts and emotions are conditioned by something else—how a thought is connected to another thought and whence come our attitudes, not all of which may be supportive of the creative life.

Level three.   This is the level of understanding change and impermanence. Buddhists do not see any unchanging absolutes—just impermanence and our attachments to things, places, and people which keep us from enjoying life. Attachment to things, which makes us fight change, is to Buddhists the basic cause of suffering in this world. When Buddhists speak of nonattachment as release from suffering, they do not mean we should not love and cherish. But, they say, and we find out in meditation, all is change. We will only suffer if we try to capture and freeze the present. Go with the flow! is the modern vernacular for this Buddhist concept.

It is at Level Three that the Buddhist philosophy speaks most to the condition of the modern scientific community—particularly to the modern theoretical physicist who lives daily in a world of flux, where outward forms cease to be meaningful—and to the psychiatrist or psychologist who sees clearly how patients suffer by trying desperately to hold onto a relationship that doesn't exist any more— and to the systems analyst who keeps decomposing whole systems to levels, and levels within levels, and observes (as we have pointed out with examples in this book) the increasing volatility and instability of a system as we go down from level to lower level.

Level four. Level Four is the level of nonverbal insight free of forms and concepts. We see patterns and we see the results of shifting the patterns. At this level we are free to be free. We can do closure and postpone closure. We can love without becoming overly attached. We can invent without becoming ego-involved with our first attempts. We can give others credit. We come as close as we ever can to seeing things as they really are. This is what the Buddhists mean by attaining enlightenment or continuous insight.

Level Four is an exciting concept for a knowledge worker, such as the systems analyst, to contemplate. Are we likely to achieve this state of consciousness? The answer is that very few of us—almost none of us (certainly including the author)— are likely to achieve this. That is why we cherish those who do.

But—all of us, with or without such a methodology as Insight Meditation, will achieve insight at least some of the time, and so we will wait for it and be ready to recognize and record it when it is given to us.

Besides, there seems to be some anecdotal evidence that people achieve just as much insight without any methodology whatsoever, just as there is anecdotal evidence that meditation does help insight. There is no proof that meditation is the sole handmaiden of invention. (There is some proof that it will lower blood pressure, stop headaches, and so forth . . .). Whatever is the truth, it is still worthwhile to have this discussion, because we are, and will continue to be, very curious about the mysterious anatomy of insight, the essential material of invention.

SIMPLE AND COMPLEX DUALITY

Nearly every problem can be broken into simple problems which can be stated in terms of some point between two extremes. In fact this is one of the main ways we have of judging solutions to problems. We go to take a shower and the water temperature ranges from very hot to very cold. Either extreme is unpleasant, so we mix the hot and cold together until we arrive at the water temperature which we feel to be right for us at this moment in this place. This can be called "finding our perfect place," which is to say finding our right place on the continuum between opposites. We do this all the time with every decision that requires analysis. This analysis of duality is very important to the process of invention. Some people call it having good taste or good judgement or a good analytical sense.

The ancient Chinese made a philosophy or methodology out of this process, calling the opposites Yin and Yang (pronounced "Yan"). Finding the perfect place was called being in the Tao (pronounced "Dow"). This is the famous Taoist philosophy which has influenced people such as Confucius and Carl Jung and many others. Also involved in this philosophy is the concept of keeping in the flow of the changing continuum. It is not enough to find our perfect place, we must also hold to it (flow with it) as everything changes. This fits nicely with the Buddhist concept of change which we have discussed. Zen Buddhism is the merging of Buddha's Dharma (teaching) with Taoism. The great sacred book of Taoist philosophy is the much misunderstood I Ching[8] or *Book of Changes*.

In a number of examples in this text we have shown a two-dimensional graph consisting of an X and a Y axis, each calibrated from one polarity or extreme to another. In this case we are showing sets of points along two continuums, so that we may vary one to see the effect on the other. Figure 15-7 in Chapter 15 is an example. In this example response time (from awful to ideal) is set against cost (from ideal to awful) and we arrive at a "perfect place" which rests on both continuums. An acceptable response time at an acceptable cost.

Sometimes problems cannot be decomposed into such nice subproblems. The reason is that when we decompose the whole into its parts, we sometimes lose something of great value—because the whole is almost always greater than the sum of its parts. The quality of a human being cannot be determined by looking at the liver or the heart or the brain. The quality of an organization cannot be determined by looking at the qualifications of all the individuals. Sometimes average people work together splendidly, and sometimes exceptional people do not. So we have complex duality.

Complex duality resists solution by normal methodologies. Creative insight is needed to *sense* the complex whole. This sensing is vital. An example of a complex whole was the Kobler Bakery case study presented in Chapter 17. At one point in that study the organization was ready to collapse, and a whole view was needed as a basis for a solution.

This does not mean that the main methodology covered in this book—specifying problems and solutions by decomposition into levels and partitions—is invalid. Far from it. It means that in addition to doing this we must also analyze the whole system.

CAUSATION AND CHANCE

We ring a doorbell and someone comes to open the door or doesn't come to open the door because:

---

[8] Richard Wilhelm and Cary F. Baynes, *The I Ching or Book Of Changes* (Princeton, N.J.: Princeton University Press, 1950). This is a much-recommended I *Ching* translation.

- they are home or they are not home
- they did or did not hear the ring
- they prefer to open or not to open the door for anyone
- they prefer to open or not to open the door for us

These are all *causation* (which is to say cause and effect). If we knew the condition of the party on the other side of the door, we would be able to predict every time whether the door would open in response to the ring. We could make a Law of Door Openings.

If we apply heat to an uncovered pan of water at sea level, the water will boil at 212 degrees Fahrenheit. We can count on it. These causal relationships are (or used to be) the basis for science and the scientific method.

However:

If we miss a plane and take a later plane and meet someone who offers us a job and we go to work for this person and meet our beloved life partner in the office elevator (we usually walk up the stairs), this is *noncausal coincidence*.

It may be true that statistically we have a better chance of being offered a job on an airplane than in a movie house and a better chance of meeting our beloved riding in an office elevator than playing ice hockey. We could call this *statistical causation*, which is not predictive in any single case event.

Or:

We dream of a yellow room where we feel a sense of great happiness and well-being. The song "I'll be seeing you in all the old familiar places . . . " goes through our head throughout that day. At lunch a Chinese fortune cookie tell us: "It pays to see old friends and loved ones."

That night we decide to fulfill an obligation to attend a local-area college reunion party at an elegant nearby inn. We go to the inn, and the common room we are directed to is decorated in a faded and delicate yellow.

Unexpectedly, she/he is there, our closest companion from college days, who was supposed to be working in Hong Kong. At one time we were inseparable and brought out the best in each other. We broke up over a misunderstanding and went different ways.

That was three years ago and now there he/she is, looking wonderful and receptive, across the room with the pale yellow walls.

A meaningless joining of events or a meaningful coincidence? Our hearts tell us that every event in this cluster of happenings is significant. We move toward her (or him) with a sense of surety that this was meant to be.

This little encounter is a "meaningful coincidence" or an example of what Carl Jung called *synchronicity*.[9] He wrote extensively about it. Arthur Koestler wrote a book on the subject[10] calling it *seriality*.

---

[9] C.G. Jung, Foreword to the Wilhelm–Baynes *I Ching* cited in footnote 7. Zurich, 1949, pp xxi–xxxix. Other works of Jung address synchronicity at greater length. The virtue of this reference is the conciseness of the presentation.

[10] Arthur Koestler, *The Roots of Coincidence* (New York: Vintage Books, 1973), pp. 82–104.

It is a subject of great interest to theoretical physicists who deal in subatomic particles/energy forms—forms whose behavior is more meaningful coincidence than it is cause and effect. Their concern is documented in books by the physicist, Fritjof Capra.[11]

What does this phenomenon of meaningful coincidence mean to us who analyze and develop information systems (besides the fact that it explains the content and direction of our personal lives more satisfactorily than any examination of cause and effect)?

It means to us that we should pay attention to these coincidences and follow them when we are caught up in the process of invention, as inconsistent and unpredictable as they may often be—because cause and effect will often let us down as a predictor when we are dealing with humans and human-based organizations. This, also, is what it means to be an intuitive person, capable of catching a truth that is "in the air."

THE UNITY BETWEEN METHOD AND INVENTION

We have come a long way together, those of us who have stayed together for the entire journey through the various methods of analysis and design, and now this perhaps strange discussion about invention and its sources. We may have missed some valuable methods and paddled up the wrong river branch on the track of invention through creativity and insight. If we have failed, it is not because we did not give it a good try. One fact we can rely on as the very truth: Method alone without invention or invention alone without method is the way to failure at systems analysis. The way to success is method with invention.

[11] Fritjof Capra, *The Turning Point* (New York: Simon and Schuster, 1982), pp. 75–97.

QUESTIONS
AND EXERCISES

CHAPTER 1

1. What five classes of physical components make up a natural system such as a social organization? If these are the physical components, what are the metaphysical components?

2. What are the two kinds of system specification?

3. What dangers are connected with omission of a sufficient systems analysis?

4. What is the systems development life cycle? Give some checkpoints on this cycle. What are the role names for the major participants in systems development?

5. Why is a system analyzed?

6. What is a current systems study? What purpose does it serve?

7. What is "chunking"? Give some examples of "chunking" from personal experience.

8. Describe decomposition of a system through leveling and partitioning. Define leveling and partitioning.

9. Give examples of a current system, first from the point of view of process flow and then from the point of view of data flow. For instance: Unloading a truck at a warehouse and storing the goods is a process. The data or information is checking the packing slip, making out a receiving slip, getting a storage location address, and updating the master inventory record. Do your best without doing additional research or looking ahead in the book. Don't get too detailed.

CHAPTER 2

1. What is the importance of requirements analysis? How does requirements analysis relate to current and new system studies?
2. Give an example of the difference between a process (function) requirement and an information (data) requirement.
3. Give an example of a new system specification from the points of view of process and information flow. This exercise should be a follow-up to Exercise 9, Chapter 1. What were the new requirements? Don't go into much detail.
4. Why would we want to use a graphic, nonlinear approach to specifying systems? In other words, why charts? Why not text only?
5. Describe "phasing" as it applies to system development.
6. Describe the elements of an implementation plan.
7. What are the major design tasks in systems development?
8. How do analysis, design, and implementation differ from one another?

CHAPTER 3

1. What are the symbols used in flowcharting a system specification? (Do not mention any symbols which relate to physical solutions, such as use of terminals or storage devices.)
2. Discuss how these symbols are used, particularly the subroutine.
3. What is the relationship between between levels, partitions, and the subroutine charting symbol?
4. What is a base level?
5. What other information is needed along with a process flow chart to fully model a system? What is a systems resource? Describe a few.
6. Flow chart the process by which a college sports team is managed (a current system study). Select any team. Show the input and output data, the triggers, and the resources involved. Do not do much research. Make assumptions if you don't know. Go into detail in one instance of the system, decomposing this instance down to the base level and charting the base level.
7. Give any advantages you can think of which are gained by using process flow charts as a method for specifying a system.

CHAPTER 4

1. What symbols are used in data flow diagramming a system specification?
2. Discuss how these symbols are used.
3. Discuss the specifying of the in-line data store as related to the moving data flow.

4. Why do we want to draw the data name on the line before we fill in the process transform bubble?

5. Discuss the differences between the symbols for the process flow chart and the data flow diagram.

6. Data flow diagram the same current system study done in Exercise 6, Chapter 3. Be sure to think of data (transaction) flow rather than process flow. Do the data flow first and then fill in the name of the process bubble. Do one base level showing the internals of the base level in text format.

7. Discuss the different methods of specifying base-level internal serial flow in this methodology. Name and defend your preference.

8. Give the advantages (in your opinion) of data flow diagrams over process flow charts. Show specific examples from the exercises you have already done in both methods.

CHAPTER 5

1. Why do we show all requirements, both those accepted for inclusion in the new system and those rejected (the YES's and the NO's)?

2. Should we show all the requirements of a new system or just those processes and that data to be automated?

3. Do a new requirements analysis for the current system study done in Exercise 6 of Chapter 3 and Chapter 4. Make up the new requirements.

4. Using the requirements done in Exercise 3, do a new system specification from the current system study done in Exercise 6 of Chapter 3 and 4. Use either the flow chart or the data flow diagram method but not both. Do not do extensive research. The point of these exercises is to understand the methodology, not to specify the best system.

CHAPTER 6

1. Describe data-in-motion and data-at-rest.

2. What do we commonly call data-in-motion; data-at-rest? Give examples of each.

3. What about temporary stores of data, such as the output of a sort program in a batch system which will soon be input into a file maintenance update—what kind of data is this?

4. Discuss the differences between data and information. What do relationship, apprehension, and surprise mean in terms of this difference?

5. Discuss the difference between information and knowledge.

6. What is meant by the data "code"?

7. Give an example of information showing the data, the relationships between the data, and the data code system. Explain how this information becomes knowledge.

8. What are the different kinds of data ownership?

9. Discuss the difference between local application data stores and integrated common data stores. Speculate on which is best and when.

10. Discuss the movement of data through a system from transmission of inputs into the system to output reporting.

CHAPTER 7

1. What is the difference between a transaction, a message, and an information stream?

2. What are the steps involved in transforming a "raw" input transaction into a "refined" input transaction ready for file update? How is batch different from on-line in this respect?

3. What is the function of data stores in the development of a refined transaction? Give an example.

4. Give the eight verbs that cover the internal processing by which a raw transaction is refined. (See the discussion associated with Figure 7-4 in the text.)

5. What is the difference between implicit and explicit relationship? Give an example of each.

6. Give examples of messages and streams.

7. What is a report? What is a query? How do they differ? How are they alike?

8. What are the steps involved in processing output retrievals?

CHAPTER 8

1. What are the major components of a physical system in support of an automated information system?

2. Name some terminal devices.

3. What types of devices are connected by a data transmission system (not including devices within the data transmission system itself, such as the modem)?

4. What two questions about data determine the physical-system specification?

5. Name some examples of a personal information resource.

6. Name three basic work needs of operators using terminals.

7. Discuss the tradeoffs between centralized and distributive computing.

8. What are the four geographical links in data transmission? What types of data transmission links are available?

9. What is a data transmission network? Give examples of different networks.

10. What are the tradeoffs in designing a physical system?

11. What is line speed?

12. What is the relationship between link, path, and session?

13. What is the difference between synchronous and asynchronous data transmission?

14. How do the multiplexor and the modem work together? Give examples of different types of multiplexors.
15. What is a LAN? What is a Value-Added LAN?
16. How should the systems analyst, who is a generalist, handle the problem of the specialized knowledge needed to specify, purchase, install, and manage a data transmission system?

CHAPTER 9

1. How does data in storage get there? Give an example. How does stored data get to be reported?
2. How do we arrive at data storage (database) requirements?
3. What are "user views"?
4. What is the relationship between systems analysis and data analysis? Do your best. There may not be a precise answer.
5. Who owns data?
6. Name some data storage physical media.
7. What are a DMS and a DBMS, and what is the difference between them?
8. Discuss the pros and cons of using data on primary databases as opposed to extracted data-bases. What do "primary" and "extracted" mean in this context?
9. What is the definition of "database"?
10. What do we mean by data dependence and independence?
11. What are the different kinds of databases? What are their characteristics? What do you think are the advantages of each kind of database?

CHAPTER 10

1. What is a *table*?
2. What is meant in this book by a *chunk*?
3. Give an example of a table, showing the table name, the primary key, and the dependent attributes.
4. What other terms are often used in place of "table"?
5. What is a *data group*? What other terms are often used instead?
6. What is a *data item*? Also known as what?
7. What is a *domain*? Also known as what?
8. What is an *attribute* (data item)? Also known as what?
9. What expressions are used as adjectives for the word *key*? Explain what each term means.
10. Describe the three-level database using the terminology suggested in Chapter 10.

11. Describe user-view bubble charting. What is the purpose of this methodology? Does this method need software support to work?

12. What is normalization? What is first normal form? Second normal form? Third normal form? Give an example of a table, first un-normalized, then in first NF, then in second NF, and finally in third NF. State the definition of a fully normalized table.

13. Draw a user-view bubble chart of all the data on your driver's license (or another document in your wallet).

14. Draw a user view of all the money in your wallet.

15. Integrate these two user views to show one user view of "partial wallet contents." In the same manner add a credit card user view to form a new combined user view.

16. Try to draw a hierarchical, multilevel, multipath database from this last combined user view. Call it the WALLET database.

17. Discuss the problems associated with building and maintaining a common database as opposed to those associated with maintaining multiple application databases.

18. How would you solve the problems raised in Exercise 17?

CHAPTER 11

1. What is a strategic plan for an organization information system?

2. What does SPIRACIS stand for? Discuss each element. What one word might describe the sum of all SPIRACIS elements? (*Hint*: Unscramble this word–"FIEBSNET.")

3. What is a *data entity*? Give examples. How many data entities are there in the average large organization? How many information groups or user views?

4. What are some of the necessary steps in strategic planning for information systems study?

5. Referring to the contents of your wallet used in the exercises in the last chapter to construct a WALLET database, try now to do a PROCESS/DATA ENTITY matrix chart from this source.

6. Can you think of other examples where an organization could strengthen competitive position (the SP of SPIRACIS)?

CHAPTER 12

1. What is *metadata*?

2. What are the possible responsibilities of the data administrator?

3. Discuss the two types of data dictionaries and the three functions which identify each type further. Describe these functions in some detail.

4. What are some of the standard categories one might find on a data dictionary?

5. What is meant by an "extended" data dictionary? Give examples of extended categories one might want to keep on a data dictionary.

6. What types of documentation are necessary to support the development and ongoing use of an information system?

7. Some complain that extensive documentation cannot be maintained, slows down development, cannot be accessed, and is always substantially inaccurate. Others complain of severe trouble in developing and maintaining systems because of insufficient documentation. What is your opinion concerning what type of documentation is appropriate? Discuss your position in some detail.

CHAPTER 13

1. What two main forms of research are used when doing systems analysis?

2. What are the steps to be covered when doing research for systems analysis?

3. What are some "types" of interviews? What are the three stages to interviewing?

4. What is dialectics? What is dialectical interviewing?

5. How do group meetings differ from one-on-one interviews?

6. Discuss methods for checking the validity of interview material.

7. How do we deal with the hostile interviewee?

8. Why do we make a point of requesting all the documentation connected with the study area?

9. What types of documents, concerning the study area, exist outside the study area?

10. What are the three main types of documentation? Discuss how the systems analyst relates these documents to the systems specification.

11. What are some practices we wish to avoid when dealing with users? Do you feel you should have a personal position on this subject? If so, what is it?

CHAPTER 14

1. Why does a systems analyst need to be skillful at selecting and using software? What kinds of software are we discussing in this chapter?

2. What is a basic DBMS? What is an extended DBMS? Discuss the elements or functional modules of the DBMS. What is the integrated DBMS? Compare the extended and the integrated DBMS.

3. What are some examples of application software? How do we integrate application software with the rest of the information system?

4. Discuss the steps by which we evaluate contending software products and make our selection of the best product.

5. How do we present the results of a software evaluation. What topics must be covered?

6. After acquiring software, what can we do to insure it is properly utilized?

CHAPTER 15

1. What problem do senior managers have with information systems as an investment choice? What other investment choices might these managers have as alternatives?
2. What problem do systems analysts have with the perspective of these senior managers regarding the nature of an information system?
3. Discuss the differences between tangible and intangible benefits. How do they differ?
4. Define cost-benefit analysis.
5. How does cost-benefit analysis relate to the new system specification? To the strategic plan effort?
6. What is the "quality of information"? Give a number of examples.
7. What is meant by "sensitivity testing"?
8. Name some of the sources useful for arriving at benefits and costs.
9. Give examples of each of the four kinds of SPIRACIS benefits.
10. What is "Bayesian analysis"?
11. Discuss risk assessment. How is it done? What are the three critical factors and how do they affect each other?
12. Discuss "quality" cost estimating. Give an example.
13. Discuss how we arrive at the net benefit.
14. What are the development cost centers?
15. Discuss how we arrive at "How long will it take" (to develop system software). Discuss the two methods given in this chapter for doing this. As you see them, what are the strengths and weaknesses of *COCOMO* and *function points* as estimation methods? Can you think of any other way of estimating development work?

CHAPTER 16

1. Relative to the boundary of the system under analysis, what are the two kinds of design?
2. What is the main input to the design phase? What is the main output from the design phase?
3. What is the transformation that converts system specifications into design specifications? What is the generic computer design solution?
4. What are the elements of the design package which is delivered to the implementation team and other concerned parties?
5. Draw a system chart for a batch system. Make it up. Draw a program chart for the file-maintenance update program in this system.
6. Draw a system chart for an on-line information system which has some batch reporting. Make it the same system that was done in batch in Exercise 5.
7. Draw an on-line query system from an extracted database for the batch system drawn in Exercise 5.

8. Draw a panel flow diagram for an automatic banking window. Your system allows deposit, withdrawal, and current-balance transactions on entry of a banking card and a secret four-digit number. Withdrawals are allowed in $10 increments up to $300 in any banking day. In addition to the panel flow do a writeup for each transaction including a transaction layout. Show on the panel flow diagram where you think a program will be needed to do validation.

9. Paint the panels necessary to make a cash withdrawal.

10. Discuss module characteristics as defined in the structured design technique.

11. Which do you prefer for design charting: flow charts or structure charts. Why?

12. What is top-down design and implementation? Give an example. What are the benefits?

13. Which do you prefer regarding the module internal design specification: pseudotext or flow charts? Why?

14. Discuss the use of tables in systems design. What is a table?

15. Can you give an example of a design standard that would apply to the appearance of a panel? One, for instance, might be that the selected response character always be placed in column 1, row 1, and that when the panel appears, the cursor always be in this position.

CHAPTER 17

1. Do you think that poor information quality was really a major factor in the downfall of Kobler's bakery? Defend your opinion.

2. Does automation cause a deterioration in the quality of a product or a service? Defend your opinion.

3. What do you think would be involved in the total automation of the bake process at Kobler's? What would be the role of the information about the route orders and the recipes in such a system?

4. Do you think it was really possible to meet the new requirements established for the Kobler's operation with the information system that was developed in this case? How would you have met the requirements?

5. Do you think the systems analyst was exceeding authority and stepping into some other type of consulting work by the scope of the analysis, or do you think this was a proper role for a consulting systems analyst? What is the limit to the role of systems analyst? Does it cover only information systems or can it, as in this case, get into such things as labor relations and personnel requirements?

6. Describe the relationship between the salesperson's route order, the master bake order, and the delivery ticket that went into each truck with the baked goods.

7. When the routeperson enters a route order onto the panel, what type of validation is provided?

CHAPTER 18

1. Discuss the relationship between insight, creativity, and invention. What is the connection?
2. Discuss right-side, left-side brain functioning. What does this mean? Does it make any difference whether this is a metaphor or a physiological fact?
3. What is the difference between concentrative consciousness and superconsciousness? Does this relate to left-right brain functioning?
4. What are the steps involved in the creative process?
5. What does it mean to "sign up"?
6. Recall and discuss an instance when you had an insight into the solution to a problem.
7. Have you ever received an insight and then lost it because you did not write it down?
8. In a classroom situation: As a group—shut your eyes, breathe in and out through your nose, feeling the air entering your body cool and leaving warm. Focus on this breathing. Avoid thinking without repressing thought. Just brush the thoughts away, if you can. Do this for five minutes, trying to achieve a space where the mind is completely quiet. Open your eyes and discuss as a group how you feel now compared to how you felt before the exercise began. Did you experience "an altered state of consciousness"?
9. Have you experienced "meaningful coincidence" in your life? Describe what happened.
10. What role has "the chance event" played in your life? Has a chance event had a major effect on the course of your life?
11. Is it possible to plan for the chance event?
12. In light of the material presented in Chapter 18, what do you think about systems analysis that so precisely defines data and process as to leave nothing to chance? In this context discuss the mission of the systems analyst in specifying systems.

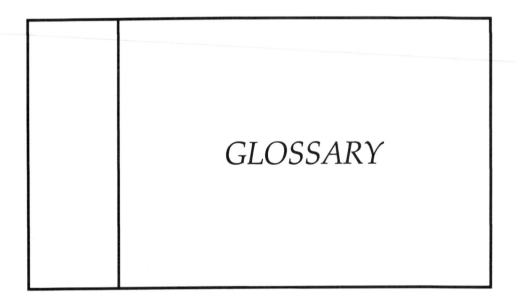

GLOSSARY

Active data dictionary: Provides documentation and is active in application program preparation but not at execution time.

Activity: See *process*.

Add: Insert a new logical record into a data store or database.

ANSI: American National Standards Institute.

Any-point-to-any-point network: See *network*. All end points connected to all other end points.

Application: A phased development chunk in a system which supports a particular organization function. See *chunk*.

Asynchronous transmission: One byte or character at a time, preceded and followed by a control bit signifying start and stop.

Attribute: A dependent data item of a *primary key*. Also known as *dependent item, dependent attribute, owned item, and owned element*. In another context attribute is also a quality that specifies a data subject. Example: field length is nine alphanumeric characters.

Bandwidth: The potential of a "line" to carry a given BPS rate. See *BPS*.

Base level: That lowest level of a system hierarchy below which it is irrelevant to decompose. This is the action level where the work of the system, such as data moves and calculations, is done.

Basic DBMS: Allows calls to the database and supports database administration. See *DBMS*.

Batch: To collect by like characteristics into temporary stores until triggered into motion.

Batch system: A system with collection points and delays to allow accumulation, where the operators enter transactions at the front end of the system and receive reports at the back end. The operator is not involved interactively with the process on a one-at-a-time basis.

BLI: Block-length indicator.

Blocking factor: The number of logical records in a physical block of data.

BPS: Bits per second.

Branching table: Multiple decisions shown in graphic form. Usually a series of "OR" decisions that distribute one type of input to many different processing modules in a program according to some distribution code, such as a transaction code.

Browse: To search through an information subset of database records and possibly select one record to see the full data view.

BSP: Business System Planning—an IBM methodology used for strategic planning of information systems. Makes heavy use of matrix charting methodologies. See *strategic planning* and *matrix analysis*.

Business function: Department or leader with jurisdiction to control and operate certain processes.

Candidate key: A possible choice for a *primary key* found when modelling a *table*.

Canonical schema: See *conceptual schema*.

Category of data dictionary: A group of subjects, such as all the data elements or all the programs.

Causation: Predictable cause and effect.

Child chunks: Chunks of data on a hierarchical database which have a parent (or parents) at the next higher level. See *chunks*.

Chunks: Elements of a system working in groups. Example: On a chess board this might be a queen, bishop, and pawn working together to accomplish an objective of the chess player. In terms of databases a chunk is the moveable unit of data, such as the segment on the IBM-IMS database.

Closure: Actualizing the insightful moment. Deciding an insight is right and signing off on the search for this particular answer to a problem.

CPU: Central processing unit of a computer.

Cohesion: Degree of association within a module.

Communication: In reference to data communication: successful information transmission.

Concatenated: See *concatenated key*.

Concatenated key: A primary key made up of more than one data item. The position of each data item in the concatenation shows the data hierarchy.

Concentrative thought: Analytical, conscious thought. Left-hemisphere brain processing that deals with language.

Conceptual schema: The logical database system. A layer between the external schema and the internal schema which serves to reduce internal-external dependency problems. Also known as *schema, canonical schema*, and *IMS logical DBD*.

Conditioned line: Special line enhancements that improve the quality of transmission. These lines have grade codes, each with its own tariff schedule.

Conversational mode: Regarding the user terminal dialogue of an on-line system: locking out all access to a data group while updating is in progress. Also known as "running long".

Cost-benefit analysis: The business case for investing in the development of an information system. What are the development and operating costs? What is the net benefit? What are the tangible and intangible benefits? How long will it take?

Coupling: Degree of dependence between modules.

Creativity: Originality of thought and performance—using insight.

Current system study: A formal analysis of the existing system under study in graphic and text format.

Customizing: Getting vendor application software to perform as required for a particular application.

DA. Data administration: The function responsible for the logical use of data in the organization: what the data means, where it originates, who maintains it; data security, data independence from application programming, lack of harmful redundancy and so forth.

DASD: Disk access storage device.

Data (singular **datum**): Some kind of a fact which has a code and attributes and is capable of relationship. Example: Coded in decimal number system; always is a two-digit numeric between 01 and 99.

Data-at-rest: Data stores, permanent and temporary. Databases and files on and off the computer system.

Database: "Collection of shared data used for multiple purposes." See Chapter 9—James Martin quote.

Data class: See *data entity*.

Data dependency: Situation where we cannot change data structures without reviewing all programs that reference the data and probably making major changes to these programs.

Data dictionary: Information about data collected into an organized data store that shows subjects, attributes of subjects, and relationships between subjects. Could also involve support of programming by providing input-output working storage definitions and the like. Could also be used in-line to validate raw input transactions. See also *passive, integrated, stand-alone, active,* and *in-line data dictionary*.

Data element: See *data item*.

Data entity: Table information about a subject, such as Employee, Client, Vendor, Machine. There are relatively few data entities, even in a large organization. These few data entities are used over and over again in various permutations to form many user views or information groups. It is essential to strategic planning to identify these data entities. Also known as *data classes* and *data groups*.

Data flow: See *information flow*.

Data flow diagram: A charting method used to specify a current or new system.

Data group: A table row. One instance of a table. Also known as a *record*, an *IMS segment*, a *tuple*, an *entity*, and a *logical record*. The moveable chunk in a database.

Data-in-motion: The data inputs and outputs of a system.

Data independence: Situation where we can add or rearrange data without changing all the programs that use the data.

Data item: One data element or field in a data group. Minimum addressable data in a macro language (like COBOL) or a nonprocedural language. Also known as *data element, field, key* or *attribute, group* or *elementary level item* (COBOL).

Data modelling: The showing of relationship between key data items and attribute data items in all possible combinations that represent user views of organization data. Can be done with user-view bubble charts. Used to define the conceptual schema or subject database.

Data ownership: Private, public, common (corporate), and limited sharing of data. Some individual, department, organization, or public holder owns the information and is therefore responsible for its maintenance.

Data set: See *table*.

Data store: Database; file.

Data transmission: The movement of data between people, computers, data storage devices, and terminals in all combinations by means of a delivery system that retains intact the data characteristics and code structure.

DBA: Database administration. The person or group responsible for the effective performance (and construction) of databases.

DBMS: Database management system. See *basic DBMS*, *extended DBMS* and *integrated DBMS*.

Decision models: Models where we can predict the sensitivity of one model function as we vary other model functions. See *models*.

Delete: To remove a logical record or a chunk of a logical record from a database. There can be no more access of this data.

Dependent attribute: See *attribute*.

Dependent item: See *attribute*.

Derived data: Data off-loaded from a primary database. Usually refreshed periodically.

Design: The specification from which the actual production system is implemented. The realization of the analyst's logical model into a solution-dependent "blueprint" that includes the physical-system design.

Design boundaries: Separation of a system into manual and automated transforms. The analyst has already defined the whole system apart from the suprasystem around it.

Dialectics: Tension between Thesis and Antithesis producing a qualitative change called the Synthesis. In this book, a research interview technique where opposing ideas clash and result in a creative third-path resolution.

DMS: Data management system. Works with major databases, has none of its own. Provides such services as report/query.

Domain: One column in a table. Also known as a *table column set* or a *field values set*. All data items in all data groups that pertain to this column of the table are included.

Download: Data off-loaded from a common or public database to a shared or private database.

DSS: Decision support system. Its primary purpose is information storage and retrieval for decision support, as opposed to an operational system, where the primary purpose is to drive the actual processes of the organization. In an operational system the decision support is there but is secondary in importance.

Elementary level item: See *data item*.

Entity: See *data group*.

Entity set: See *table*.

Extended data dictionary: See *extensibility*.

Extended DBMS: System that provides, in addition to basic DBMS functions, a long list of functions such as report/query, screen or panel generation, program library, security, dictionary, interfaces to other systems and services, and so on. See *DBMS* and *basic DBMS*.

Extensibility: Regarding data dictionaries: the ability to add user defined categories over and above the standard categories provided. See *categories*.

External schema: The application view of the common database system. Also known as *subschema*, *information group*, and *external subschema*.

External subschema: See *external schema*.

Extracted database: A database which is periodically refreshed from a primary database and used for retrievals. Seldom used for updating.

Feasibility study: An overview current and new system study specifying just enough to attempt to arrive at a go–no go decision. Usually includes a rough guesstimate cost-benefit analysis.

Field: See *data item*.

Field values set: See *domain*.

File: See *table*.

First normal form: Given an unnormalized table, remove repeating data items in domains where they exist.

Fourth-generation language: See *non-procedural language*.

Genial (congenial): User-friendly.

Global data: Available to many programs. See also *local data*.

Gross benefit: Benefit calculation before deducting development costs.

Group-level item: See *data item*.

Hard methodology: In information-system terms: a charting technique which is sufficiently rigorous and detailed to be directly applicable to the building of an information system. A soft methodology makes observations, provides brief examples and history, but is not a direct problem-solving tool.

HIPO: Hierarchical input process output. The methodology presented in chapter three as the neoclassical specification incorporates and augments the HIPO approach. See *neoclassical method*.

Host computer: One that runs organization applications as a primary function. Not a computer used solely for data transmission.

I−1 data integration: The building of common databases one level below the whole system level. Could be a division of a corporation, a branch of government, or a university school. If a department is the $i-1$ level, we would call the database shared rather than common.

IMS logical DBD (database description): See *conceptual schema*.

IMS segment: See *data group*.

Inclusion/exclusion: Setting the boundaries of the system under study—what operations are included and excluded. Sometimes quite arbitrary.

Information: Data which has use value because it has relationship to other data, can be apprehended, and is a surprise.

Information flow: The movement of the system data in and out of process transforms.

Example: information about automobile assembly necessary to change car parts into a finished car.

Information group: A user view such as a document or transaction. See *user view*.

In-line data dictionary: Documentation and perhaps application development services plus execution-time services, such as validation of raw input and checking of syntax.

Inquire: To look up in a database.

Insight: Perception into the inner nature or real character of a thing or situation.

Integrated data dictionary: Part of an automated DBMS. See *DBMS*.

Integrated database system: A connected system of databases that serves almost all the application processing needs of the organization. Connections could be predefined or available at data collection time during execution.

Integrated DBMS: Extended DBMS provided by one vendor. See *extended DBMS*. Alternative is to create an extended DBMS using software from a number of vendors which have to be made to work together.

Internal schema: The physical database system.

Invention: The product of the creative process.

IRG: Interrecord gap between blocks of data on a physical device such as magnetic tape.

IRM: Information resource management. The particular but not exclusive province of data administration.

Key: See *primary key*, *secondary key*, *candidate key*, and *concatenated key*.

Knowledge: Information which is perceived, understood, retained, elaborated upon, evaluated, formed into principles, and made retrievable for problem solving.

Knowledge database: A database of information which has enhanced characteristics, particularly in regard to the relationships of the data. A relationship database.

LAN: Local-area network, as opposed to a remote or global type network. See *network*.

Leveling: Decomposition of a system by hierarchical layers—along with partitioning, the basic method involved in doing systems analysis.

Line speed: BPS rate of both wired and wireless transmission links. See *BPS* and *links*.

Links: Data transmission connections within a workplace, between locally clustered workplaces, between remote workplaces, and to the outside environment. In data transmission parlance, links are part of the data transmission system which are made into paths during a session. An optimum path could be selected, which would vary according to circumstances.

Local data: Data for a single program. See *global data*.

Logical record: The basic unit of information which may traverse the physical chunks of a database or a database system. See *chunks* and *data group*.

Mainframe: The highest level of a system or a program. In a well-made program all other modules are a hierarchy of subroutines hanging off the mainframe. An important design concept.

Matrix analysis: A charting technique which is useful in strategic planning. Allows us to see relationships and patterns between system pairs, such as function/process, process/data entity.

Messages: Data, voice, and video information which has some precise transaction-type fields but is mostly an unstructured information flow. See *transactions* and *streams* for contrast.

Metadata: Data about data.

Models: Symbolic representations of a real process. Also miniature representations of a larger process. See *decision models*.

Modem: A modulating and demodulating device that converts signals from data to voice and back again so that voice-grade lines can be used for data transmission.

Module: A cohesive set of serial processing steps that transform inputs into outputs. Can be called or performed. Some modules are command modules, whose main function is to manage other modules, and some modules are mission worker modules, whose main function is to do serial work. See *subroutine*.

Moveable chunk: See *data group*.

Multidrop network: Every location or "stop" on the network is a potential drop point. The data transmission is so routed that it will be accepted only by the proper destination point, passing other points without making a "drop." See *network*, also *ring* and *tree networks*.

Multiplexing: Use of a multiplexor to do line sharing. Gathers several transmissions at a lower BPS into one transmission at a higher BPS for a faster outgoing line, preserving message identity. Reverse multiplexing is the opposite. Three methods are in use: time-division (TDM), frequency-division (FDM), and statistical multiplexing.

Neoclassical method: A method of specifying systems that places emphasis on process tracking through the use of flow charts.

Net benefit: Gross benefit minus development costs. See *gross benefit*.

Networks: Complex linking systems (see *links*) that allow data transmission between computer systems and terminals at multiple locations either local or global.

New system study: A formal analysis of a proposed new system that meets new requirements. A specification.

Noncausal: Not a cause-and-effect happening.

Nonprocedural language: The specification of requirements for a report, a query, an input transaction, or a database management routine which picks up predefined object-code modules to form a program out of the input-language syntax and then executes the task.

Normalization: The process by which complex data structures are decomposed into flat tables or relations. These tables are further simplified and decomposed by getting rid of repeating data items in a domain and making attribute data items dependent on the whole primary key and only the primary key.

On-line system: A system where each transaction involves a separate user-terminal dialogue and where a database is a functioning part of the dialogue.

Operating cost: The ongoing costs of resource usage.

Owned element: See *attribute*.

Owned item: See *attribute*.

Packages: Application packages. Software which is purchased rather than developed.

Painting a panel: Showing what the user will see on the functioning screen.

Panel flow diagram: The flow of screens the user sees in the user-terminal dialogue.

Paper stores: Manually stored data.

Parent chunks: Chunks of data on a hierarchical database which have dependent children chunks at the next lower level of the database. See *chunks*.

Partitioning: Horizontal decomposition. While leveling takes us top-to-bottom vertically as we decompose a system, partitioning takes each level and breaks up the activity into its processing parts.

Passive data dictionary: One that does the documenting function only.

PBX: Private branch exchange. A connecting switch between (many) internal building lines and the (few) external lines going out of the building.

Performance standard chart: A chart showing acceptable boundaries of performance for an operation. The actual current performance point is placed on this continuum, where it is either acceptable, unacceptable, below average, or above average.

Phasing: "Top-down left-right" approach to systems development. Top-down is leveling, left-right is partitioning. Allows a system to be tested and delivered in parts rather than all at once.

Physical database: See *internal schema*.

The Plan: The Strategic Plan for the development of the information system.

Plex database: Also known as *network* database. One in which a lower-level data chunk can be the child of more than one parent at the next higher level.

Primary database.: The database which is maintained by adds, changes, and deletes.

Primary key: The data item or concatenated data item set by which a table data group is accessed and moved. See *data group*.

Process: The transform of inputs into outputs.

Process (in strategic planning): A major task done by an organization function. Also known in this book as *activity*. Sometimes the term *process group* is used to indicate a very high level of *process*.

Process flow: The movement of the system inputs through transforms. Example: car parts to finished car assembly.

Program chart: Shows graphically the main program mandate (Example: validate) along with the program inputs and outputs.

Program data stores: Data stored as literals and constants in application programs rather than as external database tables.

Program dependency: Situation where, if data items are changed as to attributes, every program using such a data item has to be reviewed.

Pseudoconversational mode: Allowing access to a data group during most of the time a user-terminal dialogue to update a data group is going on. Also known as "running short."

Quality of information: A single term for all such attributes as accuracy, timeliness, clarity, relevancy, and safety.

Real-time: On-line interactive.

Record: See *data group*.

Recycle: Make an input out of an output.

Redundancy: Usually refers to the storing of identical data elements in multiple places, perhaps under different names.

Refreshed: Reloaded.

Relation: See *table* and *normalization*.

Relational database: A database where the chunks of data are not predefined concerning linkage and are joined only at process time into a hierarchical structure. These information chunks or relations are therefore more independent than predefined hierarchical database structures and can be joined for applications in a more flexible manner.

Requirements: *Gross* requirements are all the things the user might want a system to do. *Net* requirements are those requirements which are actually selected to be accomplished. *New* requirements are requirements that change a current system into a new system. Requirements are usually divided into *data* and *process* requirements.

Resources: Data stores, people, facilities, finances, materials, etc.

Response time: Usually the time the operator waits at the terminal after entering a request until the time the response information is returned. The turnaround time from request to reply.

Ring Network: See *network*. An example of a multidrop network. See *multidrop network*.

Risk assessment: Relating the size of project, complexity of project, and resource capability to arrive at a scored probability of success.

RLI: Record-length indicator.

ROI: Return on investment.

Root chunk: The highest-level chunk in a hierarchical database. The root has no parent.

Running long: See *conversational mode*.

Running short: See *pseudoconversational mode*.

Schema: See *conceptual schema*.

Second normal form: Given a table in first normal form, remove conditions where attributes are not fully dependent on the primary key.

Secondary key: An attribute which points to a primary key.

Segment: See *chunk* and *data group*. Chunks in an IMS database. Per James Martin: ". . . the basic quantum of data . . . passed to and from . . . application programs."

Span of control: Number of modules "managed" by a control module. "Managing" involves calls or performs plus switch and data passing.

Specification of a system: A model that shows by charts or text how something works now or how something will work after it is made.

SPIRACIS: An acronym standing for: Strengthen Position, Increase Revenue, Avoid Costs, Improve Service.

Stand-alone data dictionary: One that is sold and maintained as a separate product. Not integrated into a DBMS. See *DBMS*.

Star network: See *network*. All transmission traffic goes through one main link point.

Strategic plan: A high-level current and new system study where decomposition stops two or three levels deep and data is analyzed at a high level. Inferences regarding development priorities and cost-benefit are derived from the study. Outstanding opportunities and problems are highlighted. A tactical plan for development is decided on. The study is usually organization-wide.

Strategic planning for information systems: Determining application systems that need to be developed for the whole organization and the data entities that are needed in support of the applications. See *data entities* and *strategic plan*.

Streams: Information-in-motion whose content is unpredictable except in general terms. Example: newswire reports from a public database.

Structure chart: A design chart at the module level showing module flow with data and switch movement between modules.

Structured English: A method of specifying the internal processing of base modules.

Subject databases: Those formed from one or more data entities. This term is used in strategic planning data research. See *database*, *conceptual schema*, and *data entities*. High-level gues-stimates concerning eventual database needs.

Subjects: Pertaining to subjects of a data dictionary. One member of a category. See *category* of a data dictionary.

Subroutine: A predefined, self-contained module which is usually called or performed by one or more higher-level subroutines.

Subschema: See *external schema*.

Superconscious: The unconscious. Right-hemisphere brain processing that often deals in nonlanguage symbols.

Synchronicity: Meaningful coincidence.

Synchronous transmission: Block transmission of data rather than transmission of one character at a time. See *asynchronous*.

System chart: Graphic design showing programs and application packages in the system and major inputs, outputs, and data stores.

System description: Text to cover what the system does.

System specification: Top-down decomposition of a system through leveling and partitioning. Usually shown in chart form.

Table: A file with a primary data item key and a set of attribute data items related to the primary key. Also known as a *file*, an *entity set*, a *data set*, and a *relation*.

Table column: See *domain*.

Table-driven systems: Systems where variable reoccurring information used by programs is maintained on tables rather than hard-coded into programs.

Table row: See *data group*.

Tactical plan: Implementation of the results of the strategic plan. A plan for applications scheduling and development based on the findings of the strategic plan.

TCC: Terminal cluster controller. The control unit between a group of terminals and the rest of the data transmission system.

Terminal: Any device which interfaces between a user and an information system. Example: display screen, telephone, printer.

Third normal form: Given a table in second normal form, remove attribute dependencies on other attributes which are not the primary key.

Tight English: See *structured English*.

Transactions: Input data-in-motion where the data consists of a collection of data items which can be validated, often against a set or range of precise values. These fields in the transaction have implicit relationships to each other. Example: Employee #, Employee name, and Employee initials are all related implicitly.

Transforms: Partitioned processes which change input data to output data.

Tree database: A predefined hierarchical database with multiple levels and usually multiple paths. There is a child-parent relationship between chunks at different levels. Any child can have only one parent.

Tree network: See *network*. An example of a multidrop network. See *multidrop network*.

Triggers: Predefined times and events that set off a process. Example: Stop batching input and enter into computer system at 8:00 A.M.

Tuple: See *data group*.

Update: Change the values of an existing database record.

Upload: Data off-loaded from a local, private, or shared data store to the common database. Movement of data is from local to organization-wide.

User-view bubble charts: See *data modelling*.

User views: System inputs and outputs such as transactions and report documents.

Validate: Check data-in-motion against data-at-rest.

Validation: Assure, insofar as possible, the accuracy of raw input into the system.

VAN: Value-added network. See *network*. Offers an added service, such as synchronous–asynchronous conversion.

VHL systems analysis: Very high level systems analysis.

RECOMMENDED READING LIST

These books are considered by the author to be necessary supplemental reading for any individual planning to be a practicing systems analyst.

1. Tom DeMarco, *Structured Analysis and System Specification* (New York: Yourdon Press, 1978).
2. James Martin, *Computer Data-Base Organization*, 2d ed., (Englewood Cliffs, N.J.: Prentice-Hall, Inc., 1977).
3. Chris Gane and Trish Sarson, *Structured Systems Analysis: Tools And Techniques* (Englewood Cliffs, N.J.: Prentice-Hall, Inc., 1979).
4. James Martin, *Systems Analysis for Data Transmission* (Englewood Cliffs, N.J.: Prentice-Hall, Inc., 1972). Needs updating with current developments but still is the best general book on the subject.
5. IBM, Information Systems Planning Guides (Business Systems Planning, GE20-0527-3, 3d ed., July 1981).
6. Barry W. Boehm, *Software Engineering Economics* (Englewood Cliffs, N.J.: Prentice-Hall, Inc., 1981). Not as much for the main topic, the COCOMO model, but rather for the excellent documentation of productivity "cost drivers." A well-researched book.
7. Tracy Kidder, *The Soul of a New Machine* (New York: Avon Books, 1981).
8. Fritjof Capra, *The Turning Point* (New York: Bantam Books, 1981). Only Chapter 9 is relevant—concerns general systems theory.

9. Ed Yourdon and Larry L. Constantine, *Structured Design* (New York: Yourdon Inc., 1975).

10. Ervin Laszlo, *The Systems View of the World* (New York: George Braziller, Inc., 1972). A brilliant work on general systems theory. Essential reading for any systems analyst.

NOTE: The author wished to include recently published books on this list but, alas, could not find suitable selections worthy of being listed with Martin, DeMarco, Gane and Sarson, Laszlo, et al. Part of the problem may be the speed at which development is moving. An up-to-date volume on data transmission, for instance, is not current at all just two years after writing commences.

A

A CASE STUDY SHOWING A NEW SYSTEM SPECIFICATION FOR A LARGE BUSINESS ENTERPRISE

STATEMENT OF THE PROBLEM

United Consumer Products Inc. developed through the years as a loosely structured conglomerate of subsidiary corporations, each becoming a division in the corporate structure. These divisions were all involved in the manufacture and sale of consumer products. Taken all together, the United family was well up on the list of so-called Fortune 500 companies. Each division had its own national salesforce, which consisted of salespeople working out of their homes in territories remote to the division headquarters and selling to distributors and very large end users. The operations of one divisional salesforce were much the same as those of the others, which is also to say that their information requirements were very much the same. The size of each salesforce did vary from under fifty to over five hundred.

As time went on, many of the corporate information needs were automated. A central data processing staff aided each division in the development of computerized information systems for accounting, materials and manufacturing requirements planning, customer service, marketing analysis, and so forth. This central service was also responsible for pulling the information from each subsidiary together into a corporate management information system.

As is typical of many corporations with large outside salesforces, no attempt had been made to provide a similar automated information system for this highly individualistic group. To the last division, the sales operation was a manual, work-out-of-a-shoebox affair. This, if one thinks about it, is quite strange, since no functional group in such a business organization is more critical to profit than the sales staff.

This shoebox operation was surprising, because these sales staffs were all in significant need of a good information system. Salespeople spent up to 20% of their time handling paperwork, particularly trying to supply to, and receive from, sales management the information necessary to perform their vital function. It is not surprising that much of the information developed in this environment was inaccurate and less than timely.

Complaints included late and inaccurate reporting of booked orders, the inability to adjust a wrong booked order, no information concerning which booked orders were actually shipped, unrealistic quotas, long delays in getting expense accounts settled, lack of information concerning a salesperson's weekly schedule, lack of competitiveness in bidding on large accounts, slow new account approval, lack of good information on salesperson and territory performance, and lack of good, quick information on account profiles and the product line.

Numerous complaints about the situation finally promoted the problem onto the agenda of the central corporate evaluation, design, and implementation group. This central group assigned a systems analyst to look into the feasibility of developing a salesforce information support system that would work for all subsidiary divisions. The analyst, following the research techniques developed in Chapter 13 pulled together the relevant information from interviews and document analysis.

Next the analyst used the data-flow-diagram technique to chart the current system and also, working closely with the potential users, determined the requirements for a new system. The analyst also wrote a feasibility report, giving the business case for several new systems. One of them was selected by the management as the best solution of the problem, and the analyst, working with a junior, was given the go-ahead to produce the new system specification from which detailed design specifications could be made.

The new system specification would consist of the following elements:

1. data flow diagrams
2. data store field definitions
3. transaction definitions
4. report definitions
5. panel flow diagrams
6. a passive data dictionary

This specification package, with the option of the data dictionary, was the irreducible and minimal logical system specification needed by the design and implement team to build the SALESFORCE information system.

The sales management placed one restriction on the analyst. The system physical design must be a stand-alone system, giving the sales function full autonomy. There was a preference for a solution which could be maintained and enhanced by the users with minimum DP support. The senior analyst determined that this design requirement would have no impact on the logical systems analysis.

DATA FLOW DIAGRAMS

The systems analysts now returned to the data flow diagramming technique to chart the new specification, using as a basis the current system specification. Figure A-1 shows the first diagram, which was the highest (context) level. The new sales support system was to be known as <u>SALESFORCE</u>. The context diagram shows a system of databases (note the store symbols) in support of four process bubbles: CHECK SECURITY, PROCESS INPUT TRANSACTIONS, UPDATE DATABASES, MAKE REPORTS and QUERIES. The data flow in this data flow diagram is transactions to refined transactions to database changes to raw reporting to refined reports. The external source which gives and receives data is, in this example, the sales staff, which includes field sales reps and sales management.

This brief context diagram uses all the symbols needed to do data flow diagramming. For a refresher on these symbols and their use turn back to Chapter 4.

This context diagram sets the boundary on the system. We are interested only in input from the sales staff (with the addition of shipped orders downloaded from HQ operations on the mainframe). We are interested (for now) only in providing information to the sales staff.

We go now, in Figure A-2, to the full exposition of the SALESFORCE system, which is the level one data flow diagram (DFD). All further DFDs will serve to show the internals of the bubbles shown on level one. We will do only as many levels as are necessary to get to the base level. The base levels will become, for the part of the new system that is automated, the process modules which are designed and programmed. In the example shown here the entire chart is to be automated.

External Sources of Input

The DFD in Figure A-2 shows, starting at the left, the four external sources of input.

The Field Sales Staff.   In all divisions the field sales teams are organized into geographically based hierarchies. At the base of the pyramid is the TERRITORY, which is usually, but not always, identical with one salesperson. These territories are grouped into REGIONS headed up by a regional sales manager. REGIONS are grouped into AREAS, which are grouped into DISTRICTS, which

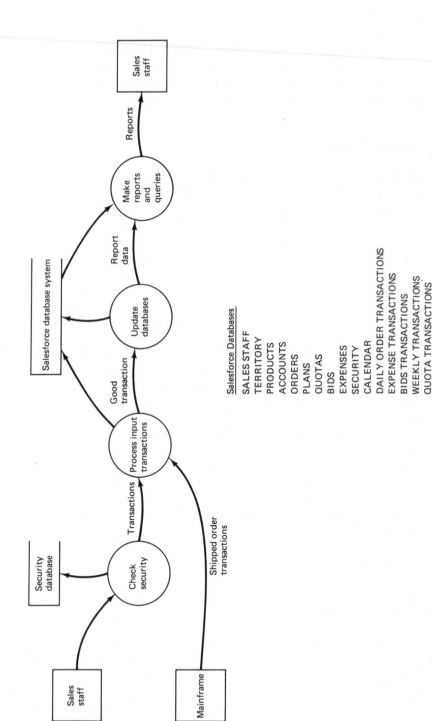

Figure A-1 Context Diagram of the New System.

Sales staff

Reports

Make reports and queries

Salesforce database system

Report data

Update databases

Good transaction

Process input transactions

Transactions

Shipped order transactions

Check security

Security database

Sales staff

Mainframe

Salesforce Databases

SALES STAFF
TERRITORY
PRODUCTS
ACCOUNTS
ORDERS
PLANS
QUOTAS
BIDS
EXPENSES
SECURITY
CALENDAR
DAILY ORDER TRANSACTIONS
EXPENSE TRANSACTIONS
BIDS TRANSACTIONS
WEEKLY TRANSACTIONS
QUOTA TRANSACTIONS

in turn are part of the NATIONAL SALESFORCE. INTERNATIONAL sales teams are also hierarchically arranged. There are sales managers at each level. With all this hierarchy the salesperson is still a highly individualistic person, working out of a home, making his or her own hours, and having contact with the regional sales manager mostly by telephone. It is for these people, the sales reps (and the managers), that SALESFORCE is to provide a better way to get and receive information.

Sales reps have three kinds of information that must be input to the system. These are:

- expense items daily
- booked orders daily
- next week's schedule by day

Following the flow in the upper right hand corner, we see that these three items are entered into batched transaction files after being carefully validated. Only completely valid transactions are accepted onto the transaction file, which at some time trigger is posted to the relevant master file (database). These three files make up most of the volume of the database system. For instance, booked orders may come in at the rate of 1000 per day for, say, 240 days a year, and each record is about 55 bytes in length.

Validation of the order transaction file consists of checking the date, looking up the sales rep's identification on the sales rep file, looking up the territory number on the territory file, looking up the customer account number on the account file, looking up the product number on the product file, checking the order-number numeric range, checking the type order, the activity and status codes, and the price code for valid string values, and checking the quantity and price for reasonableness. These are all the fields on the order file and they are all validated, many of the fields with 100% accuracy! So we can see that the bubble RECEIVE AND VAL-IDATE contains a good deal of process logic as the incoming unvalidated transaction becomes the outgoing validated transaction.

We can see, still moving from left to right across the top tier, that expense, order, and schedule transactions lose their transaction identity in the middle part of the data flow when they are posted to their respective databases. Then we see the output reporting system coming out of the database system as a third part of this input, update, report processing. This is a very classical flow. Almost all information systems look just like this.

Please note, before we move on, that parts of the database system came into play as lookup files during the input processing, and this is why the relevant data stores are shown associated with the validation bubbles. Also note that we are tracking input and output data through various transform bubbles. That is what data flow diagrams are all about.

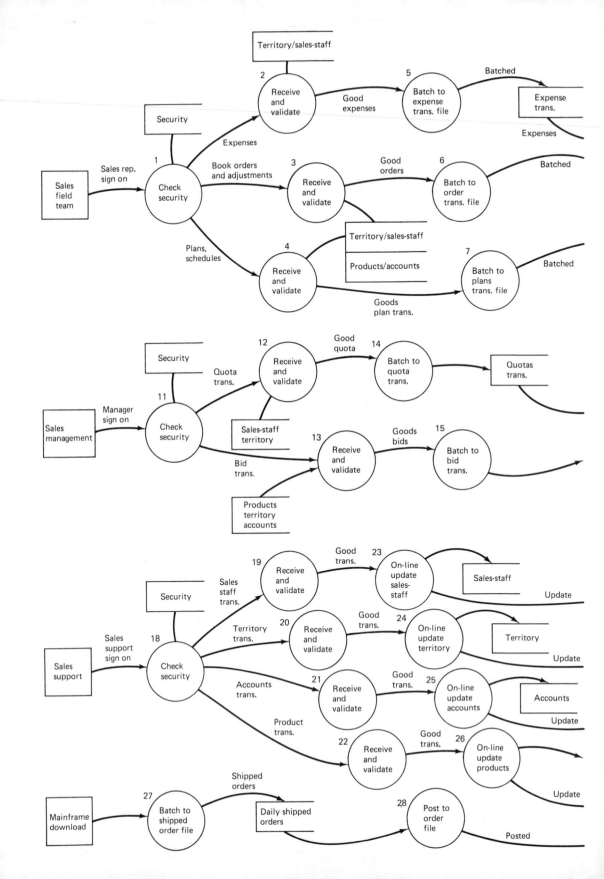

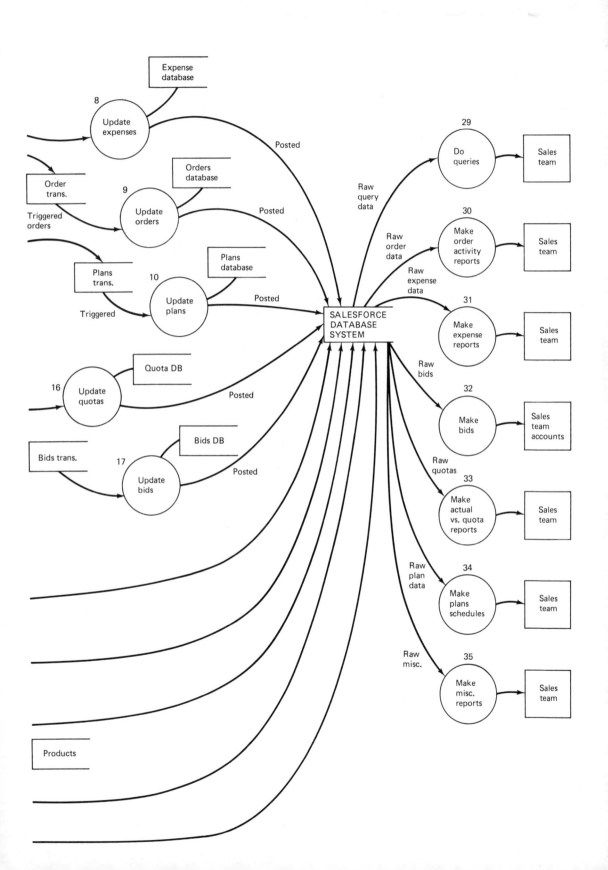

The Headquarters Staff: Sales Management and Sales Support.   We can see on Figure A-2 the file management controlled by the sales manager and support staff. This could be regional management, national management, or any level in between, depending on the size of the organization. The management is going to want to control QUOTAS (forecasts) through the quota database and to control account and product bidding through the BIDS database. These files are updated as batch jobs from transaction files created and validated on-line. The data flow diagram shows this flow.

The sales support function, besides managing the physical system and answering phoned in communications with the field sales staff, is responsible for maintaining the databases that support all the other input transaction validation. These are the following files:

- SALES-STAFF (one record for each salesperson)
- TERRITORY (one record for each territory)
- ACCOUNTS (one record for each customer or prospect)
- PRODUCTS (one record for each product sold by the sales staff)

These four databases, besides supporting validation, are joined with the other databases to create the various reports and queries in the SALESFORCE output system. They are also useful sources of stored information in their own right. For instance, if a salesperson has a question about the product line, the PRODUCT database is the place to look for the information. These databases, which are small in volume, are maintained entirely online.

Mainframe download.   Finally, there are the SHIPPED ORDERS, which are an output of the material management system on the mainframe. These shipped-order transactions are matched to the booked orders entered previously by the salesperson.

Security

Everyone must pass a security test to get into the system. Each external source has a different security level.

Reports Produced

Figure A-2 also shows an overview of the reports produced from the database system. These reports are broken down on this first level into

- ad hoc queries and browses
- order activity by any time period from daily to year-to-date
  - by salesperson
  - by geographical unit, such as region

* by account and product
* by product
* expenses by time period, especially weekly
* bids by account and product
* quota versus actual results reporting—especially year-to-date, using last year's actual and this year's actual to date
* salesperson advance weekly schedule plan

These are the first reports delivered by the development group. It won't be long after "going on the air" that this system will have many more reports. With this much information stored on nine databases, many more needed reports can be retrieved.

HOW THE DIAGRAM CLARIFIES THE SYSTEM

The important fact for us to see here is how easy it is to understand this system when a data flow diagram is drawn. We see the flow of input, update, and report as it applies specifically to the SALESFORCE system. Figure A-2 gives concreteness to Figure A-1. It also partitions the task of data acquisition, storage, and retrieval to many transform bubbles. Now each of these bubbles can in turn be exploded into more detail, down as many levels as we need to go to reach bubbles that will make up nice little modules of programming.

Examples of Bubble Explosion

An example of this bubble explosion (decomposition through levelling and partitioning) is shown in Figure A-3, where the bubble numbered 1 on Figure A-2 is exploded into 1.1 through 1.11. We are dealing with security as it applies to the field sales staff and are now already (at level 2) at the base level. All the bubbles in Figure A-3 are suitable for modules and need no further decomposition. As a matter of fact, if working in a high-level language, a very good case can be made that level one is the base level (the lowest level of decomposition necessary to program).

NOTE: See Chapters 16 and 17 on design for discussion concerning the transfer of the systems analysis to the systems design.

Our last example of the data flow diagram technique is Figure A-4. On this chart we explode bubble 3 of Figure A-2 into 3.1 through 3.20 to get a complete picture of the refinement of a raw order transaction into a valid order transaction ready for batching. Notice again the relationship between the data store and the transform bubble. For instance, in order for bubble 3.8 to do its job of checking the account number for validity, it has to use the order transaction account number as a search argument to do a lookup to the account database. A hit on this database is the best possible validation of this field.

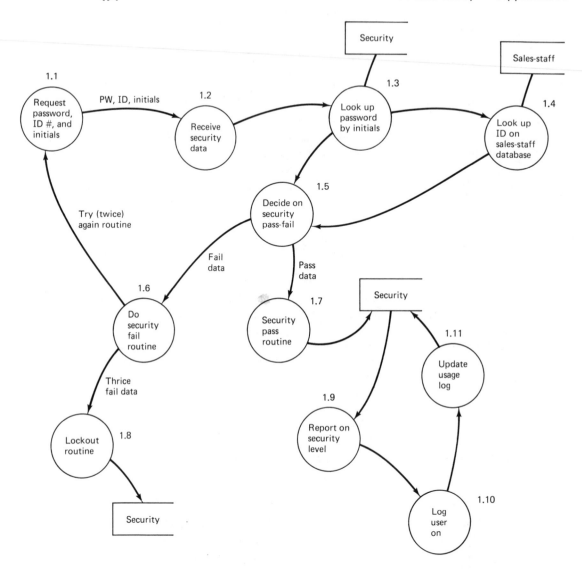

Figure A-3   Check Security.

Return of Invalid Fields

The data flow diagram in Figure A-4 also shows the method by which we return invalid fields to the operator. In this case the entire transaction is checked, and probably the invalid fields will all be returned highlighted in some way. Whatever approach is used will usually be standardized.

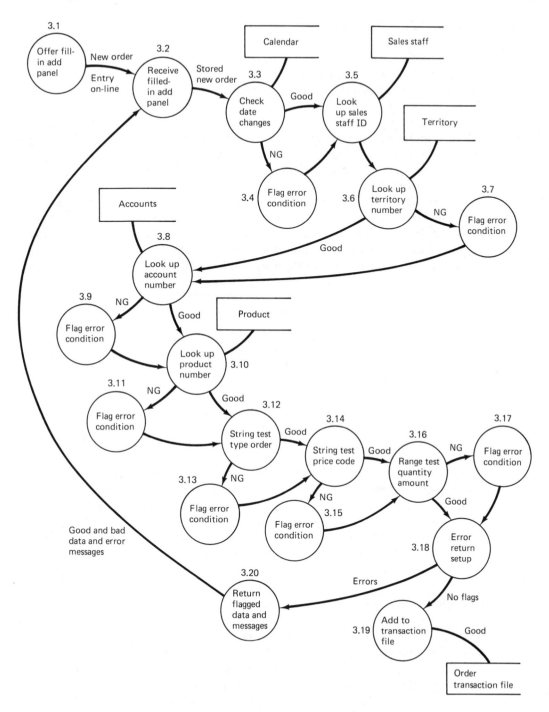

Figure A-4 Receive and Validate Orders Transactions.

DATA STORE FIELD DEFINITIONS

Before, during, and after finishing the data flow diagrams the systems analyst will be working on the data stores (databases or data files). It is hard to imagine doing data flows without having a pretty good idea already of the nature of the data to be kept in storage.

Chapters 9 and 10 were concerned with the data modelling techniques by which complex common databases are now modelled out of the user views of the system (such as reporting).

No new system specification can possibly be useful without specifying, along with the moving data flow, the stored data.

SALESFORCE has nine databases and a number of other temporary data files and small stored tables. By whatever means we arrived at it, whether by scientific data modelling or by a mix of common sense, servicing the output report requirements, and using some intuition, here is the definition of the SALES-STAFF database. It is relational in the third normal form (Chapter 9).

SALES-STAFF DATABASE

#.	NAME	TYPE	SIZE	DEC
01	sales-id	ch	04	
02	repname	ch	24	
03	address	ch	24	
04	auxaddr	ch	18	
05	city-st-zip	ch	24	
06	phone	ch	12	
07	title-code	ch	02	
08	sales-ytd	nm	09	2

When we say this file is in third normal form (see Chapter 10), we are assuming one address, one phone per sales rep. Field #8 is defined as a nine position numeric amount—with six whole numbers, a decimal point, and two numbers to the right of the decimal point making up the nine positions.

Definitions like the one shown here are done for all data stores (even humble little tables) and are part of the specification of the new system.

THE DATA DICTIONARY

In Chapter 12 we discussed passive, active, and in-line data dictionaries, emphasizing the more sophisticated forms. Here, just to keep the example in its simplest form, is the SALESFORCE file data shown as part of a home-grown passive data dictionary. Shown here are the kinds of information about available data that users want to see in the dictionary. When all the data stores are in the dictionary, the users can know what kinds of information it is possible to retrieve by making permutations and combinations and derivations of dictionary data. They also know what it is necessary to do to present the data to the system so that it won't be rejected as invalid. More sophisticated dictionaries will do the validation from the dictionary rules, construct file definitions for copy into computer programs, and much more.

Here is a data element (field) from the SALES-STAFF database:

SALESFORCE DATA DICTIONARY

standard name:	SALES IDENTIFICATION NUMBER
computer name:	SALES-ID
file name:	SALES-STAFF
key field:	YES, PRIMARY KEY
type:	APHANUMERIC
size:	4
decimals:	–
entry source:	REGIONAL SALES MANAGER
description:	LEADING ALPHA, FOLLOWED BY THREE NUMERICS. ALPHA DESIGNATES DIVISION, NUMERIC IS SALESPERSON #.
values:	B231, D73
validation rules:	A-K BUT NOT I; 002-600

Here is another example:

standard name:	SALES STAFF TITLE CODE
computer name:	TITLE-CODE
file name:	SALES–STAFF
key field:	NO
type:	ALPHANUMERIC
size:	2
decimals:	–
entry source:	NATIONAL SALES MANAGER
description:	CORPORATE JOB CLASSIFICATION
values:	PS (probationary), QS (qualified rep), SS (senior rep), RM (regional manager), AM (area manager), DM (divisional manager), NM (national manager)
validation rules:	VALIDATE VALUE STRING AS SHOWN

TRANSACTION DEFINITIONS

Every data store starts out as a transaction. One or several transactions may create a data store record; also one transaction could create more than one data store record. After creation, the data store record will probably be updated with field changes and record deletes. No system specification can be called complete without reference to the transactions which make up the data flow on the data flow diagrams. Here is an example of the transaction form that updates the QUOTA file in the SALESFORCE system. There are four transaction codes:

01Q   add a new quota record
02Q   change a quota record
03Q   delete a quota record
04Q   update quota record with last year's actuals

QUOTA TRANSACTION

##	FIELD	TYPE	SIZE	DEC
01	MONTH	CH	03	
02	TERRITORY	CH	04	
03	PRODNUM	CH	07	
04	TRANCODE	CH	03	
05	FCAST-QTY	NM	07	
06	LASTACTQTY	NM	07	
07	FCASTAMT	NM	07	
08	LASTACTAMT	NM	07	

LASTACTAMT, of course, stands for last actual amount. The reason there are no decimal places is that quota forecasts are rounded off to the nearest dollar.

Twice a year the national sales manager creates the quotas for each sales rep and sales manager. A terminal operator enters this transaction set to a temporary transaction file, the quota transaction file. Then the transaction file is sent to the quota database for matching (mostly insertions). Next, a special run of the SALESFORCE system recycles last year's actual sales from the order file to the quota file, using the 04Q transaction. Then the file is printed out for careful examination by all concerned, each individual seeing only that portion of the data which is of direct concern. This leads us into the next essential part of the specification: the reporting.

REPORT DEFINITIONS

Without question, the part of the specification that most interests the user is the reporting. Reporting is the user's main view of the system and, really, is what the user is buying. What we want to do to specify a report is show a picture of what the report will look like. Other standard information is associated with reporting, such as when the report is to be run and who gets it and so forth, but this is usually dealt with during the design phase.

We show here, as an example of a report in the SALESFORCE system, the weekly EXPENSE report. The sales rep sends in daily transactions, and the system produces weekly and other time-frame reports for sales reps, regions, and other geographical units, including national. All the expense reports, whatever the time frame or geography, look like this one. This is known as a *full-screen* or *row-oriented* report, in contrast to the familiar table report where there are column headers and the data in any column is identified by the header.

```
            WEEKLY EXPENSE REPORT FOR WEEK OF MM/DD/YY
            SALES REP NAME: _____
            SALES-ID#: _____TERRITORY #: _____


       ITEM     SUN   MON   TUE   WED   THU   FRI   SAT   CASH  CHG
  MAJOR MINOR
  _____   _____ _____ _____ _____ _____ _____ _____ _____ _____


  CAR
    TOLLS
    PARKING
    GAS
    REPAIRS    SUN-SAT, CASH AND CHARGE FIELDS ARE AMOUNTS
    FEES       WITH THIS TEMPLATE: XXXX.XX. ZERO SUPPRESSION
    RENTALS    TO LEFT OF DECIMAL LEAVING ONE ZERO. FOR
  ENTERTAIN    INSTANCE, 0.50 FOR A TOLL.
    BFST
    LUNCH
    DINNER
    OTHER
  OVERNITE
    HOTEL
    BFST
    LUNCH
    DINNER
```

. . . and so forth, you get the idea.

PANEL FLOW DIAGRAMS

The final element of the new system specification is the panel flow diagram. Almost all modern information systems rely on the so-called user-terminal dialogue to confront the system. Even the batch-type system, except for unusually large input files, accepts transactions on-line. Many systems are almost entirely on-line except for batch reporting. The panel flow diagram gives the user (and the designer) a very satisfactory idea of what the system is trying to do. A panel flow consists of menus from which the operator makes choices, which finally bring the operator to a panel where a record may be added, changed, deleted, or browsed or some combination of these functions. Showing this dialogue laid out on a big sheet of paper goes a long way toward letting the user/designer understand the nature of the system, particularly a mostly on-line system.

Figure A-5 shows some of the panel flow diagram for the SALESFORCE

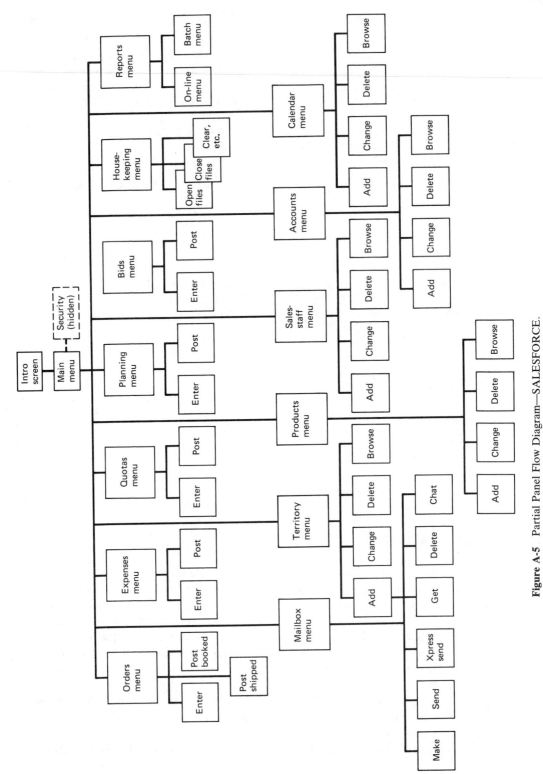

Figure A-5 Partial Panel Flow Diagram—SALESFORCE.

information system. For a really satisfactory workup of this panel flow diagram all menus should be shown as well as all action panels hanging off the menus. Then we might show the boxes associated with each action panel and menu, which are the supporting programs. For a really complete layout, each panel should then be displayed as a separate page, where the detail of each is laid out. If this is accompanied by a dummy prototype panel system on-line, where the user can actually tap through the panels, the analyst gets a gold star and a medal.

SUMMARY

DFDs, data stores, data dictionary entries, transactions, reports, and panel flows—these, taken together, are the minimum definition of a systems analysis for a new system. It doesn't matter whether we are dealing with a home mail-order business in the basement, a big system for a major corporation, or a monitor system for all the ships in all the navies of the world—we show the same basic specification. It doesn't make any difference whether the application goes eventually onto a personal computer or a network of mainframes and micros; the logical system problem remains the same. Thus it has been, thus it is now, thus it will continue to be.

Meanwhile, as we have shown in this book, there are other considerations involved with all levels of analysis—such as cost-benefit, strategic planning, selecting software, and so forth. The point is, we cannot understand what is expected of us on every level of systems analysis until we have mastered the core concerns addressed in this appendix.

INDEX

A

Activities, 24
Activity:
in strategic plan, 216
AD/M (application development maintenance):
the function point method, 333
Albrecht, Allen, 338
Analysis:
basis problem of, 18–21
data, 5
definition, 348
domain, 49
level of trust in, 34
process, 5
Ann Arbor connection, 215
Ann Arbor method, 217
ANSI X3 SPARC, 171
Antientropic, 88
Application:
implementation, 92
Application derived database, 204
Application development:
buy or build, 29–30
phased, 29–30
Application packages, 5
Application problems:
data problems in disguise, 93
Application software, 287
customizing, 307
customizing costs, 307
hooked into DBMS, 287
integration with data stores, 299–301
Application systems, 91–92
Application views, 92
Apprehension of information, 86
Approvals, 23
Architecture (SNA), 143–44
Artificial intelligence, 59
Asynchronous transmission, 139
Attribute, 98–99
data item, 173

B

Bandwidth, 138
Base level, 10, 30, 43–45, 85
Base level specification:
case example, 74–78
decent outline, 73, 75
flowcharting, 73, 78
structured English, 72–76
through data flow diagrams, 72
tight English, 72–75
Basic DBMS, 287–88
appending for full service, 288
Batch system, 355
input flow, 355–57
update and output flow, 357–59
Bell Labs, 88
Benefits:
avoid costs, 316–17
identification of, 313–18
increase revenue, 316
and investment choices, 321
improve service, 317–18
multiple benefit calculation, 320
strengthen position, 314–16
Benefits sources, 311–12
Benefits spreadsheet, 319
Boehm, Barry W., 264, 333–38
Brain functioning:
right-left sides, 424
Branching, 45
Brassboard, 28
BSP (business system planning), 215, 217
Buddha's Dharma, 439
Business function:
in strategic planning, 215

C

Candidate key.
See Key
Canonical, 181

Capra, Fritjof, 441
Case studies:
bakery analysis problem, 64–70
bakery design problem, 389–419
large business enterprise, 465–81
major league baseball team, 52–56
Category:
as used in data dictionary, 238
Causation, statistical, 440
Causation and chance, 439–41
examples, 439–40
Centralization of data, 132–33
Centralized computing, 132–33
Charting systems, 56–58
Chunks and Chunking, 15, 19, 165, 170
Cincom systems, 246
Classical methodology, 57
Closure, 428
COBOL, 59
COBOLese, 58
COCOMO, 264–65, 333–38
basic, 336–38
cost drivers, 336
model, 334–36
CODASYL, 171
Code, 88
as antientropic, 88
system, 88
tests for transaction validation, 112
Coded data:
evolutionary development, 90
Common data, 91
different views of, 152–153
Common data base:
as primary, 198
Common data store, 155
from the i-1 level, 203
uploaded from application data stores, 201
Common database, 190
alternative strategies, 201
path lockout, 199

Common database (*cont.*)
  political problems, 199–200
  skeptical view, 198–201
  technical problems, 198, 199
Communicating analysis, 9
Communication, 86
  definition, 89
  informal, 115
Communication systems, 125–26
Complex duality, 439
Computer system design:
  generic solution, 349–51
  mix of on-line and batch, 349
  package, 351–53
  reasons for doing it, 347
  transactions, 349–51
  transformation from specification, 347–49
Computing:
  centralized, 131
  distributed, 131
Concatenated key.
  See Key
Concentrative consciousness, 423
Conceptual schema, 177
  credibility of data modelling, 195
  defined by user views, 189
Constantine, Larry L., 373
Conversational mode, 199
Core specification, 49
Corporate data, 91
Cost analysis, 333
Cost of development, 327–31
Cost sources, 311–12
Cost-benefit:
  definition of intangible costs, 309
  definition of tangible costs, 309
  ratio, 29
  using historical information, 312
  using the technical expert, 312
Cost-benefit analysis, 5
  cash flow over time, 341
  definition, 308–10
  minimum delivered requirement, 343–44
  putting it all together, 341–44
  ROI (return on investment), 341–43
  and the strategic plan, 309
  two levels at which conducted, 310, 311
  uncertainties, 308
Creative process, 428–35
Creative process model, 424–28
  absorption, 424
  closure, 428
  confirmation, 427
  frustration, 426
  incorporation, 427
  insight, 426–27
  introduction to problem, 424
  letting go, 426
  logical, concentrative analysis, 426
  recognizing necessity, 424
  signing up, 424
  testing, 427
  using, 427
Creativity, 5
  definition, 423
  and the urgent problem, 428
Cross reference checks (when validating transactions), 112
Cullinet software corporation, 246
Current system and requirements study overview, 81
Current system specification:
  conversion to logical from physical, 81

Current system study, 17–21
  data flow programs, 21
  data resource, 21
  data store evaluation, 150–52
  entropy evaluation, 150–52
  Kobler's bakery, 395–98
  process flow charts, 21
  user sign off, 21
Cylinder (DASD), 161

D

DASD, 162
Data, 86
  as potential fact, 86
  at rest, 5
  common, 91
  corporate, 91
  development alternatives, 91–93
  dictionary, 5
  fully defined, 92
  integrated, 91
  in motion, 5
  ownership, 90
  private, 91
  shared, 90
  structure of, 85
  transmission, 89
  validation rules, 92
  well controlled, 85
Data, information, knowledge:
  understanding relationships, 87
Data, input and output:
  as part of core specification, 46
Data administration, 240–43
  and management of information, 241–43
Data administrator, 237
Data analysis, 149–50, 170
  introduction, 85–97
  subjects, 90
  theory and principles, 86
  two kinds of data, 85
Data and system disorganization, 85
Data applications, 92
Data-at-rest, 85, 140–68
  as available data store, 89
  data storage mediums, 157–62
  relationship to data-in-motion, 146–52
  stored and retrieved, 90
Data attributes, 71
Data automation:
  generic computer solution, 90
Data bank, 167
Database, 30–31, 164–68
  complexity levels, 165–68
  conceptual schema, 177
  definition, 164
  design, 5
  external schema, 177
  external view, 30
  hierarchical, 167
  integrated, 30
  internal schema, 176–77
  logical design, 169
  management systems, 5
  multilevel, multipath, 165–68
  multilevel, single path, 166–67
  physical, 30
  plex, 166
  problems, 165
  redundancy of data elements, 164
  relational, 165–67
  scope of use, 168
  simple, single level, 165

  subject, 92, 169
  subject databases, 30
  3-level ansi/x3/sparc, 31–32
  tree, 166
Database administrator, 174–76
Database design inc.:
  a Doll-Martin company, 215
Database design terminology:
  table expressions, 170–74
Database management system, 31
  congenial, 162
  minimal, 162
Database system:
  as different from database, 167–68
  for Kobler bakery case, 404–7
Database system organization, 174
  ANSI X3 3-LEVEL, 178
Database systems:
  and dasd, 174
Database terminology, 170–79
  American nat'l standards institute, 176–78
Data communication network system, 132–36
Data confusion, 200
Data development:
  alternatives, 91
  application systems, 91–92
  integrated organization systems, 92
Data dictionary, 30–31, 92, 202, 237
  active, 246
  and artifical intelligence systems, 239–40
  as central organization directory, 251–52
  definition, 206
  documenting master organization plan, 93
  example of extended service, 251–53
  full service extended, 248–53
  IBM extended categories, 250
  IBM OS/VS DB/DC dictionary, 248–50
  in-line, 246–47
  integrated, 245
  internals, 247–53
  minimum, 247–48
  passive, 246
  stand-alone, 245
  types and functions, 245–47
Data entities, 92
  in strategic plan study, 213
Data entity, 211
  examples, 212
  in strategic plan, 216–17
Data evolution, 89
  from raw to refined, 148–50
Data flow diagram, 60–80
  advantages, 78–79
  as center of whole specification, 79–80
  breaking up wordy bubbles, 70
  case example: the bakery, 64–72
  charting tools, 61–64
  compared to text, 68
  determining base level, 68
  nonlinear comprehension, 68
  nonserial flow, 79
  transaction to process ratio, 79
  at whole system level, 66–67
Data flow diagram add-ons:
  base level process specification, 70–78
  data specification, 70–72
  physical system specification, 70

Data flow diagram add-on (*cont.*)
    terminal dialogue specification, 70
Data flow diagram symbols, 62
    data store open rectangle, 63
    exclusive or, 62
    external source or receiver box, 63
    inclusive or, 62
    line arrow, 62–63
    transform bubble, 61–62
Data group, 172
Data independence, 151
Data-in-motion, 85
    becoming part of knowledge base, 95
    built and transmitted, 90
    identify input transaction, 104
    input refining process, 95
    making knowledge out of, 105
    receive, 103–04
    as system output, 116–24
Data item, 172–73
Data Language/1, 174
    examples, 179
Data management system, 162
    functions, 162
Data model:
    as minimal sum of combined user views, 189–98
    relationships, 255
Data modelling, 24, 30, 92, 148, 201
    steps, 180–81
Data ownership:
    private data, 90
    workstation data, 90
Datapro, 247
Data relations, 92
Data relationships, 71–72
Data requirements, 82
Data sources, 247
Data storage mediums, 157
    consciousness, 158
    paper, 158
    programs, 158–59
    tape, 159–60
Data storage ownership:
    shared, 155–57
Data storage resource:
    as part of core specification, 46–48
Data store, 85
    contention for, 153
    data flowing out of store one way, 63
    data flowing to and from store, 63
    extracted, 151
    need for real-time access, 163
    primary, 151
    primary and secondary storage, 154
    as resource, 3
    store between two transforms, 63
    three approaches to organization, 152
Data store approaches:
    application driven, 152
    I-1, 152
    integrated common, 152
Data store definition:
    through user views, 148
Data store development:
    basic questions to be answered, 151–52
    controls and triggers, 147
    input requirements, 147
    output requirements, 147
    synthesizing inputs and outputs, 148
Data store evolution:

from motion data to resting data, 148–50
Data store ownership:
    common, 152–55
    private, 147
    public, 157
Data tracking method, 36–37
Data tracking to arrive at process flow:
    rationale for, 60–61
Data transmission, 95, 97, 125–45
    characteristics, 137–42
    front ends, 140
    line speed, 137–38
    lines and links, 138–39
    links, 134–35
    local area network, 141–42
    models, 139
    multiplexing, 139–40
    node operations, 140
    PBX, 141
    standards, 142–44
    systems analysis and design of, 136–37
    tradeoffs and compromises, 129–30
    units, 139
    value added network, 141–42
Data transmission system:
    any computer to any computer, 133
    autonomous-centralized, 133
    definition, 125
    host to satellite, 133
    load sharing, 133
Data views, 92
Datum, 86
DBA:
    database administrator, 168
    as physical database custodian, 168
DBMS, 362
    basic, 287–88
    calls, 288
    comparisons between types, 298
    components of a "roll-your-own" system, 288–90
    data dictionary/directory function, 294
    download handling of extracts, 291
    extended, 288–94
    integrated system, 294–99
    other utilities and functions, 293–94
    procedural Executive manager, 292–93
    procedural language support, 292
    procedures editor/library function, 293
    records (input transaction) management, 291–92
    report and query, 290
DBMS-DMS comparison, 163
DBMS types:
    criteria for making comparisons, 299
Decentralization of resources, 131
Decision diamond:
    compound relational expression, 37
    multiple simple form, 37
    precise form, 37
    question form, 37
    simple relational expression, 37
    switch form, 37
Decision models and sensitivity testing, 311–12

Decision support system, 71, 117, 146
Decomposition, 20, 52, 85
De Marco, Tom, 322
Design:
    and implementation, 346–47
    and systems analysis, 346
    elements, 34–35
    manual versus computer system, 345–47
    manual-computer connections, 346
Design deliverables package, 352–53
Design documentation, 260–61
Design of computer systems:
    batch, 5
    charting, 5
    on-line, 5
Design of information systems, 345–88
Design package:
    cost-benefit reanalysis, 353
    data dictionary updates, 353
    panel flow diagrams, 353
    panel paintings, 353
    physical system, 353
    program chart, 352
    program descriptions, 352
    program modules, 352
    system chart, 352
    system description, 352
    transactions, 352
    users manual, 353
Design phase main output, 346–47
Design team concerns, 34
Development costs, 327–31
    by work type, 331
    contract labor, 329
    how long will it take?, 327
    physical system, 329–31
    resident labor, 329–30
    vendor software, 329–30
Device as small number system, 309
Device system, 1
Direct access storage device (DASD), 160–62
Disk pack (of DASD), 161
Disorganization, 88
DL/1, 168
DNA code, 88
Document analysis:
    and conceptual database schema, 281
    and process flow diagrams, 281
    and reports, 281–82
    and study research area, 280–82
    and user-view diagrams, 281
Documentation, 256–63
    appropriate for information systems, 256–57
    as critical path, 256
    for implementation, 259
    importance to organization, 279
    integration on dictionary, 263
    graphics, 254
    methodology, 254
    need to see, 262–63
    of new system, 259
    program source code, 257
    run book, 257–58
    for system maintenance and use, 257–59
    for systems analysis, 261–62
    for systems design, 259–61
    for users, 259
    using extended data dictionary, 253–56
    where to keep, 262–63
Document research techniques, 278–82

Documents:
  invisible documents, 281–82
  reading for decomposition, 52
  three types, 280
DO loop, 41
Domain, 173
Domain of analysis, 49
Downloading, 29
Duality:
  example, 438
  simple and complex, 438–39
Dynamic Equilibrium, 1

E

Ehrenberg, Miriam, 433
Ehrenberg, Otto, 433
Elaborated knowledge, 89
Electromechanical component of
  DASD, 161
Electronic mailbox, 115
Entropic, 88.
  See also Antientropic
Entropy, 88
Equilibrium, 7.
  See also Dynamic equilbrium
Equipment, facilities and materials,
  48–49
Ergonomics, 127–28
Estimating gross benefit, 318–23
Evaluated knowledge, 89
Evolutionary potential of a system, 8
Executive letter of authorization, 81
Extended DBMS, 288–94
External Schema, 177

F

Feasibility studies, 16
Financial resources:
  as part of core specification, 48
First normal form, 182, 183
Flow charting, 37–59
  appropriate leveling, 42
  combining forms to specify sys-
    tem, 40–42
  loops, 39
  physical solution independence, 39
Flow charting base levels, 78
Flow charting process flow, 45
Flow chart symbols, 37–45
  branching table, 45
  connector, 45
  decision diamond, 37–39
  line arrow, 45
  operations box, 39–42
  subroutine box, 42–45
FOCUS DBMS, 202
Fourth normal form, 182
Functional and data requirements, 82,
  83
Function point method, 338–40
  elements, 338
  list of adjusting characteristics,
    339
Functions in strategic plan study, 213

G

Gane and Sarson, 382
Gane, Christopher, 373
General systems theory, 3, 309

Generic information system, 93–97,
  265–67
  data transmission, 93
  input processing, 95
  knowledge processing, 96
  output data transmission, 97
  output delivery, 96
  processing, 95
  retrieval processing, 96
Glans, Thomas, 57
Goals, 7
Grad, Burton, 57
Gross benefit:
  estimating, 23, 318
  requirements level, 320
  system level, 320
Group meeting dynamics, 273–74

H

Hard methodology, 218
Hierarchy, 2
HIPO (Hierarchical input process out-
  put) 57
Holland System Corporation, 215
Holstein, David, 57
Horizontal partitioning.
  See Partitioning
Host computers, 131
How long will it take?, 331–40, 335
Human resource:
  in core specification, 48
Human social organization:
  compared to a device, 309

I

I Ching (the book of changes), 439
i-1 data integration, 92–93
i-1 level, 203–4
IBM:
  CICS, 143
  DB2 database, 202
  DMS (application development
    system), 362
  IMS database system, 174
  system network architecture,
    143–44
  3380 disk drive and control unit,
    161
IDMS database system, 182
Immediate access, 154
Implementation elements, 35
Implementation plan, 31
  elements, 31
Implementation team, 34
IMS database, 167–68, 182
Inclusion-exclusion, 18–19, 49
Inference, 87
Information, 86
  apprehension, 86
  attributes of, 86
  batched, 29
  centralized, 29
  code-based, 88
  comprehension, 88
  definition, 87
  distributed, 29
  integration into knowlege base, 89
  and knowledge, 88
  made from data, 89
  on-line, 29
  private, 29
  relationship, 86
  suprasystem, 86

surprise, 86
  validation from databases, 89
Information as meaning:
  in biologic and social systems, 88
Information attributes, 96
Information automation:
  overall computer solution model,
    350
Information chunks:
  use of keys for addressability, 106
Information flow:
  related to process flow, 20
Information group:
  in strategic plan, 216
Information groups:
  in strategic plan study, 213
Information quality:
  defined, 311
Information resource management:
  and data administration, 241–43
  objectives, 240
  responsibilities, 240
  role players, 240–44
Information retrieval:
  from primary databases, 129
Information systems development:
  importance of inventive skills, 421
  lack of inventiveness as main
    problem, 422
Input information processing system,
  95
Inputs, 2
Insight, 5, 426–27
  anatomy of—an in depth discus-
    sion, 435–38
  Buddhist approach, 435–38
  definition, 423
  level four achievements, 438
  level one achievement, 436–37
  level three achievements, 437–38
  level two achievements, 437
  through fantasy, 432–33
  through meditation, 433
  through relaxation, 432
  through sleep, 432
Integrated database:
  cost of, 92
Integrated DBMS, 294–99
  problems, 297–98
  resolution of confusion, 294
Integrated organization data, 91
Integrated organization systems, 92
Integrated systems, 92
Internal schema, 176–77
Interview, 52
  confirmation of findings, 274–76
  dialectics, 272
  gathering information, 269–71
  I/O, 275–76
  negative situation, 276–78
  the Platonic dialogue, 271–72
  stages, 269–70
  techniques, 268–78
  three-level, 275
  types, 268–69
  understanding what has been
    gathered, 271–73
  validity checking of findings,
    275–78
Invention, 5, 26, 264
  being prepared, 435
  the creative society, 428–29
  definition, 423
  the family background, 429–30
  the mysterious, 431–32
  peak times, 434–35
  personal commitment, 431
  power source for running method-
    ology, 421

Invention (*cont.*)
  the schools, 430–31
  understanding duality, 438–39
Investment options, 308
Irreducibility:
  in a system, 8

J

Jung, Carl G., 440

K

Key, 172
  candidate, 174
  concatenated, 106, 174
  data-in-motion, 106
  primary, 173
  secondary, 173–74
  in transactions, 113
Knowledge, 87, 89, 239
  accumulation into culture, 90
  attributes of, 89
  base, 89
  formed into principles, 89
  from information, 88
  perceived, 89
  retained, 89
  retrievable, 89
  used to solve problems, 90
Knowledge base, 89
Knowledge processing system, 96
Kobler's bakery:
  case study in design, 389–419
Kobler's bakery problem:
  description of the business prob-
    lem, 391–93
Koestler, Arthur, 440

L

Language use, 147–48
Languages:
  user-genial, 151
Level and partition numbering
    scheme, 49
Leveling, 10, 19–20, 42
  finding proper level, 49
  sports teams example, 51
Levels, 85
Line arrows:
  limitations for showing data, 59
Line conditioning, 138
Lines and links:
  coax, 138
  fiber optics, 138
  in-plant cable, 138
  microwave, 139
  satellite, 139
  twisted wire, 138
Line speed:
  high, 138
  medium, 138
  slow, 137
Links:
  dial up, 135
  group shared leased, 135
  leased private, 135
  own private, 135
Lisp, 59
Load sharing, 132
Logical current system specification,
    81

M

Management:
  concerns, 33–34
Mandates, 7
Martin, James, 164
Mathematica Products Group, Inc:
  RamisII, 167
Matrix analysis, 226–35
  arriving at subject databases, 229
  bounding affinity groups, 229
  case example—the university,
    228–35
  examining relationships, 227
  function/process, 227
  other paired relationships, 227
  process/data entity, 227
  process/problem, 227
Meaningful coincidence, 440
Meetings:
  strategic seating, 273–74
Messages, 95
  compared to streams, 114
  as data, 98
  informal, 115
  input, 114–16
  precise fields, 114
Metadata, 237
Method and invention:
  unity between, 441
Methodologies, 264
  data tracking, 36–37
  hard, 1, 4
  process tracking, 36–37
Modelling:
  quick nonparametric method, 312
Models, 311–12
Module:
  construction of subroutine mod-
    ule, 44
Module internal design:
  flow chart versus COBOL-link
    text, 379
Multiplexors:
  frequency-division, 140
  statistical, 140
  time-division, 140

N

Natural system, 1
Negative thinking;
  the positive value of, 433–34
Neoclassical, 85
Neoclassical methodology:
  advantages, 58–59
  for system specification, 58
Net benefit calculation, 327
Networks:
  definition, 135
  dial up and leased line, 135–36
  dial up to public line, 135–36
  exclusive leased line, 135–36
  global private line, 135–36
  local private line clusters, 135–36
  private line permanent, 135–36
  shared leased line, 135–36
  workplace private line, 135
Network topology:
  any point to any point, 135–36
  ring, 135–36
  star, 135–36
  tree, 135–36
New requirements, 81
  general statement, 82
  input and output workload statis-
    tics, 84

New requirements analysis:
  base level processing logic, 82
  net, 82
  rate and/or volume, 82
  yes or no marking, 82
New system requirements:
  from logical current system, 81
New system specification, 18, 25–34
  acceptances, 33
  alternatives, 26
  baseball team case study, 56
  boundaries, 30
  case study—large enterprise,
    465–81
  diagram or chart, 30
  feasibility, 33–34
  from current system plus require-
    ments, 25–26
  implementation plan, 31
  package components, 33
  phasing, 26–27
  role players, 33–34
  statement of problem, 465–67
  summary, 481
  text, 30
  transactions, 30
  when necessary, 150–52
New system specification case study:
  data flow diagrams, 466–75
  data store definitions, 476
  examples of bubble explosion,
    473–74
  panel flow diagrams, 479–81
  report definitions, 478–79
  specification elements, 466
  the data dictionary, 476–77
  transaction definitions, 477–78
New system specification particulars,
    30–31
Noncausal coincidence, 440
Normalization, 171–72
  decomposition, 182–85
  definition, 181–82
  first to third normal form, 184
  steps, 182–85
  summed up, 185
  of user views, 195
  value of, 182

O

Observations, 264
On-line in real time, 154
On-line system, 355
Operating-cost centers, 325
Operating costs:
  identifying, 323–27
Operating plan, 207
Operational system, 71, 118, 146
Operations, 24
Operations box:
  computations, 39–40
  logical action, 40
  moving data, 39
Operator:
  needs while using terminals, 130
  work patterns, 128
Operator productivity:
  information entry, 127–28
Operators and terminals, 126–30
Organizations:
  human social, 3
  mechanisms for downfall, 214
Output:
  of an information system, 96
  used to stimulate an input, 117
Output data transmission system, 97

Output delivery processing system, 96
Output systems message:
    used in AI systems, 117
Outputs, 2
    of information systems, 116–24
Overall computer solution:
    generic information processing, 93
Oxford:
    UFO application development system, 362

P

Panel flow diagram:
    elements, 366–67
Panel painting:
    elements and problems, 368–69
    for Kobler bakery design, 418
    for Kobler design case, 410–12
Partitioning, 20, 42
    baseball example, 54–55
Partitions, 85
Personal information on-line:
    types of resources, 127
Phased development model:
    mainframe level, 29
Phasing, 26–30
    application development, 29–30
    phased development model, 27
    prototyping, 27–28
Phenomena:
    of process, 36
Physical database, 174–77
Physical system, 371–72
    batch environment, 127–28
    data-at-rest, 125
    data-in-motion, 125
    first conceptual view, 125
    operators and terminals, 126–30
    parts of, 125
    systems analyst's role, 144–45
Physical system analysis, 149–50
Physical system specification:
    where and how to present data,
        126
    where to store data, 126
Plex:
    database, 171
Prebuilt software:
    increasing use of, 59
Premature closure.
    See also Closure, 427
Presentation:
    alternative solutions, 275
Primary data storage, 154
Primary key, 195.
    See Key
Private ownership:
    of data store, 157
Process:
    in strategic planning, 215–16
Process analysis, 149–50
Process charting methodology, 50
Processes, 24
    in strategic plan study, 213
Process flow:
    related to information, 36–37
    related to information flow, 20
Process flow charts:
    core specification, 45–49
    incorporation into complete methodology, 45–48
Process-group, 216
Process tracking method, 36–37
Productivity, 214
Program chart, 363

Program dependence:
    upon data, 164–65
Program design:
    classical batch file maintenance,
        375–76
    global and local data, 380–81
    modules and subroutines, 374–78
    structure charts, 381–86
Program example:
    Kobler bakery design, 412–14
Programmer productivity, 264–65
Program module design, 372–78
    cohesion, 373
    coupling, 373
    flow chart method, 372–78
    scope of control, 373
    span of control, 373
    structure chart method, 372–86
Programs:
    cognitive and inferencing, 96
    maintenance, 96
Prototype, 27–28
Pseudoconversational mode, 199
Public ownership:
    of data stores, 157

Q

Quality of information:
    cost, 324
    defined, 311
Quality/cost estimating:
    quick estimation method, 325
Quanta of data, 237
Queries, 2
    characteristics, 118–19
    as output data, 98
Quick fixes, 15–16

R

RamisII, 59, 167, 202
Range tests:
    transaction validation, 112
Read-write mechanism:
    of DASD, 161
Real time, 154
Redundancy, 164
    elimination, 195
Reenterent, 110
Relation, 24
    chunk of stored data, 165
Relational database, 170–71
    limitations, 182
Relationship, 7, 52
    attributes, 238
    between information subjects, 86
    explicit, 87, 98–99
    explicit and implicit, 239
    expressed as mathematical notation, 88
    forward and reverse, 238
    implicit, 87, 98–99
    invention, 424
Relationship lists:
    used in strategic plan, 223–24
Replication:
    in a system, 8
Report, 2
    characteristics, 118–19
    design specifications, 123
    Kobler bakery case design, 419
    output data, 98
Reporting examples:
    from Kobler bakery design case,
        414–17

Report/Query:
    accumulation, 122
    compared, 117–23
    data collection, 120–22
    extension, 122
    formatting, 122–23
    ordering, 122
    presentation, 123
    reduction of data, 122
    syntax checking, 120
    use of 4th generation languages,
        123
    validation, 123
Requirements, 23–25
    data, 24
    document, 25
    net, 24
    new, 24
    partial, 24
    process, 24
    related to goals, 24–25
    statement of, 82
    total, 24
Requirements analysis, 81–84
    automated versus manual, 84
    data, 82
    functional, 82
    new requirements for Kobler's
        bakery, 397–400
    scope of analyst participation, 400
Research, 264
    invention, 265
    steps to be covered, 268
    through document analysis, 268
    through interviewing, 268
    two kinds, 265
Research documents:
    from outside study area, 278–80
Research techniques, 5
    documents analysis, 5
    interviewing, 5
    used in strategic plan study,
        221–25
    working with users, 5
Resource, 3, 7
Resource requirements, 84
Retrieval, 96
Retrieval processing system, 96
Risk assessment, 321–23
    and team performance, 323
ROI (return on investment):
    complicating factors, 341–43
    some well known methods, 343
Running long, 199
Running short, 199

S

Sarson, Trish, 373
Schema:
    canonical, 181
Secondary key, 195.
    See also Key
Second normal form, 183–84
Segment, 165
Sensitivity models, 311–12
Seriality, 440
Session, 143
Shannon, Claude E., 88
Shared data, 90–91, 155–57
Software:
    customizing application software,
        307
    DBMS, 287–99
    getting the best, 301–7
    passing control of product to
        users, 306–7

488

Software (*cont.*)
off the shelf, easy to customize, 287
two areas of getting and using, 287
Software acquisition:
case study of requirements analysis, 302–4
dealing with vendors, 301
evaluating and scoring, 304
interview form, 301–2
logical steps involved, 301
making final decision, 305–6
presenting analysis results, 306
requirement weight, 302
scoring products against requirement, 302–4
testing on computer, 305–6
Software solutions:
during logical system specification, 286–87
Solution dependence in analysis, 61
SOP (study organization plan), 57
Specification responsibility, 73
Specifying current system: 9
Specifying new system, 9
Specifying system data:
for computer system, 71
data dictionary, 71
data modelling, 71
outside the computer system, 71
SPIRACIS, 207
SPIRACIS benefits:
AC—avoid costs, 314
IR—increase revenue, 313
IS—improve service, 314
SP—strengthen position, 313
Standards:
ANSI, 142
ISO, 142–43
Statistical causation, 440
Strategic plan, 26–30
activity, 216
data entity, 216–17
downward control, 28–29
example of research interview, 223
goals, 214
information group, 216
resulting in subject databases, 217
same methods as systems specifying, 218
Strategic planning, 5
business case for doing it, 208–10
business function, 215
case studies, 208–10
the data dictionary, 206, 253–56
data model, 206
examples of the processes, 207
inductive, 205
for information systems, 205–36, 210–11
the operating plan, 207
an organization function, 207
part of planning and control, 207
process, 215–16
process model, 206
specification for information systems, 206
SPIRACIS, 207–8
to strengthen position, 236
top down meets bottom up, 206
Strategic planning documentation:
examples from IBM Systems Journal, 254–56
Strategic planning for information:
data modelling, 205
selecting applications, 205

Strategic planning methodology, 217–21
Strategic planning study:
duplication of processes, 212
Strategic planning systems analysis, 205
Strategic planning terminology, 211, 214–17
Strategic plan study:
activities, 213
conducting the research, 222
data entities, 213
document analysis, 225
functions, 213
information groups, 213
list relationships, 223–24
matrix analysis, 226–35
multiple interview techniques, 222–23
necessary steps, 226
overview, 214
presentation techniques, 226–35
processes, 213
research techniques, 221–25
skimming top vertical levels, 219
user's role, 221–22
who does the research, 221–22
Strategic plan study elements, 219
Streams, 95, 116–17
compared to messages, 114
as data, 98
public media, 95
teleconferencing, 95
String tests, 112
Structure charting:
batch system, 385–86
on-line system, 384–85
Structure charts, 381–86
examples, 382–86
symbols used, 383
Structured, 85
Structured analysis and design, 57
Structured English, 59, 72–76
Study Organization Plan (SOP), 57
Subject, 86
in data dictionary, 238
Subject databases:
in strategic plan, 217
Subroutine box, 42–45
defining base level modules, 43
for leveling and partitioning, 42–45
Subroutine module:
beginning and ending module, 44
Subsystem, 2, 24
Superconscious, 423–24
Suprasystem, 2
Surprise in information, 86
Synchronicity, 440
Synchronous, transmission, 139
Synergy, 2
Synthesizing a new system, 49
System:
definition, 2
design, 5
development life cycle, 5
device, 1
evolution, 2, 8
hierarchy, 2
human, 1, 2
irreducibility, 8
law of synergy applied to, 2
natural, 1, 2
organizational rules, 3
physical, 3
random, unstructured, 1
replication, 2, 8
resources, 1, 2

specifying, 1
subsystem, 1, 2
suprasystem, 1
triggers, 2
unvarying organization of, 2
whole, 49
System and program chart symbols, 353–55
System chart and program chart symbols, 353–63
System charts, 355–63
for batch type system, 355–59
for new Kobler bakery system design, 401–4
for on-line system, 360–63
System components:
inputs in motion, 7
outputs in motion, 7
processes that transform, 7
resources, 7
triggers, 7
System design:
for command level COBOL systems, 362
data dictionary and database system, 369–70
defining transactions, 365
elements or major tasks, 34–35
system and program text descriptions, 363–65
System designer:
communicating with, 9
System development tasks, 35
System development life cycle:
design elements, 35
implementation, 35
System interface problems, 294
Systems analysis:
analysis of systems data, 85
applied, 4
charting, 4, 5
and the communications specialist, 144
costs, 9
definition, 9
designer and programmer, 35
documentation, 261–62
effect of excellence on project, 265
including understanding of data, 85
models, 4
political skills, 200
very high level (VHL), 205
VHL grade, 221
Systems design:
application software requirement, 371–72
bakery case study, 389
cost-benefit reforecast, 370–71
design standards manual, 387–88
mixed batch and on-line, 363
panel flow diagram, 365–68
panel screen painting, 365–68
physical system requirement, 371–72
program charts, 363
project scheduling and tracking, 388
responsibility for data store definition, 379–80
table driven systems, 387
top down, 378–79
transactions for Kobler system, 407–10
Systems development:
implementing team, 351
maintenance team, 351

Systems development life cycle, 10–35
  checkpoints, 10
  current systems study, 17
  does sufficient executive support
    exist? 13–14
  elements, 10–12
  executive letter of authorization,
    14
  fast overview analysis, 14–15
  feasibility study, 16–17
  getting approval, 23
  is analysis needed?
  quick fixes, 15
  requirements, 23–25
  role of chief executive officer, 14
  roles, 10
Systems model generalized, 169
Systems process flow, 36
System specification:
  for new system, 400–401
  role of data analysis, 150–52
System specifying:
  data level, 149–50
  device level, 149–50
  process level, 149–50
Systems specifying, 264

T

Tables, 148, 170–72
  as chunks of information, 170
  different ways of showing, 172
  flat, 182
  normalization of, 171–72
Tactical plan, 207
Tactical planning and the data administrator, 243
Taoist philosophy, 439
Tape storage:
  drives, 159
  fixed length record, 159
  inter-record gap, 159
  logical records, 159
  method of record insertion,
    159–60
  physical records, 159
  tape reels, 159
  variable length record, 159–60
Team performance:
  importance of, 322
Terminal and operators, 126–30
Terminals, 125–6, 134
Terminology clarification, 267
Third normal form, 24–25, 185, 195
Tight English, 72–75
TLIF table system, 387
Tools.
  See Methodologies
TOTAL database system, 182
Track (on a DASD), 161
Tracking data to arrive at process,
  60–61
Transaction:
  acceptance, 110
  add, 105
  alignment, 111
  batch input, 107
  batch machining, 104
  browse, 105
  code system, 105
  combination, 111
  concept of, 98–102
  as data, 98
  decomposition, 111
  definitions, 98–99
  delete, 105

  derivation, 111
  distribution, 104
  enhancement, 111
  entry complexity, 114
  entry techniques, 107
  feedbacks, 117
  field characteristics, 98–102
  fully refined, 113
  as information, 98–99
  inquiry, 105
  internal processing, 106, 110–11
  key, 96
  machine entry options, 104–14
  motion path from input to stor-
    age, 102–6
  on-line entry types, 105–6
  on-line machining, 104
  as output, 117
  precise fields, 98
  preparation for machining, 104
  presentation, 111
  real time on-line, 109
  sorting and merging, 96
  terminal entry on-line, 107
  update, 105
  validation, 96, 110
  videotex, 105
Transaction entry on-line system:
  readout only, 126
  two-way, 126–27
Transaction input processing, 106–14
Transactions, 95
  combined, split, enhanced, 96
  man-machined interface, 96
  unrefined, 3
Transaction tracking:
  relationship to process, 60–61
Transaction validation by lookup, 146
Transforms, 20
Transmission modes:
  full duplex, 139
  half duplex, 139
  simplex, 139
Tree database, 171
Trigger, 2, 3, 147
  part of core specification, 45
Typing and word processing, 128–29

U

Understanding data importance, 85
Un-normalized table, 182
Uploading, 29, 202
Uploading from the i-1, 203
User:
  communications with, 9
  concerns, 33
User-analyst problem solving team,
  284
User relations, 282–85
  analyst as teacher, 283–84
  encouraging user independence,
    283–85
  new role of the analyst, 283–84
  professional ethics for the analyst,
    283–84
  unsavory practices, 282–83
User-terminal dialogue, 5
User view bubble charting.
  See User view diagramming
User view diagramming, 180–98
  advantages listed, 192
  arriving at conceptual schema, 180
  arrow, 185–86
  bubble, 185
  bubble-arrow combinations,
    186–87

  charting elements, 185–87
  combining user views, 181,
    189–97
  and data-in-motion, 180
  examples, 188–98
  and normalization, 180
  purposes, 185–86
User views, 24, 30, 148

V

Victorian novel specification, 21–22
Validation of transactions, 112
Vendor developed application system,
  162
Vertical leveling.
  See Leveling
Very high level systems analysis, 205
VHL systems analysis.
  See Very high level systems analy-
    sis
Videotex, 105
Voice mailbox, 114–15

W

Weinberg, Victor, 373
Workplace data, 90–91

Y

Yourdon, Ed, 373
Yourdon-Constantine:
  on structure charting, 382